T0260642

Information and Exponential Families

Information and Exponential Families

In Statistical Theory

O. BARNDORFF-NIELSEN
Matematisk Institut
Aarhus Universitet

John Wiley & Sons Chichester · New York · Brisbane · Toronto

This edition first published 2014

Registered office
John Wiley & Sons Ltd, The Atrium, Southern Gate, Chichester, West Sussex, PO19 8SQ, United Kingdom

For details of our global editorial offices, for customer services and for information about how to apply for permission to reuse the copyright material in this book please see our website at www.wiley.com.

Library of Congress Cataloging-in-Publication Data

Barndorff-Neilsen, Ole
Information and exponential families

(Wiley series in probability and mathematical statistics – tracts)

Includes bibliographical references and index.

1. Sufficient statistics. 2. Distribution (Probability theory).
3. Functions, Exponential. I. Title.

QA276.B2847 519.5 77-9943
ISBN 0 471 99545 2

A catalogue record for this book is available from the British Library.

ISBN: 978-1-118-85750-2

2 2014

A Note from the Author

This book - a reprint of the original from 1978 - provides a systematic discussion of the basic principles of statistical inference. It was written at the time near the end of the forty years period where, starting with R. A. Fisher's seminal work, the debate about such principles was very active. Since then there have been no major developments, reflecting the fact that the core principles are accepted as valid and seem effectively exhaustive. These core principles are likelihood, sufficiency and ancillarity, and various aspects of these.

The theory is illustrated with numerous examples, of both theoretical and applied interest, some of them arising from concrete questions in other fields of science.

Extensive comments and references to the relevant literature are given in notes at the end of the separate chapters.

The account of the principles of statistical inference constitutes Part I of the book. Part II presents some technical material, needed in Part III for the exposition of the exact, as opposed to asymptotic, theory of exponential families, as it was known at the time. Some discussion is also given of transformation families, a subject that was still in its early stages then. The exact properties are related to the inference principles discussed in Part I.

Much of the later related work has been concerned with approximate versions of the core principles and of associated results, for instance about the distribution of the maximum likelihood estimator. A few references to related subsequent work are given below.

Exponential transformation models. (1982) Proc. Roy. Soc. London A 379, 41-65; coauthors Blæsild, P., Jemsen, J.L. and Jørgensen, B..; On a formula for the distribution of the maximum likelihood estimator. (1983) Biometrika 73, 307-322.; Likelihood Theory. Chapter 10 in D.V. Hinkley, N. Reid and Snell, E.J. Statistical Theory and Modelling. (1983) London; Chapman and Hall.; Inference on full and partial parameters, based on the standardized log likelihood ratio. (1983) Biometrika 73, 307-322.; Parametric Statistical Models and Likelihood. (1988). Springer Lecture Notes in Statistics. Heidelberg: Springer-Verlag; Inference and Asymptotics .London: Chapman and Hall.(1994); coauthor Cox, D.R; General exponential families. Encyclopedia of Statistical Sciences. (1997) Update Volume 1, 256-261.

October 2013
Ole E. Barndorff-Nielsen

Preface

This treatise brings together results on aspects of statistical information, notably concerning likelihood functions, plausibility functions, ancillarity, and sufficiency, and on exponential families of probability distributions. A brief outline of the contents and structure of the book is given in the beginning of the introductory chapter.

Much of the material presented is of fairly recent origin, and some of it is new. The book constitutes a further development of my Sc.D. thesis from the University of Copenhagen (Barndorff-Nielsen 1973a) and includes results from a number of my later papers as well as from papers by many other authors. References to the literature are given partly in the text proper, partly in the Notes sections at the ends of Chapters 2–4 and 8–10.

The roots of the book lie in the writings of R. A. Fisher both as concerns results and the general stance to statistical inference, and this stance has been a determining factor in the selection of topics.

Figures 2.1 and 10.1 are reproduced from Barndorff-Nielsen (1976a and b) by permission of the Royal Statistical Society, and Figures 10.2 and 10.3 are reproduced from Barndorff-Nielsen (1973c) by permission of the Biometrika Trustees. The results from R. T. Rockafellar's book *Convex Analysis* (copyright © 1970 by Princeton University Press) quoted in Chapter 5 are reproduced by permission of Princeton University Press.

In the work I have benefited greatly from discussions with colleagues and students. Adding to the acknowledgements in my Sc.D. thesis, I wish here particularly to express my warm gratitude to Preben Blæsild, David R. Cox, Jørgen G. Pedersen, Helge Gydesen, Geert Schou, and especially Anders H. Andersen for critical readings of the manuscript, to David G. Kendall for helpful and stimulating comments, and to Anne Reinert for unfailingly excellent and patient secretarial assistance. A substantial part of the manuscript was prepared in the period August 1974–January 1975 which I spent in Cambridge, at Churchill College and the Statistical Laboratory of the University. I am most grateful to my colleagues at the Department of Theoretical Statistics, Aarhus University, and the Statistical Laboratory, Cambridge University, and to the Fellows of Churchill for making this stay possible.

Aarhus, May 1977 O. B. -N.

Contents

CHAPTER 1 INTRODUCTION 1

1.1 *Introductory remarks and outline* 1
1.2 *Some mathematical prerequisites* 2
1.3 *Parametric models* 7

Part 1
Lods functions and inferential separation

CHAPTER 2 LIKELIHOOD AND PLAUSIBILITY 11

2.1 *Universality* 11
2.2 *Likelihood functions and plausibility functions* 12
2.3 *Complements* 16
2.4 *Notes* 16

CHAPTER 3 SAMPLE-HYPOTHESIS DUALITY AND LODS FUNCTIONS 19

3.1 *Lods functions* 20
3.2 *Prediction functions* 23
3.3 *Independence* 26
3.4 *Complements* 30
3.5 *Notes* 31

CHAPTER 4 LOGIC OF INFERENTIAL SEPARATION. ANCILLARITY AND SUFFICIENCY 33

4.1 *On inferential separation. Ancillarity and sufficiency* 33
4.2 *B-sufficiency and B-ancillarity* 38
4.3 *Nonformation* 46
4.4 *S-, G-, and M-ancillarity and -sufficiency* 49
4.5 *Quasi-ancillarity and Quasi-sufficiency* 57
4.6 *Conditional and unconditional plausibility functions* 58
4.7 *Complements* 62
4.8 *Notes* 68

Part II
Convex analysis, unimodality, and Laplace transforms

CHAPTER 5 CONVEX ANALYSIS 73

5.1 *Convex sets* 73
5.2 *Convex functions* 76
5.3 *Conjugate convex functions* 80
5.4 *Differential theory* 84
5.5 *Complements* 89

CHAPTER 6 LOG-CONCAVITY AND UNIMODALITY 93

6.1 *Log-concavity* 93
6.2 *Unimodality of continuous-type distributions* 96
6.3 *Unimodality of discrete-type distributions* 98
6.4 *Complements* 100

CHAPTER 7 LAPLACE TRANSFORMS 103

7.1 *The Laplace transform* 103
7.2 *Complements* 107

Part III
Exponential families

CHAPTER 8 INTRODUCTORY THEORY OF EXPONENTIAL FAMILIES 111

8.1 *First properties* 111
8.2 *Derived families* 125
8.3 *Complements* 133
8.4 *Notes* 136

CHAPTER 9 DUALITY AND EXPONENTIAL FAMILIES 139

9.1 *Convex duality and exponential families* 140
9.2 *Independence and exponential families* 147
9.3 *Likelihood functions for full exponential families* 150
9.4 *Likelihood functions for convex exponential families* 158
9.5 *Probability functions for exponential families* 164
9.6 *Plausibility functions for full exponential families* 168
9.7 *Prediction functions for full exponential families* 170
9.8 *Complements* 173
9.9 *Notes* 190

CHAPTER 10 INFERENTIAL SEPARATION AND EXPONENTIAL FAMILIES 191

10.1 *Quasi-ancillarity and exponential families* 191
10.2 *Cuts in general exponential families* 196
10.3 *Cuts in discrete-type exponential families* 202
10.4 *S-ancillarity and exponential families* 208
10.5 *M-ancillarity and exponential families* 211
10.6 *Complement* 218
10.7 *Notes* 219

References 221

Author index 231
Subject index 233

CHAPTER 1

Introduction

1.1 INTRODUCTORY REMARKS AND OUTLINE

The main kinds of task in statistics are the construction or choice of a statistical model for a given set of data, and the assessment and charting of statistical information in model and data.

This book is concerned with certain questions of statistical information thought to be of interest for purposes of scientific inference. It also contains an account of the theory of exponential families of probability measures, with particular reference to those questions. Besides exponential families, the most important type of statistical models are the group families, i.e. families of probability measures generated by a unitary group of transformations on the sample space. However, only the most basic facts on group families will be referred to. (Some further introductory remarks on these two types of models are given in Section 1.3.) Another limitation is that asymptotic problems are not discussed, except for a few remarks.

The reader is supposed to have a fairly broad, basic knowledge of statistical inference, and in particular to be familiar with the more conceptual aspects of likelihood and plausibility, such as are discussed in Birnbaum (1969) and Barndorff-Nielsen (1976b), respectively.

Probability functions, likelihood functions, and plausibility functions are charts of different types of statistical information. They are the three prominent instances of the concept of ods functions, due to Barnard (1949). An ods function is a real function on the space of possible experimental outcomes or on the space of hypotheses, which expresses the relative 'credibility' of the points of the space in question. It is often convenient to work with the logarithms of such functions and these are termed lods functions. For the objectives of this treatise the interest in lods (or ods) functions lies mainly in the very concept which is instrumental in bringing to the fore the duality between the sample aspect and the parameter aspect of statistical models, and in constructing prediction functions. Thus, although the concept of lods function will be referred to at a number of places, the theoretical developments relating to lods functions and presented in Barnard (1949) are not of direct relevance in the present context and will only be indicated briefly (in Section 3.1).

Generally, only part of the statistical information contained in the model and the data is pertinent to a given question, and one is then faced with the problem of separating out that part. The key procedures for such separations are margining to a sufficient statistic and conditioning on an ancillary statistic. Basic here is the concept of nonformation, i.e. the concept that a certain submodel and the corresponding part of the data contain no (accessible) pertinent or relevant information in respect of the question of interest.

A general treatment of the topics of statistical information indicated above is given in Part I, while the theory of exponential families is developed in Part III. Properties of convex sets and functions, in particular convex duality relations, are of great importance for the study of exponential families. Since much of convex analysis is of fairly recent origin and is not common knowledge, a compendious account of the relevant results is given in Part II, together with properties of unimodality and Laplace transforms. A reader primarily interested in lods functions and exponential families may concentrate on Chapters 2, 3, 8, and 9, just referring to Part II, which consists of Chapters 5–7, as need arises. Inferential separation, hereunder notably nonformation, ancillarity, and sufficiency, is discussed in Chapters 4 and 10. The chapters of Parts I and III contain Complements sections where miscellaneous results which did not fit into the mainstream of the text have been collected.

Each known methodological approach, of any inclusiveness, to the questions of statistical inference is hampered by various difficulties of logical or epistemic character, and applications of these approaches must therefore be tempered by independent judgement. The merits of any one approach depend on the extent to which it yields sensible and useful answers as well as on the cogency of its fundamental ideas.

The difficulties, of the kind mentioned, connected with likelihood, plausibility, ancillarity, and sufficiency have been discussed in Birnbaum (1969), Barndorff-Nielsen (1976b), and numerous other papers. Many of these papers will be referred to in the course of this treatise, but a comprehensive exposition of the arguments adduced will not be given. One of the difficulties, whose seriousness seems to have been overestimated, is that different applications of ancillarity and sufficiency, to the same model and data, may lead to different inferential conclusions (cf. Section 4.7(**vi**)). However, as has been stressed and well illustrated by Barnard (1974b), it is in general impossible to obtain unequivocal conclusions on the basis of statistical information. It is therefore not surprising that if uniqueness in conclusions is presupposed as a requirement of inference then paradoxical results turn up, such as is the case with Birnbaum's Theorem (Section 4.7(v)).

1.2 SOME MATHEMATICAL PREREQUISITES

Let M be a subset of a space $\mathfrak{M}$. The *indicator* of M is the function 1_M defined by

$$1_M(x) = \begin{cases} 1 & \text{for } x \in M \\ 0 & \text{for } x \in M^c \end{cases}$$

where M^c is the complement $\mathfrak{M} \backslash M$ of M. If $\mathfrak{M}$ is a product space, $\mathfrak{M} = \mathfrak{M}_1 \times \mathfrak{M}_2$, and if $x_1 \in \mathfrak{M}_1$ then M_{x_1} is the section of M at x_1, i.e. $M_{x_1} = \{x_2 : (x_1, x_2) \in M\}$. When $\mathfrak{M}$ is a topological space the interior, closure, and boundary of M are denoted by int M, cl M, and bd M, respectively. Suppose $\mathfrak{M} = R^k$. The affine hull of M is written aff M, and dim M is the dimension of aff M. An affine subset of M is a set of the form $M \cap L$ where L is an affine subspace of R^k.

For any mapping f the notations domain f and range f will be used, respectively, for the domain of definition of f and the range of f, and f is said to be a mapping *on* domain f.

If x is a real number then $[x\}$ will stand for $x - 1$ or x provided x is an integer and for $[x]$, the integer part of x, otherwise. Furthermore, the notations $N = \{1, 2, \ldots\}$, $N_0 = \{0, 1, 2, \ldots\}$, and $Z = \{\ldots, -2, -1, 0, 1, 2, \ldots\}$ are adopted.

All vectors are considered basically as row vectors, and the length of a vector x is indicated by $|x|$. A set of vectors in R^k are said to be *affinely independent* provided their endpoints do not belong to an affine subspace of R^k. The transpose of a matrix $\mathbf{A}$ is denoted by $\mathbf{A}'$ and, for $\mathbf{A}$ quadratic, $|\mathbf{A}|$ is the determinant and tr $\mathbf{A}$ is the trace of $\mathbf{A}$. The symbols $\mathbf{I}$ or $\mathbf{I}_r$ are used for the $r \times r$ unit matrix ($r = 1, 2, \ldots$). Occasionally an $r \times r$ symmetric matrix $\mathbf{A}$ with elements a_{ij}, say, will be interpreted either as a point in R^{r^2} or as the point in $R^{\binom{r+1}{2}}$ whose coordinates are given by $(a_{11}, a_{22}, \ldots, a_{rr}, a_{12}, \ldots, a_{1r}, a_{23}, \ldots, a_{2r}, \ldots, a_{r-1r})$. Let $\mathbf{\Sigma}$ be a positive definite matrix, set $\mathbf{\Delta} = \mathbf{\Sigma}^{-1}$, and let

$$\mathbf{\Sigma} = \begin{pmatrix} \mathbf{\Sigma}_{11} & \mathbf{\Sigma}_{12} \\ \mathbf{\Sigma}_{21} & \mathbf{\Sigma}_{22} \end{pmatrix} \quad \text{and} \quad \mathbf{\Delta} = \begin{pmatrix} \mathbf{\Delta}_{11} & \mathbf{\Delta}_{12} \\ \mathbf{\Delta}_{21} & \mathbf{\Delta}_{22} \end{pmatrix}$$

be similar partitions of $\mathbf{\Sigma}$ and $\mathbf{\Delta}$. Then, as is well known,

$$\mathbf{\Delta}_{22}^{-1} = \mathbf{\Sigma}_{22} - \mathbf{\Sigma}_{21}\mathbf{\Sigma}_{11}^{-1}\mathbf{\Sigma}_{12} \tag{1}$$

$$-\mathbf{\Sigma}_{11}^{-1}\mathbf{\Sigma}_{12} = \mathbf{\Delta}_{12}\mathbf{\Delta}_{22}^{-1}. \tag{2}$$

When indexed variables, as for example $x_i, i = 1, \ldots, m$, or $x_{ij}, i = 1, \ldots, m$; $j = 1, \ldots, n$, are considered the substitution of a dot for an index variable signifies summation over that variable. Furthermore, the vector $(x_1, \ldots, x_m)$ will be denoted by x_*, the vector $(x_{i1}, \ldots, x_{in})$ by x_{i*}, etc.

Consider a real-valued function f defined on a subset $\mathfrak{X}$ of R^k. The notations $Df = \partial f/\partial x$ and $\partial^2 f/\partial x'\partial x$ are used for the gradient and the matrix of second order derivatives of f, respectively, while $D^i f$, where $i = (i_1, \ldots, i_k)$ is a vector of non-negative integers, indicates a mixed derivative of f. (Thus $Df = D^{(1,\ldots,1)}f$.) In the case where a partition $(x^{(1)}, \ldots, x^{(m)})$ of $x (\in \mathfrak{X})$ is given then the (i, j)th matrix component of the corresponding partition of $\partial^2 f/\partial x'\partial x$ is denoted by $\partial^2 f/\partial x^{(i)\prime}\partial x^{(j)}$. Let h be a twice continuously differentiable mapping on an open subset $\mathfrak{Y}$ of R^k and onto $\mathfrak{X}$, also assumed open, and set

$$\frac{\partial x}{\partial y'} = \frac{\partial h}{\partial y'} = \begin{pmatrix} \frac{\partial h_1}{\partial y_1} & \cdots & \frac{\partial h_k}{\partial y_1} \\ \cdot & & \cdot \\ \cdot & & \cdot \\ \cdot & & \cdot \\ \frac{\partial h_1}{\partial y_k} & \cdots & \frac{\partial h_k}{\partial y_k} \end{pmatrix}$$

the Jacobian matrix of h. Moreover, set

$$\frac{\partial^2 x}{\partial y' \partial y} = \frac{\partial^2 h}{\partial y' \partial y} = \left(\frac{\partial^2 h_1}{\partial y' \partial y}, \ldots, \frac{\partial^2 h_k}{\partial y' \partial y} \right).$$

If f is twice continuously differentiable then, writing $\tilde{f}$ for the composition $f \circ h$ of f and h, one has

$$\frac{\partial^2 \tilde{f}}{\partial y' \partial y} = \frac{\partial x}{\partial y'} \frac{\partial^2 f}{\partial x' \partial x} \frac{\partial x'}{\partial y} + \frac{\partial f}{\partial x} \cdot \frac{\partial^2 x}{\partial y' \partial y} \tag{3}$$

where $\cdot$ is a matrix multiplication symbol defined in the following way. For a $1 \times k$ vector $v = (v_1, \ldots, v_k)$ and an $m \times nk$ matrix $\mathbf{A} = [\mathbf{A}_1, \ldots, \mathbf{A}_k]$, $\mathbf{A}_i$ being $m \times n$ $(i = 1, \ldots, k)$, the product $v \cdot \mathbf{A}$ is given by

$$v \cdot \mathbf{A} = v_1 \mathbf{A}_1 + \cdots + v_k \mathbf{A}_k.$$

(Thus the operation $\cdot$ is a generalization of the ordinary inner product of two k-dimensional vectors.)

Measure-theoretic questions concerning null sets, measurability of mappings, etc., will largely be bypassed. (Section 4.2, however, forms something of an exception to this.) The mathematical gaps left thereby may be filled out by standard reasoning.

Lebesgue measure will be denoted by λ, counting measure by ν. (The domains of these measures vary from case to case but it will be apparent from the context what the domain is.)

Let H be a class of transformations on a space $\mathfrak{X}$, i.e. the elements of H are one-to-one mappings of $\mathfrak{X}$ onto itself. The class H is *unitary*, respectively *transitive*, if for every pair of points x and $\tilde{x}$ in $\mathfrak{X}$ the equation $\tilde{x} = h(x)$ has at most, respectively at least, one solution h in H. In the case where H is transitive, the set $H(x) = \{h(x): h \in H\}$ is equal to $\mathfrak{X}$. A measure μ on a σ-algebra $\mathfrak{A}$ of $\mathfrak{X}$ is transformation invariant under H if $\mu h = \mu$ for every $h \in H$, where μh is defined by $\mu h(A) = \mu(h^{-1}(A))$, $A \in \mathfrak{A}$. Suppose H is a group under the operation $\circ$ of composition of mappings. Then $H(x)$ is called the *orbit* of x and the orbits form a partition of $\mathfrak{X}$. If, in addition, H is unitary then each orbit can be brought into one-to-one correspondence with H, and thus $\mathfrak{X}$ can be represented as a product space

$\mathfrak{U} \times \mathfrak{V}$ of points (u, v) such that u determines the orbit and v the position on that orbit of the point x in $\mathfrak{X}$ corresponding to (u, v). As is well known (see e.g. Nachbin 1965), if H is a locally compact, topological group then there exist left invariant as well as right invariant measures on H. For H unitary and transitive, these measures can, by the above identification of $\mathfrak{X}$ and H, also be viewed as transformation invariant measures on $\mathfrak{X}$.

The sample spaces to be considered are exclusively *Euclidean*, i.e. they are Borel subsets of Euclidean spaces, and the associated σ-algebras of events are the classes of Borel subsets of the sample spaces. Moreover, all random variables and statistics take values in Euclidean spaces. Generally, the letter $\mathfrak{X}$ will be used to denote the sample space, and x is a point in $\mathfrak{X}$.

Ordinarily, the same notation—a lower case italic letter—will be used for a random variable or statistic and for its value, the appropriate interpretation being determined by the context. In cases where clarity demands a distinction the mapping is denoted by the capital version of the letter.

Let $\mathfrak{X}$ ($\subset R^k$) be a sample space, $\mathfrak{A}$ the σ-algebra of Borel subsets of $\mathfrak{X}$, and $\mathfrak{P}$ a family of probability measures on $\mathfrak{X}$. The triplet $(\mathfrak{X}, \mathfrak{A}, \mathfrak{P})$ is termed a *statistical field*. Let P be a member of $\mathfrak{P}$ and let t (also T) be a statistic.

The marginal distribution of t under P has probability measure Pt given by $Pt(B) = P(t^{-1}(B))$ for Borel sets B. Further, $E_P t$ and $V_P t$ stand for the mean value (vector) and the variance (matrix) of t. For an event A with $P(A) > 0$ the conditional probability measure given A is denoted by $P(\cdot|A)$ or P^A. If $\mathfrak{B}$ is a sub-σ-algebra of $\mathfrak{A}$ then $P_{\mathfrak{B}}$ denotes the restriction of P to $\mathfrak{B}$ and $P^{\mathfrak{B}}$ is the Markov kernel of the conditional distribution given $\mathfrak{B}$ under P. The conditional mean value given $\mathfrak{B}$ under P of a random variable y is written $E_P^{\mathfrak{B}} y$. When $\mathfrak{B}$ is the σ-algebra generated by a statistic t the notations P_t, P^t or $P(\cdot|t)$, and $E_P^t y$ are normally used instead of $P_{\mathfrak{B}}$, $P^{\mathfrak{B}}$, and $E_P^{\mathfrak{B}} y$. For any measure μ on $\mathfrak{X}$, let $\mu^{(n)}$ indicate the measure on the product space $\mathfrak{X}^n$ which is the n-fold product of μ with itself, and let $\mu^{(*n)}$ be the n-fold convolution of μ (provided it exists). Set $\mathfrak{P}_t - \{P_t : P \in \mathfrak{P}\}$, $\mathfrak{P}^t - \{P^t : P \in \mathfrak{P}\}$, $\mathfrak{P}^{(n)} : P \in \mathfrak{P}\}$, etc.

If P and Q are two probability measures on $\mathfrak{X}$ having common support then

$$\frac{dP_t}{dQ_t} = E_Q^t \frac{dP}{dQ} \tag{4}$$

and

$$\frac{dP(\cdot|t)}{dQ(\cdot|t)} = \frac{dP/dQ}{dP_t/dQ_t}. \tag{5}$$

A distribution on R^k is *singular* if its affine support (i.e. the affine hull of its support) is a proper subset of R^k. Let u and v be statistics. The conditional distribution of u given v and under P is *singular* provided that the marginal distribution of u under the conditional probability measure given v is singular with probability 1, i.e.

$$P\{Pu(\cdot|v) \text{ is singular}\} = 1.$$

The probability measure P is said to be of *discrete type* if the support S of P has no accumulation points, of *c-discrete type* if S equals the intersection of the set Z^k and some convex set, and of *continuous type* if P is absolutely continuous with respect to Lebesgue measure λ on $\mathfrak{X}$. The same terms are applied to $\mathfrak{P}$ provided each member of $\mathfrak{P}$ has the property in question.

A function ψ defined on $\mathfrak{P}$ and taking values in some Euclidean space is called a *parameter function*. As with random variables and statistics, the same notation ψ will generally be used for the function and its value, but when it seems necessary to distinguish explicitly the function is indicated by $\psi(\cdot)$. Suppose $\mathfrak{P}$ is given as an indexed set, $\mathfrak{P} = \{P_\omega : \omega \in \Omega\}$, then $\mathfrak{P}$ is called *parametrized* provided Ω is a subset of a Euclidean space and the mapping $\omega \to P_\omega$ is one-to-one. Any parameter function ψ on $\mathfrak{P}$ may be viewed as a function of ω, and its values will be denoted, freely, by $\psi(\omega)$ as well as by ψ or $\psi(P_\omega)$. Similarly for other kinds of mappings.

The family $\mathfrak{P}$ is said to be generated by a class H of transformations on $\mathfrak{X}$ if for some member P of $\mathfrak{P}$ one has $\mathfrak{P} = \{Ph : h \in H\}$. In the case where H is a unitary group the family $\mathfrak{P}$ will be called a *group family*. Suppose that $\mathfrak{P}$ is a group family and that u is a statistic which is constant on the orbits under H but takes different values on different orbits (thus u is a maximal invariant). Then u is said to *index* the orbits, and the marginal distribution of u is the same under all the elements of $\mathfrak{P}$. It is also to be noticed that if $\mathfrak{P}$ is a transitive group family (i.e. a group family with H transitive) and if μ is a left or right invariant measure on $\mathfrak{X}$ which, when interpreted as a transformation invariant measure on $\mathfrak{X}$, dominates $\mathfrak{P}$ then the family $\mathfrak{p}$ of probability functions or densities of $\mathfrak{P}$ relative to μ is of the form

$$\mathfrak{p} = \{p(h^{-1}(\cdot)) : h \in H\} \tag{6}$$

where $p = dP/d\mu$.

For the discussions in Parts I and III (except Section 3.1) it is presupposed that a statistical model, with sample space $\mathfrak{X}$ and family of probability measures $\mathfrak{P}$, has been formulated. Unless explicitly stated otherwise, it is moreover supposed that $\mathfrak{P}$ is parametrized, $\mathfrak{P} = \{P_\omega : \omega \in \Omega\}$, and determined by a family of probability functions $\mathfrak{p} = \{p(\cdot; \omega) : \omega \in \Omega\}$, i.e. $p(\cdot; \omega)$ is the density of P_ω with respect to a certain σ-finite measure μ which dominates $\mathfrak{P}$. For discrete-type distributions this dominating measure is always taken to be counting measure, so that $p(x; \omega)$ is the probability of x. (In subsequent chapters certain topics in plausibility inference will be considered. Whenever this is the case, it is—for non-discrete distributions—presupposed that $\sup_x p(x; \omega) < \infty$ for every $\omega \in \Omega$.) In the case $\mathfrak{p}$ is of the form (6), for some probability function p with respect to μ and some class H of transformations on $\mathfrak{X}$, then $\mathfrak{p}$ is said to be generated by H. The points x in $\mathfrak{X}$ for which $p(x; \omega) > 0$ for some $\omega \in \Omega$ are called *realizable*, and the *realizable values* of a statistic are the values corresponding to realizable sample points x.

Viewed as a function on $\mathfrak{X} \times \Omega$, $p(\cdot;\cdot)$ is referred to as the *model function.* The notation $p(u;\, \omega|t)$ is used for the value of the conditional probability function of a statistic u given t and under ω.

From the previous discussion it is apparent that if the parametrized family $\mathfrak{P} = \{P_\omega \in \Omega\}$ is a group family under a group H of transformations on the sample $\mathfrak{X}$ then, under mild regularity assumptions, $\mathfrak{X}$ can be viewed as a product space $\mathfrak{U} \times \mathfrak{V}$, the spaces $\mathfrak{V}$, H, and Ω may be identified, and $\mathfrak{P}$ has a model function of the form

$$p(x;\, \omega) = p(u)p(\omega^{-1}(v)|u)$$

in an obvious notation.

The r-dimensional normal distribution with mean (vector) ξ and variance (matrix) Σ will be indicated by $N_r(\xi, \Sigma)$, and $\mathfrak{N}_r$ will stand for the class of these distributions. (The index r will be suppressed when $r = 1$.) The *precision* (matrix) Δ for $N_r(\xi, \Sigma)$ is the inverse of the variance, i.e. $\Delta = \Sigma^{-1}$. The probability measure of $N_r(\xi, \Sigma)$ will be denoted by $P_{(\xi,\Sigma)}$ or $P_{(\xi,\Delta)}$ according as the parametrization of $\mathfrak{N}_r$ by (ξ, Σ) or by (ξ, Δ) is the one of interest.

The symbol ▶ designates the end of proofs and examples.

1.3 PARAMETRIC MODELS

The statistical models considered in this tract are nearly all parametric and determined by a model function $p(x;\, \omega)$. Rather more attention than is usual will be given to the parametric aspect of the models, i.e. to the variation domains of the parameters and subparameters involved and to the structure of $p(x:\, \omega)$ as a function of ω. Thus the observation aspect and the parameter aspect are treated on a fairly equal footing. There are several reasons for this. The most substantial is that the logic of inferential separation cannot be built without certain precise specifications of the role of the parameters. Secondly, it is natural in connection with a comparative discussion of likelihood functions and plausibility functions to give an exposition of Barnard's theory of lods functions, and in a considerable and fundamental portion of that theory observations and parameters occur in a formally equivalent, or completely dual, way. Finally, the stressing of the similarity or duality of the observation and parameter aspects, as far as is statistically meaningful, leads to a certain unification and complementation of the theoretical developments.

There are two main classes of parametric models: the exponential families and the group families. The exponential families, the exact theory of which is a main topic of this book, are determined by model functions of form

$$p(x;\, \omega) = a(\omega)b(x)\,\mathrm{e}^{\theta\cdot t}$$

where θ is a k-dimensional parameter (function) and t is a k-dimensional statistic.

Group families typically have model functions which may be written

$$p(x; \omega) = p(u)p(\omega^{-1}(v)|u),$$

as explained in Section 1.1. A theory of group families—the theory of structural inference—has been developed by Fraser (1968, 1976) (see also Dawid, Stone and Zidek 1973) from Fisher's ideas on fiducial inference. Although the core of fiducial/structural inference is a notion of induced probability distributions for parameters which few persons have found acceptable, the theory comprises many results that are highly useful in the handling of group families along more conventional lines.

The overlap between the two classes of families is very little; thus in the case ω is one-dimensional, the only notable instances of families which belong to both classes appear to be provided by the normal distributions with a known variance and the gamma distributions with a known shape parameter (cf. Lindley 1958, Pfanzagl 1972, and Hipp 1975). Moreover, essential distinctions exist between the mathematical–statistical analyses which are appropriate for each of the two classes. It is remarkable indeed that both classes and the basic difference in their nature were first indicated in a single paper by Fisher (1934).

Each class covers a multitude of important statistical models and allows for a powerful general theory. This strongly motivates studying these classes *per se* and choosing the model for a given data set from one of the two classes, when feasible. Once this is realized, it seems of secondary interest only that one may be led to consider, for instance, exponential families by arguing from various viewpoints of a principled character, such as sufficiency, maximum likelihood, statistical mechanics, etc. (see the references in Section 8.4), especially since each of these viewpoints and its consequences only encompass a fraction of what is of importance in statistics.

PART I

Lods Functions and Inferential Separation

Log-probability functions, log-likelihood functions and log-plausibility functions are the three main instances of lods functions. It is an essential feature of the theory of lods functions that it incorporates a considerable part of the statistically relevant duality relations which exist between the sample aspect and the parameter aspect of statistical models.

Separate inference is inference on a parameter of interest from a part of the original model and data. Margining to a sufficient statistic and conditioning on an ancillary statistic are key procedures for inferential separation.

PART I

CHAPTER 2

Likelihood and Plausibility

In this short chapter important basic properties of likelihood functions and plausibility functions are discussed, with particular reference to similarities and differences between these two kinds of function. As a preliminary, the definition and some properties of universality are presented. Universality will also be of significance in the discussions, in subsequent chapters, of prediction, inferential separation, and unimodality.

2.1 UNIVERSALITY

The concept of universality is of significance in the discussions, given later in the book, on plausibility, M-ancillarity, prediction and unimodality.

The probability function $p(\cdot; \omega)$ is said to have a point x as mode point if

$$p(x; \omega) = \sup_{x} p(x; \omega).$$

and the set of mode points of $p(\cdot; \omega)$ will be denoted by $\check{x}(\omega)$. More generally, x will be called a *mode point for the family* $\mathfrak{p}$ provided that for all $\varepsilon > 0$ there exists an $\omega \in \Omega$ such that

$$(1) \qquad (1 + \varepsilon)\, p(x; \omega) \geq \sup_{x} p(x; \omega).$$

With this designation *universality* of the family $\mathfrak{p}$ is defined as the property that every realizable x is a mode point for $\mathfrak{p}$. If, in fact, every realizable x is a mode point for some member of $\mathfrak{p}$ then $\mathfrak{p}$ is called *strictly universal.*

For convenience in formulation, universality and strict universality will occasionally be spoken of as if they were possible properties of the family of probability measures $\mathfrak{P}$ rather than of $\mathfrak{p}$. Thus, for instance, '$\mathfrak{P}$ is universal' will mean that the family $\mathfrak{p}$ of probability functions determining $\mathfrak{P}$ is universal.

A family $\mathfrak{p}$ for which $\sup_x p(x; \omega)$ is independent of ω will be said to have *constant mode size.*

Most of the standard families of densities are universal, and many examples of universal families will be mentioned later on. Clearly, one has:

Lemma 2.1. *Let H be a class of transformations on $\mathfrak{X}$. Suppose H is transitive and* *that for some* $\omega_0 \in \Omega$

$$\mathfrak{p} = \{p(h^{-1}(\cdot); \omega_0): h \in H\}.$$

Then $\mathfrak{p}$ *is universal with constant mode size.*

As a simple consequence of the definition of mode point one finds:

Theorem 2.1. *Let t be a statistic and let*

$$p(x; \omega) = p(t; \omega)p(x; \omega | t)$$

be the factorization of the probability function of x into the marginal probability function for t and the conditional probability function for x given t.

Suppose x_0 *is a mode point of* $\mathfrak{p}$ *and set* $t_0 = t(x_0)$. *Then* x_0 *is a mode point of the family of conditional probability functions*

$$\{p(\cdot; \omega | t_0): \omega \in \Omega\}.$$

Corollary 2.1. *If* $\mathfrak{p}$ *is universal then for any given value of t the family of conditional probability functions*

$$\{p(\cdot; \omega | t): \omega \in \Omega\}$$

is also universal.

Furthermore, it is trivial that if $\mathfrak{p}$ has only a single member p, say, then $\mathfrak{p}$ is universal if and only if p is constant, i.e. the density is uniform.

The family $\mathfrak{p}$ will be said to *distinguish between the values of x* if for every pair x' and x'' of values of x there exists an $\omega \in \Omega$ such that

$$p(x'; \omega) \neq p(x''; \omega).$$

If $\mathfrak{p}$ is universal and distinguishes between the values of x then, under very mild regularity conditions, x is minimal sufficient. To see this, let x' and x'' be realizable points of $\mathfrak{X}$ and suppose that

$$c'p(x'; \omega) = c''p(x''; \omega) \quad \text{for every } \omega \in \Omega$$

where c' and c'' are constants (which may depend, respectively, on x' and x''). By the universality of $\mathfrak{p}$, the ratio c''/c' must be 1, and this implies $x' = x''$ since $\mathfrak{p}$ distinguishes between values of x. In other words, the partition of $\mathfrak{X}$ induced by the likelihood function is (equivalent to) the full partition into single points; the result now follows from Corollary 4.3.

2.2 LIKELIHOOD FUNCTIONS AND PLAUSIBILITY FUNCTIONS

A brief, comparative discussion of basic properties of likelihood and plausibility functions is given here.

Both likelihood and plausibility functions are considered as determined only up to a factor which does not depend on the parameter of the model. However, unless otherwise stated, the notations L and Π will stand for the particular choices

$$L(\omega) = L(\omega; x) = p(x; \omega)$$
$$\Pi(\omega) = \Pi(\omega; x) = p(x; \omega)/\sup_x p(x; \omega)$$

of the likelihood and plausibility functions, based on an observation x.

It is important to note that L and Π differ only by a factor

$$s(\omega) = \sup_x p(x; \omega)$$

which is independent of x.

This implies that $\ln \Pi(\omega; x)$, as well as $\ln L(\omega; x)$, is a b-lods function corresponding to the f-lods function $\ln p(x; \omega)$—in the terminology of Barnard's (1949) fundamental theory of lods. Some important common properties of likelihood and plausibility functions may be derived naturally in the theory of lods functions (see Section 3.2).

The *normed* likelihood and plausibility functions will be denoted by $\bar{L}$ and $\bar{\Pi}$, i.e.

$$\bar{L}(\omega) = L(\omega)/\sup_\omega L(\omega)$$
$$\bar{\Pi}(\omega) = \Pi(\omega)/\sup_\omega \Pi(\omega).$$

Clearly,

$$\sup_\omega \Pi(\omega; x) = 1$$

if and only if x is a mode point of $\mathfrak{p}$. Hence, $\Pi = \bar{\Pi}$ for every x if and only if $\mathfrak{p}$ is universal.

For any family $\mathfrak{p}$ such that $\sup_\omega p(x; \omega) < \infty$ for every x one has

$$\bar{L}(\omega; x) = s(\omega) r(x) \bar{\Pi}(\omega; x)$$

where

$$r(x) = \sup_\omega \Pi(\omega; x)/\sup_\omega p(x; \omega).$$

If $\bar{L}$ and $\bar{\Pi}$ are equal for a given value of x then $s(\omega)$ must be independent of ω on the set $\{\omega : p(x; \omega) > 0\}$ which means that $\mathfrak{p}$ has constant mode size on that set. On the other hand, constant mode size of $\mathfrak{p}$ obviously implies that $\bar{L} = \bar{\Pi}$ for every x. In particular, $\bar{L}$ and $\bar{\Pi}$ are thus equal for every x if $\mathfrak{p}$ is generated by a transitive set of transformations.

The set of maximum points of the likelihood or plausibility function constitute respectively the maximum likelihood estimate $\hat{\omega}(x)$ and the maximum plausibility estimate $\check{\omega}(x)$ of the parameter ω, i.e.

$$\hat{\omega} = \hat{\omega}(x) = \{\omega: L(\omega) = \sup_{\omega} L(\omega)\}$$

$$\check{\omega} = \check{\omega}(x) = \{\omega: \Pi(\omega) = \sup_{\omega} \Pi(\omega)\}.$$

Example 2.1. Figure 2.1 shows the normed likelihood and plausibility functions for the binomial model

$$p(x; \pi) = \binom{n}{x} \pi^x (1 - \pi)^{n-x} \tag{1}$$

when n equals 1 or 3 and $x = 1$.

The most prominent difference between $\bar{L}$ and $\bar{\Pi}$ ($= \Pi$) in the two cases is that $\bar{L}$ takes its maximum at one point only, $\hat{\pi} = 1$ respectively $\frac{1}{3}$, while $\bar{\Pi}$, as is typical with discrete models, is 1 on a whole set, $\check{\pi} = [\frac{1}{2}, 1]$ respectively $[\frac{1}{4}, \frac{1}{2}]$. ▶

The plausibility function is not, in contrast to the likelihood function, independent of the stopping rule. This is illustrated by the following example.

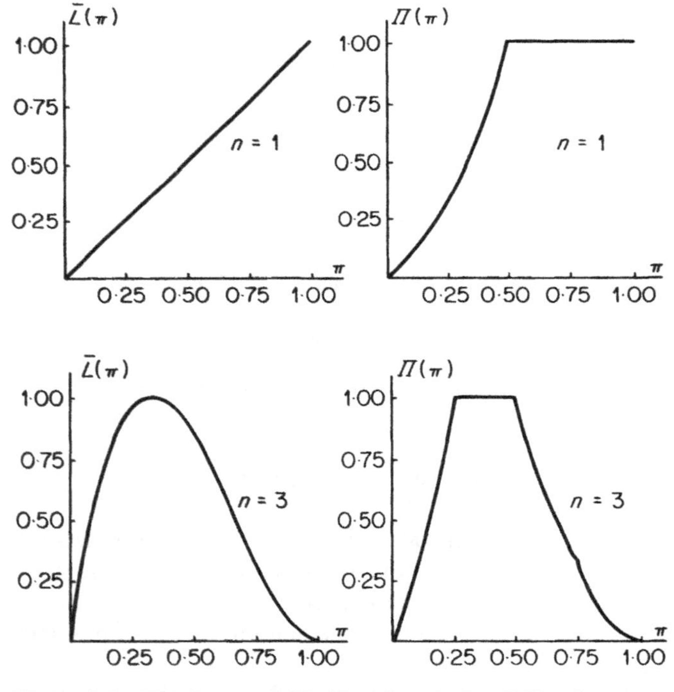

Figure 2.1 The (normed) likelihood and plausibility functions corresponding to the observation $x = 1$ of a binomial variate with trial number $n = 1$ or 3

Example 2.2. If x follows a binomial distribution with $n = 2$ and if x is observed to be 0 then the plausibility function is given by

$$\Pi(\pi) = \begin{cases} 1 & \text{for } 0 < \pi < \frac{1}{3} \\ \dfrac{1-\pi}{2\pi} & \text{for } \frac{1}{3} \leq \pi \leq \frac{2}{3} \\ \left(\dfrac{1-\pi}{\pi}\right)^2 & \text{for } \frac{2}{3} < \pi < 1. \end{cases}$$

Suppose, on the other hand, that t has the negative binomial distribution

$$p(x; \pi) = (x + 1)(1 - \pi)^2\pi^x$$

and that, again, $x = 0$. Then

$$\Pi(\pi) = (s + 1)^{-1}\pi^{-s} \quad \text{for} \quad \frac{s}{s+1} < \pi \leq \frac{s+1}{s+2}, \quad s = 0, 1, 2, \ldots .$$

However, in both instances

$$L(\pi) = (1 - \pi)^2.$$

▶

Note that the relation $x \in \check{x}(\omega)$ entails $\omega \in \check{\omega}(x)$, and that the converse implication is true if (and only if) x is a mode point for $\mathfrak{p}$.

If the maximum plausibility estimates corresponding to two different values, x' and x'', of the variate x have a point in common then $c'p(x'; \omega) = c''p(x''; \omega)$ where $c' = 1/\sup_\omega \Pi(\omega; x')$ and $c'' = 1/\sup_\omega \Pi(\omega; x'')$. Thus, provided x is minimal sufficient with respect to $\{p(\cdot; \omega): \omega \in \Omega_0\}$ for any open subset Ω_0 of Ω, the estimates $\check{\omega}(x')$ and $\check{\omega}(x'')$ will, in general, have at most boundary points in common (cf. Corollary 4.3).

For discrete data, the set of possible values of the maximum likelihood estimator is nearly always a proper, and small, subset of Ω. Thus, for instance, if x follows the binomial distribution (1) then $\hat{\pi}(\mathfrak{X}) = \{i/n: i = 1, 2, \ldots, n - 1\}$ while the domain of π is $(0, 1)$. In other words, the likelihood approach has the feature that ordinarily, for discrete models, most of the parameter values are considered not to give the best explanation of the observation x for any value of the latter. In contrast, $\check{\omega}(\mathfrak{X})$ does, whenever $\mathfrak{X}$ is finite and otherwise as a rule, equal Ω.

The maximum plausibility estimator for a truncated model is often a simple modification of the estimator for the full model.

Example 2.3. Let $\check{\pi}$ and $\check{\pi}_0$ denote, respectively, the maximum plausibility estimators for the binomial model and its zero-truncation

$$\binom{n}{x}\pi^x(1 - \pi)^{n-x}/\{1 - (1 - \pi)^n\}, \quad x = 1, \ldots, n.$$

Here $\check{\pi}_0(x) = \check{\pi}(x)$ for $x > 1$ and $\check{\pi}_0(1) = \check{\pi}(0) \cup \check{\pi}(1)$.

▶

The example also illustrates that $\tilde{\omega}$ may have a simple explicit expression in cases where no such expression exists for $\hat{\omega}$. A further illustration of this is provided by the $r \times c$ contingency table with both marginals given, cf. Example 4.19.

2.3 COMPLEMENTS

(i) Finucan (1964) has given the following characterization of mode(s) of the multinomial distribution

$$\frac{n!}{x_1!\ldots x_m!}\pi_1^{x_1}\ldots\pi_m^{x_m}. \tag{1}$$

Let $\tilde{n}$ denote a non-negative real number. Any point $([\tilde{n}\pi_1\},\ldots,[\tilde{n}\pi_m\})$, for which $[\tilde{n}\pi_1\} + \cdots + [\tilde{n}\pi_m\} = n$, is a mode of the multinomial distribution (1), and such a point exists. If none of the numbers $\tilde{n}\pi_1,\ldots,\tilde{n}\pi_m$ is an integer then the mode is unique.

A proof of this result will be given in Example 4.17.

(ii) Suppose $\mathfrak{p}$ is a transitive group family (cf. Section 1.1). For any x the normed likelihood and plausibility functions $\bar{L}$ and $\bar{\Pi}$ are then identical. Moreover, for any constant d the set $\{\omega: \bar{L}(\omega) \geqslant d\} - \{\omega: \bar{\Pi}(\omega) \geqslant d\}$ is a confidence set for ω whose confidence coefficient is given by the integral of $\bar{L}/c = \bar{\Pi}/c$ over the set in question and with respect to right invariant measure, c being the norming constant which makes the integral over all of Ω equal to 1.

This is a particular instance of the following result. If a fixed subset A of $\mathfrak{X} \equiv \Omega$ is, in repeated sampling, transformed by each observation x into xA^{-1} then the frequency of cases in which xA^{-1} contains the actual value of ω will tend to $P(A)$ as the sample size increases to infinity, and

$$P(A) = \int_A p\,d\mu$$

may be rewritten as

$$P(A) = \Delta(x)\int_{xA^{-1}} L(\cdot;x)\,dv$$

where v denotes the right invariant measure on Ω and $\Delta(x)$ is a norming constant (actually, $\Delta(\cdot)$ is the so-called modular function, which is determined by $\mu(dh) = \Delta(h)v(dh)$).

2.4 NOTES

It seems superfluous to make any bibliographical notes on likelihood here, except perhaps to draw the historically interested reader's attention to the account by

Edwards (1974) of the origination of likelihood in Fisher's writings and of earlier related ideas.

The concepts of universality and plausibility were introduced in Barndorff-Nielsen (1973a) and (1976b) respectively, and the material of the present chapter has been taken largely from the latter paper.

The inverse of the mode mapping $\check{x}$, which to any observed x assigns the values of ω such that $p(\cdot;\omega)$ has x as mode, was propounded as an estimator of ω by Höglund (1974). This estimator was termed the exact estimator by Höglund and the maximum ordinate estimator in Barndorff-Nielsen (1976b). It may be noted that when the maximum ordinate estimate exists then it equals the maximum plausibility estimate. Thus, in particular, for universal families the maximum ordinate and the maximum plausibility estimators are identical.

CHAPTER 3

Sample-Hypothesis Duality and Lods Functions

Statistical models $(\mathfrak{X}, p(\cdot;\cdot), \Omega)$ have a *sample aspect* and a *hypothesis aspect*. When considering the sample aspect one thinks of the family $\{p(\cdot;\omega): \omega \in \Omega\}$ of probability functions on $\mathfrak{X}$ and of the probabilistic properties of x embodied in this family. The hypothesis aspect concerns the evidence on ω contained in the various possible observations x, and embodiments of this evidence are given by the family $\{p(x;\cdot): x \in \mathfrak{X}\}$ of likelihood functions and the family $\{p(x;\cdot)/\sup_x p(x;\cdot): x \in \mathfrak{X}\}$ of plausibility functions.

There is a certain degree of duality between the sample aspect and the hypothesis aspect in that various notions and results in either aspect have, partly or completely, analogous counterparts in the other. This is the *sample-hypothesis duality*.

To take a primitive example, one is interested in the 'position' and 'shape' of likelihood and plausibility functions as well as of probability functions, and the position, for instance, is often indicated by the maximum point or set of maximum points for the function in question. Another primitive example is provided by the vector-valued functions on $\mathfrak{X}$ and Ω, i.e. respectively the statistics and the subparameters. The indicator functions, in particular, correspond to events and hypotheses.

More interestingly, stochastic independence has a significant analogue in the concept of likelihood independence which is discussed in Section 3.3.

The extent of the duality varies in some measure with the type of model function. For linear exponential model functions

$$p(x;\omega) = a(\omega)\, b(x)\, \mathrm{e}^{\omega \cdot x}$$

the duality is particularly rich in structure, as will become apparent in Part III. (Obviously, this form of model function in itself strongly invites mathematical duality considerations. Moreover, it so happens that for exponential models the mathematical theory of convex duality, summarized in Part II, fits closely with the sample-hypothesis duality.)

A study of how far the sample-hypothesis duality goes, or could be brought to go, at a basic, axiomatic level was made by Barnard (1949); see also Barnard (1972a, 1974a). A brief account of the theory he developed is given in Section 3.1, with the emphasis on his concept of lods functions, particular instances of which are log-probability functions, and log-likelihood and log-plausibility functions.

By a suitable combination of lods functions it is possible to obtain various types of prediction functions, as will be discussed in Section 3.2. These prediction functions have a role in predictive inference which is similar to the role of log-likelihood and log-plausibility functions in parametric inference.

3.1 LODS FUNCTIONS

Barnard (1949) established an axiomatic system which he proposed as a basic part of statistical inference theory; see also Barnard (1972a, 1974a). The notions and basic properties of probability (and likelihood) are not presupposed for this system, and the addition and multiplication rules satisfied by probabilities are not used as axioms, but are derived at a fairly late stage of the development. In the earlier stages the so-called lods functions are introduced and studied. Log-probability, log-likelihood, and log-plausibility functions are the three substantial examples of this kind of function. It is with the theory of lods functions that we shall be concerned here.

The theory deals with pairs of sets, $\mathfrak{X}$ and Ω say, and with certain types of connections between $\mathfrak{X}$ and Ω. $\mathfrak{X}$ is to be thought of as the set of possible outcomes of an experiment, while Ω is the set of hypotheses about the experiment. Each hypothesis is supposed to determine a complete ordering of the points of $\mathfrak{X}$ and, dually, each outcome of the experiment determines a complete ordering of Ω. The extra-mathematical meaning of these orderings is that they give the ranking of the points of $\mathfrak{X}$, respectively Ω, according to how 'likely' or 'plausible' the points are under the hypothesis, respectively outcome, in question. The supposition that the outcomes determine orderings amounts, in the words of Barnard (1949), 'to assuming that a theory of inductive inference is possible'. Let the orderings induced by the hypotheses and the outcomes be called *f-orderings* and *b-orderings*, respectively. (f and b stand for forward and backward.)

The ideas of independent experiments and conjunctions of such experiments together with their respective hypothesis sets are reflected in the theory as a number of axioms which specify simple consistency relations between the f-orderings of experiments and their independent conjunctions as well as the exact same consistency relations for the b-orderings.

Adding two dual (Archimedean) axioms for the orderings makes it possible to show that to each pair $\mathfrak{X}$, Ω there exist real-valued functions $f(x;\omega)$ and $b(\omega;x)$ of $x \in \mathfrak{X}$, $\omega \in \Omega$ such that the f-ordering determined by any ω is the same as the ordering of $\mathfrak{X}$ induced by $f(\cdot;\omega)$, and similarly for b. The function $f(\cdot;\omega)$ on $\mathfrak{X}$ is termed an *f-lods function* while $b(x;\cdot)$ is termed a *b-lods function*. Any such function is a *lods function*. Furthermore, f- and b-lods functions for the conjunction of a number of independent experiments are obtained by addition of the f- and b-lods functions of the component experiments.

Finally, an assumption is introduced which relates the two kinds of orderings. In effect (see Barnard 1949) it is equivalent to the requirement that there exist functions d on $\mathfrak{X}$ and h on Ω such that

$$f(x;\omega) + h(\omega) = b(\omega;x) + d(x).$$

This crucial relation, called the *inversion formula*, is the bridge which indicates how evidential rankings of the hypotheses are induced by the observed outcome x and by the family of f-functions given by the hypotheses.

The theory is exemplified, of course, by taking a family

$$\mathfrak{p} = \{p(\cdot;\omega): \omega \in \Omega\}$$

of probability functions on $\mathfrak{X}$ and setting

$$f(x;\omega) = \ln p(x;\omega) = b(\omega;x)$$

whence

$$h(\omega) = d(x) = 0.$$

The b-lods functions are then the log-likelihood functions determined by $\mathfrak{p}$. Indeed, excepting situations where ω is considered as having a prior distribution, it may fairly be said that up till recently this was the only real example, although the possibility of choosing $h(\omega)$ different from 0, with the accompanying change of $b(\omega;x)$, was touched upon in Barnard (1949, 1972a). However, another example is now furnished by plausibility inference through the choices

$$f(x;\omega) = \ln p(x;\omega),$$

$$b(\omega;x) = \ln\{p(x;\omega)/\sup_x p(x;\omega)\},$$

and

$$h(\omega) = -\ln \sup_x p(x;\omega), \qquad d(x) = 0.$$

Two lods functions are considered as equivalent if they are equal up to an additive constant. This accords with the standard practice of disregarding factors of a likelihood function which depend on the observations only. For purposes of comparison, etc. of non-equivalent lods functions it is convenient to introduce a norming of lods functions, i.e. to select a representative from each equivalence class in some suitable way. Following the usual manner in which likelihood functions are normed, a lods function will be said to be *normed* if its supremum equals 0.

If f is an f-lods function and b is a b-lods function then $F = \exp f$ is called an *f-ods function* and $B = \exp b$ is called a *b-ods function*. The ods functions F and B are normed if they have supremum 1.

When a given family of f-lods functions is to be inverted into a family of b-lods functions, two elemental procedures, which do not require introduction of acceptability functions that would have to be motivated by further principles, are possible: direct inversion, or norming and then direct inversion (where by direct inversion is meant an application of the inversion formula with $h(\omega) = d(x) = 0$). Taking $\{\ln p(\cdot; \omega): \omega \in \Omega\}$ as the family of f-lods functions, these two procedures yield the families of, respectively, log-likelihood and log-plausibility functions.

Let $f(x; \omega)$ be taken as $\ln p(x; \omega)$, let $h(\omega)$ and $d(x)$ be arbitrary, and suppose that different values of ω correspond to different probability measures on $\mathfrak{X}$. Then different values of ω do also determine different and non-equivalent f-lods functions but, in general, different x values may give equivalent b-lods functions. Here then is a lack of duality which in itself directs the attention to those cases where no two members of the family of b-lods functions are equivalent. This latter condition means that x is *minimal sufficient*. To see this, consider the partition of $\mathfrak{X}$ generated by the family of b-lods functions, i.e. the partition for which two points x and $\tilde{x}$ of $\mathfrak{X}$ belong to the same element of the partition if and only if $b(\cdot; x)$ is equivalent to $b(\cdot; \tilde{x})$. This partition is the same whatever the choice of the functions $h(\omega)$ and $d(x)$, and hence equals the likelihood function partition of $\mathfrak{X}$ which is minimal sufficient (cf. Section 4.2).

Given a family $\{b(\cdot; x): x \in \mathfrak{X}\}$ of b-lods functions let us define the *maximum b-lods estimate* of ω based on the observed outcome x as the set

$$\tilde{\omega} = \{\omega: b(\omega; x) = \sup_{\omega} b(\omega; x)\}.$$

It is clear that this procedure, which encompasses maximum likelihood and maximum plausibility estimation, has the property that the operations of estimation and reparametrization are interchangeable.

One tends to think of lods functions—and in particular of probability, likelihood, or plausibility functions, or their logarithms—as having characteristic locations (or positions) and shapes. In many concrete situations this makes good sense, and is useful in summarily describing such a function and in classifying, sometimes only in a rather rough sense, the members of a family of such functions into subfamilies.

It is often natural to indicate the position of the function by specifying that or those arguments for which the function takes its maximum. Lods functions which are quasi-concave or concave have a simple shape, and use of the maximum points as location indicators is particularly natural with such functions. Quasi-concave and especially concave log-probability, log-likelihood, and log-plausibility functions occur frequently in statistics and will play a rather prominent role in the present treatise.

A lods function of the form

$$\omega \cdot x - h(\omega) - d(x), \tag{1}$$

where $\omega \in \Omega \subset R^k$, $x \in \mathfrak{X} \subset R^k$, will be designated as *linear*. The log-probability, log-likelihood, and log-plausibility functions of an exponential model

$$p(x; \omega) = a(\omega)b(x)\, e^{\omega \cdot x}$$

are all of this type.

Clearly, if a b-lods function is linear and given by (1) then, under smoothness assumptions, the maximum b-lods estimate of ω is determined by the equation

$$Dh(\tilde{\omega}) = x.$$

Furthermore, concavity of a linear f- or b-lods function is equivalent to convexity of, respectively, d and h.

3.2 PREDICTION FUNCTIONS

Let x denote the outcome of a performed experiment and y the outcome of contemplated, independent experiment, which may be of a different type, and suppose the two experiments relate to one and the same set of hypotheses, indexed by the parameter ω. The domains of variation of x, y, and ω are denoted by $\mathfrak{X}$, $\mathfrak{Y}$, and Ω. A *prediction function* for y based on x is a non-negative function $\vec{C}(\cdot|x)$ on $\mathfrak{Y}$ which is interpreted as expressing how credible (or likely or plausible) the various possible outcomes of the contemplated experiment are relative to each other, in the light of the observation x. Thus the interpretation of prediction functions is similar to that of f-ods or b-ods functions.

Suppose that $B(\omega; x)$ is a b-ods function for ω based on x and that $F(y; \omega)$ is an f-ods function for y based on ω. The product $B(\omega; x)\bar{F}(y; \omega)$, where $\bar{F}$ is the normed version of F, may be viewed as the joint credibility of ω and y, and it is thus an immediate ideà to consider

$$\overrightarrow{BF}(y|x) = \sup_{\omega} B(\omega; x)\bar{F}(y; \omega)$$

as a prediction function for y.

Now, let $p(x; \omega)$ be a model function on $\mathfrak{X} \times \Omega$ and $p(y; \omega)$ a model function on $\mathfrak{Y} \times \Omega$, and take $F(y; \omega)$ to be equal to $p(y; \omega)$. Then

$$\bar{F}(y; \omega) = \Pi(\omega; y),$$

and choosing $B(\omega; x) = L(\omega; x)$ one obtains $\overrightarrow{BF} = \vec{L}$, say, where

$$\vec{L}(y|x) = \sup_{\omega} L(\omega; x)\Pi(\omega; y), \tag{1}$$

which will be called the *likelihood prediction function.* Similarly, with the choice $B(\omega : x) = \Pi(\omega ; x)$ the prediction function $\overrightarrow{BF}$ is denoted by $\vec{\Pi}$ and is called the *plausibility prediction function*, and one has

$$\vec{\Pi}(y|x) = \sup_{\omega} \Pi(\omega; x)\Pi(\omega; y). \tag{2}$$

The set of points ω for which the supremum on the right hand side of (1) is attained will be denoted by $\hat{\vec{\omega}}$ or $\hat{\vec{\omega}}\,(y|x)$, and the *maximum likelihood predictor* is the function $\hat{\vec{y}}$ on $\mathfrak{X}$ such that $\vec{y}(x)$ is the set of points y which maximizes $\vec{L}(\cdot|x)$. The symbols $\check{\vec{\omega}}$, $\check{\vec{\omega}}(y|x)$ and $\check{\vec{y}}$—the *maximum plausibility predictor*—are defined similarly in the plausibility case, i.e. (2).

Example 3.1. Suppose x and y are binomially distributed for trial numbers m and n, and with common probability parameter $\pi \in (0, 1)$.

The likelihood prediction function may be written

$$\vec{L}(y|x) = \binom{m}{x}\left(\frac{\hat{x}}{m}\right)^{\hat{x}}\left(1 - \frac{\hat{x}}{m}\right)^{m-\hat{x}}\binom{n}{y}\Big/\binom{n}{\check{y}} \tag{3}$$

where $\hat{x}$ is a mean value and $\check{y}$ a mode point, corresponding to a common π and determined so that $\hat{x} + \check{y} = x + y$; this is provided such π, $\hat{x}$, and $\check{y}$ exist. (If they do not exist the expression (3) still holds good in a generalized sense.) Furthermore, the plausibility prediction function is

$$\vec{\Pi}(y|x) = \binom{m}{x}\binom{n}{y}\Big/\binom{m}{\check{x}}\binom{n}{\check{y}} \tag{4}$$

where $\check{x}$ and $\check{y}$ are mode points corresponding to a common π and such that $\check{x} + \check{y} = x + y$. Formulas (3) and (4) follow simply from general results given in Section 9.7.

For $n = 1$ one finds

$$\vec{L}(1|x)/\vec{L}(0|x) = \begin{cases} \{(x+1)^{x+1}(m-x-1)^{m-x-1}\}/\{x^x(m-x)^{m-x}\} & \text{if } x \leq \tfrac{1}{2}m - 1 \\ (m/2)^m/\{x^x(m-x)^{m-x}\} & \text{if } x = \tfrac{1}{2}m - \tfrac{1}{2} \\ 1 & \text{if } x = \tfrac{1}{2}m \\ (m/2)^m/\{(x-1)^{x-1}(m-x+1)^{m-x+1}\} & \text{if } x = \tfrac{1}{2}m + \tfrac{1}{2} \\ \{x^x(m-x)^{m-x}\}/\{(x-1)^{x-1}(m-x+1)^{m-x+1}\} & \text{if } x \geq \tfrac{1}{2}m + 1 \end{cases}$$

and

$$\vec{\Pi}(1|x)/\vec{\Pi}(0|x) = \begin{cases} (x+1)/(m-x) & \text{for } x < [\tfrac{1}{2}(m+1)\} \\ 1 & \text{for } x = [\tfrac{1}{2}(m+1)\} \\ x/(m-x+1) & \text{for } x > [\tfrac{1}{2}(m+1)\} \end{cases}$$

▶

Suppose that the maximum likelihood estimate $\hat{\omega} = \hat{\omega}(x)$ of ω based on x exists and that the probability function $p(y; \omega)$ has a mode $\check{y}(\omega)$ whatever the value of ω in Ω. Then, since plausibilities are less than or equal to 1,

$$\begin{aligned} \vec{L}(y|x) &= \sup_{\omega} L(\omega; x)\Pi(\omega; y) \\ &\leq \sup_{\omega} L(\omega; x) \\ &= L(\hat{\omega}; x) \\ &= L(\hat{\omega}; x)\Pi(\hat{\omega}; \check{y}(\hat{\omega})) \end{aligned}$$

and hence

$$\check{y}(\hat{\omega}(x)) \subset \hat{\vec{y}}(x). \tag{5}$$

The same kind of argument shows that

$$\check{y}(\check{\omega}(\hat{x})) \subset \check{\vec{y}}(x). \tag{6}$$

It is obvious that when $\hat{\omega}(x)$ and $\check{\omega}(x)$ exist the inclusions in (5) and (6) are, in fact, equalities except in rather pathological instances, i.e. as a rule

$$\vec{y}(x) = \check{y}(\hat{\omega}(x))$$

and

$$\vec{y}(x) = \check{y}(\check{\omega}(x)).$$

Thus these predictates are simply the sets of mode points corresponding to the respective estimates of ω based on x.

Finally, it may be noted that if the outcome y is to be predicted but without x having been observed then it is natural to define the likelihood and plausibility prediction functions as

$$\vec{L}(y) = \vec{\Pi}(y) = \sup_{\omega} \Pi(\omega; y).$$

In the case when the family of probability functions for y is universal $\vec{L}(\cdot)$ and $\vec{\Pi}(\cdot)$ are both constant (and equal to 1).

Suppose, for example, that a coin is to be thrown n times and no information is available about the probability π that it turns up heads in any single throw, and that the aim is to predict the number of heads y. Since the distribution of y is universal, any of the possible values of y are held equally credible by likelihood prediction as well as by plausibility prediction. The same conclusion is true if the outcome y to be predicted is the actual sequence of heads and tails.

3.3 INDEPENDENCE

Let $w_1, \ldots, w_m$ be a collection of variables and let $M_1, \ldots, M_m$ denote their domains of variation. If the domain of variation M, say, of the combined variable $w = (w_1, \ldots, w_m)$ is equal to the product $M_1 \times \cdots \times M_m$ of the single domains then $w_1, \ldots, w_m$ are called *variation independent*. Furthermore, let f be a non-negative function defined on M. Then $w_1, \ldots, w_m$ are said to be *independent under f* provided

(i) $w_1, \ldots, w_m$ are variation independent.

(ii) There exist (non-negative) functions $f_1, \ldots, f_m$ defined, respectively, on $M_1, \ldots, M_m$ such that $f(w) = f_1(w_1) \ldots f_m(w_m)$, $w = (w_1, \ldots, w_m) \in M$.

The definition of independence under a function f covers the classical definition of stochastic independence of random variables in terms of densities, as well as the definition of independence of σ-algebras under a probability measure. Dual to the former is the following definition of *likelihood independence*, or, for short, *L-independence*. Let $p(x; \omega)$ be a model function and let $(\omega^{(1)}, \ldots, \omega^{(m)})$ be a partition of the parameter vector ω. Then $\omega^{(1)}, \ldots, \omega^{(m)}$ are said to be *L-independent at* $x (\in \mathfrak{X})$ provided they are independent under $p(x; \cdot)$ and, furthermore, $\omega^{(1)}, \ldots, \omega^{(m)}$ are called *L-independent* if they are L-independent at x for every $x \in \mathfrak{X}$. (Obviously, one might also introduce a concept of *plausibility independence* or *Π-independence*. However, Π-independence seems to be so rare a property as to be of no real interest.)

The remainder of this section consists of some comments on the use of L-independence, and of a number of examples of this property.

Example 3.2. Let $\mathfrak{P}$ be the family of k-dimensional multinomial distributions with fixed trial parameter n. The model function is

$$\frac{n!}{x_1! \ldots x_k!(n - x.)!} \pi_1^{x_1} \ldots \pi_k^{x_k} (1 - \pi_1 - \cdots - \pi_k)^{n - x_1 - \cdots - x_k} \tag{1}$$

and the domain of variation of the parameter $\pi = (\pi_1, \ldots, \pi_k)$ is the k-dimensional simplex ▶

$$\Pi = \{\pi : \pi_1 > 0, \ldots, \pi_k > 0, \pi_1 + \cdots + \pi_k < 1\}. \tag{2}$$

The set of equations

$$\omega_1 = \pi_1$$

$$\omega_2 = \frac{\pi_2}{1 - \pi_1}$$

$$\vdots$$

$$\omega_k = \frac{\pi_k}{1 - \pi_1 - \cdots - \pi_{k-1}}$$

determines a reparametrization of $\mathfrak{P}$ and $\omega = (\omega_1, \ldots, \omega_k)$ has domain of variation

$$\Omega = (0, 1)^k.$$

Moreover, (1) can be written

$$b(x_1; n, \omega_1) b(x_2; n - x_1; \omega_2) \ldots b(x_k; n - x_1 - \cdots - x_{k-1}, \omega_k)$$

where $b(z; v, \alpha)$ denotes the point probability at z for the binomial distribution with trial parameter v and probability parameter α.

Thus $\omega_1, \ldots, \omega_k$ are L-independent. ▶

A main consequence of the property of L-independence is that it simplifies the handling and visualization of a likelihood function. In particular, the problem of finding the maximum likelihood estimate of ω falls into m separate pieces if $\omega^{(1)}, \ldots, \omega^{(m)}$ are L-independent.

Note also that if prior likelihood in the sense of Edwards (1969) is available and if the components $\omega^{(1)}, \ldots, \omega^{(m)}$ are likelihood independent both under the prior likelihood and under the likelihood provided by the actual data then they are also independent *a posteriori*. Also, if $\omega^{(1)}, \ldots, \omega^{(m)}$ are independent at x and if they follow an *a priori* probability distribution under which they are (stochastically) independent, then they are independent too under the *a posteriori* distribution.

A special, close connection between certain cases of stochastic independence and L-independence, in exponential families, will be discussed in Section 9.2.

Besides the examples of L-independence mentioned here some instances may be found in the discussions of cuts and S-ancillarity to be given in Sections 4.4, 10.2, and 10.3.

Example 3.3. Other types of factorization of the likelihood function for the multinomial family than that mentioned in Example 3.2 are possible. For instance, when $k = 3$ the expression (1) can be recast as

$$b(x_1 + x_2; n, \alpha)\, b(x_1; x_1 + x_2, \beta)\, b(x_3; n - x_1 - x_2, \gamma)$$

where $\alpha = \pi_1 + \pi_2$, $\beta = \pi_1/(\pi_1 + \pi_2)$, $\gamma = \pi_3/(1 - \pi_1 - \pi_2)$, and (α, β, γ) varies in $(0, 1)^3$. ▶

Example 3.4. Let x_1 and x_2 be independent and Poisson distributed with mean values λ_1 and λ_2. The distribution of $x_.$ is Poisson with mean value $\lambda_.$, while conditionally on $x_.$ the variate x_1 follows the binomial distribution having trial number $x_.$ and probability parameter $\chi = \lambda_1/\lambda_.$. Thus $\lambda_.$ and χ are L-independent provided they are variation independent, which is the case not only if $\lambda = (\lambda_1, \lambda_2)$ varies freely in $(0, \infty)^2$ but also if, for instance, λ_2 is known to be less than or equal to λ_1, since then $(\lambda_., \chi)$ has domain of variation $(0, \infty) \times [\frac{1}{2}, 1)$

A concrete example of an experiment for which the latter model with $\lambda_2 \le \lambda_1$ could well be appropriate is that of readings of a Geiger counter, without and with a piece of material inserted between the counter and the radioactive source. ▶

Example 3.5. Consider a finite collection $\{x_i : i \in I\}$ of independent Poisson variates, let $I_1, \ldots, I_r$ be the elements of a partition of I and set

$$x_{\cdot}^{(k)} = \sum_{i \in I_k} x_i, \qquad k = 1, 2, \ldots, r.$$

From Examples 3.2, 3.3, and 3.4, one sees that the joint distribution of the x_i may be split into the marginal (Poisson) distribution of $x_{\cdot}$, the conditional (multinomial) distribution of $x_{\cdot}^{(\cdot)}$ given $x_{\cdot}$, and the product over k of the conditional (multinomial) distribution of $\{x_i : i \in I_k\}$ given $x_{\cdot}^{(k)}$, and that corresponding to this factorization one has L-independent subparameters.

The unconditional model for $\{x_i : i \in I\}$, the conditional model given $x_{\cdot}$, and the conditional model given $x_{\cdot}^{(\cdot)}$ are, of course, the three base models for contingency table analysis. ▶

Example 3.6. The negative multinomial distribution of order k and with shape parameter χ and probability parameter $\pi = (\pi_1, \ldots, \pi_k)$ is the distribution on N_0^k, where $N_0 = \{0, 1, \ldots\}$, having point probabilities

$$\frac{\Gamma(\chi + x_{\cdot})}{\Gamma(x_1 + 1) \ldots \Gamma(x_k + 1)\Gamma(\chi)} \pi_1^{x_1} \ldots \pi_k^{x_k} (1 - \pi_{\cdot})^{\chi}.$$

The domain of (χ, π) is $(0, \infty) \times \Pi$ where Π is given by (2).

Several types of factorizations are possible here; consider, for instance, the case $k = 2$. Let $b^-(\cdot; \cdot, \cdot)$ denote the model function for the negative binomial distribution. Conditioning on x_1 yields the factorization

$$b^-(x_1; \chi, \rho) b^-(x_2; \chi + x_1, \pi_2)$$

where $\rho = \pi_1/(1 - \pi_2)$. And conditioning on $x_{\cdot}$ gives

$$b^-(x_{\cdot}; \chi; \pi_{\cdot}) b(x_1; x_{\cdot}, \sigma)$$

where $\sigma = \pi_1/\pi_{\cdot}$. ▶

Example 3.7. Suppose $x_1, \ldots, x_n$ is a random sample from the r-dimensional normal distribution $N_r(\xi, \Sigma)$ and let $\mathfrak{P}$ be the class of probability measures of $x = (x_1, \ldots, x_n)$ for (ξ, Σ) varying freely, Σ being nonsingular. Split ξ into two components, $\xi = (\xi^{(1)}, \xi^{(2)})$, of dimensions q and $r - q$, respectively, and partition Σ correspondingly. Then $\omega^{(1)} = (\xi^{(1)}, \Sigma_{11})$ and

$$\omega^{(2)} = (\xi^{(2)} - \xi^{(1)} \Sigma_{11}^{-1} \Sigma_{12}, \Sigma_{22} - \Sigma_{21} \Sigma_{11}^{-1} \Sigma_{12}, \Sigma_{11}^{-1} \Sigma_{12})$$

are independent, (i.e. the parameters of respectively the marginal distribution of $x_i^{(1)}$ and the conditional distribution of $x_i^{(2)}$ given $x_i^{(1)}$ are L-independent. (The variation independence of $\omega^{(1)}$ and $\omega^{(2)}$ is not difficult to verify directly, but may also be obtained as an immediate consequence of Theorem 9.3, see Example 9.5). ▶

Example 3.8. In a medical study, the l persons under observation entered the study at individual times $t_1 < \cdots < t_l$ but were thence continuously monitored till a time $T(> t_l)$. All the persons were fit at the time of entrance but might, during the observation period, interchangeably be in this state and a state of disabledness, and might pass from each of these states to that of death. Supposing that the time–state records for the various persons are a set of l independent observations of a Markov process with transition intensities between states as indicated in the diagram

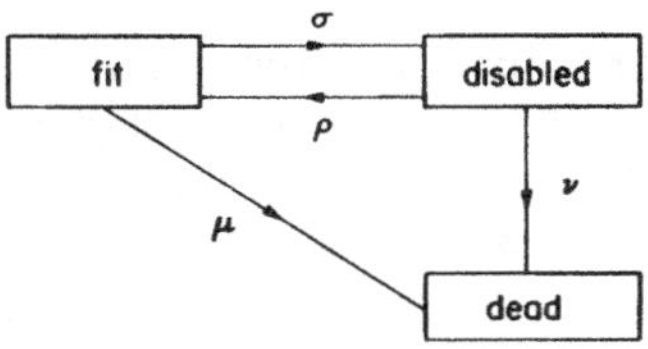

then the likelihood function becomes

$$\mu^{m_\cdot}\cdot\sigma^{s_\cdot}\cdot\nu^{n_\cdot}\cdot\rho^{r_\cdot}\cdot e^{-(\mu+\sigma)v_\cdot-(\nu+\rho)w_\cdot} \tag{3}$$

where

$m_\cdot =$ number of deaths among fit persons

$s_\cdot =$ number of disablements

$n_\cdot =$ number of deaths among disabled persons

$r_\cdot =$ number of recoveries

$v_\cdot =$ total time lived in fit state

$w_\cdot =$ total time lived in disabled state.

This model was studied by Sverdrup (1965).

It follows from (3) that the parameters μ, σ, ν, and ρ are L-independent, on the assumption that (μ, σ, ν, ρ) varies in $(0, \infty)^4$.

Moreover, the parameters $\mu + \sigma$, $\sigma/(\mu + \sigma)$, $\nu + \rho$, and $\rho/(\nu + \rho)$ are L-independent.

Similar results hold for the birth and death process. Denote the birth and death intensities by λ and μ. Then the likelihood function, based on continuous observation during a time interval $[0, T]$ of a population consisting initially of l individuals, is given by

$$\lambda^b \mu^d e^{-(\lambda+\mu)z}$$

where

$b =$ number of births

$d =$ number of deaths

$z =$ total time lived.

Hence λ and μ are L-independent, and so are $\lambda + \mu$ and $\lambda/(\lambda + \mu)$.

Indeed, it is obvious from these cases that analogous conclusions hold generally for Markov processes with continuous time and finite state space.

3.4 COMPLEMENTS

(i) The shapes of the lods functions of a given family of such functions can sometimes be made simpler or more uniform through transformation to another argument variable. The normalizing and the variance-stabilizing transformations of probability functions which have long been of standard use in statistics are of this kind, and more recently analogous transformations have been introduced for (log-) likelihood functions (see Anscombe 1964a, Sprott 1973, and Box and Tiao 1973). A treatment of these two particular types of transformations—*normalizing* and *spread-stabilizing*—will be given, for one-dimensional exponential models, in Section 9.8(**v**).

(ii) *Approximate L-independence.* Consider a model function $p(x;\omega)$ and a partition $(\omega^{(1)}, \ldots, \omega^{(m)})$ of ω. In cases where $\omega^{(1)}, \ldots, \omega^{(m)}$ are not L-independent it may still be, of course, that the likelihood function $p(x);\cdot)$ factorizes either exactly or approximately in a neighbourhood of the maximum likelihood estimate. Such a property is often helpful in the statistical analysis, both conceptually and with respect to the numerical and graphical handling of the likelihood function.

The following kind of approximate factorization of $p(x;\cdot)$ is of particular interest. Let l denote the log-likelihood function,

$$l(\cdot) = \ln p(x;\cdot).$$

The components $\omega^{(1)}, \ldots, \omega^{(m)}$ are said to be *infinitesimally L-independent* at x provided the likelihood function has a unique maximum point $\hat{\omega}$ which belongs to the interior of Ω and provided

$$\frac{\partial^2 l}{\partial\omega^{(i)\prime}, \partial\omega^{(j)}}(\hat{\omega}) = 0, \qquad i \neq j. \tag{1}$$

Anscombe (1961, 1964a,b) has discussed the possibilities of obtaining infinitesimal L-independence, as well as approximate normal shape, of the likelihood function by transformation of the parameters (see Sections 9.8(**v**) and (**vi**)).

Infinitesimal L-independence is closely related to the concept of orthogonality of parameters introduced by Jeffreys (1948).

The parameters $\omega^{(1)}, \ldots, \omega^{(m)}$ are said to be *orthogonal* under P_ω if

$$E_\omega \frac{\partial^2 l}{\partial\omega^{(i)\prime}, \partial\omega^{(j)}} = 0, \qquad i \neq j. \tag{2}$$

Under standard regularity assumptions this is equivalent to asymptotic stochastic independence of the components $\hat{\omega}^{(1)}, \ldots, \hat{\omega}^{(m)}$ of the maximum likelihood estimate $\hat{\omega}$ in the asymptotic normal distribution of $\hat{\omega}$.

If the dimension of ω is k then the problem of finding a transformation of ω such that the k one-dimensional components of the new parameter are orthogonal amounts to solving $\binom{k}{2}$ partial differential equations in k unknown functions. As mentioned by Huzurbazar (1950), it is therefore to be expected that when $k > 3$ the problem will be solvable only for very special families of distributions. For $k = 3$ and especially for $k = 2$ there is better hope of solving the equations. Huzurbazar (1950, 1956) indicated a way of doing this and applied it to a number of two-parameter distributions.

Clearly L-independence implies both infinitesimal L-independence and orthogonality. Note also that orthogonality in many cases implies approximate infinitesimal L-independence and *vice versa*, because for large sample sizes the left hand side of (1) tends to be near the left hand side of (2). For exponential families the notions of orthogonality and infinitesimal L-independence are even more closely related (see Section 9.8(vi)).

3.5 NOTES

The method of constructing prediction functions described in Section 3.2 was first proposed in Barndorff-Nielsen (1976b) and the two particular cases of plausibility and likelihood prediction have been briefly discussed there and in Barndorff-Nielsen (1977b). The general method was prompted by (4) of Section 3.2, which was originally derived as a formal plausibility analogue of Fisher's (1956) (likelihood based) prediction function for binomial experiments. More specifically, the two latter prediction functions are equal to, respectively, the plausibility ratio and the likelihood ratio test statistics for the hypothesis that the probability parameter is the same in the two experiments. The reasoning leading to likelihood and plausibility prediction functions, which is given in Section 3.3, is of a different kind, and it appears that Fisher's prediction function is not derivable as a special case of the general construction discussed in that section. For binomial experiments, both the likelihood and the plausibility prediction method of Section 3.2 have the property that if the sample size m of the observed experiment tends to infinity and the probability parameter π thus becomes known then in the limit the prediction function is proportional to the probability function for the unobserved experiment; this property is not shared by Fisher's method. Mathiasen (1977) has given a detailed study and comparison of exact and asymptotic properties of four particular prediction procedures, including plausibility prediction and the generalization of Fisher's proposal for binomial experiments. References to other approaches to prediction may be found in Mathiasen's paper.

CHAPTER 4

Logic of Inferential Separation. Ancillarity and Sufficiency

The operations of margining to a sufficient statistic and conditioning on an ancillary statistic are the primary procedures leading to separate inference, i.e. inference on a parameter of interest based on only a part of the original model and data. The logic relating to inferential separation and in particular to the general, intuitive notions of ancillarity and sufficiency is discussed in Section 4.1, and then various, mathematically defined concepts of ancillarity and sufficiency are studied in Sections 4.2, 4.4, and 4.5. A key problem in this connection is that of giving precise meaning to the notion that a certain part of the model and data does not contain any information with respect to the parameter of interest, and definitions of this notion, which is called nonformation, are given in Section 4.3. Finally, Section 4.6 contains some results on the relations between conditional and unconditional plausibility functions.

4.1 ON INFERENTIAL SEPARATION. ANCILLARITY AND SUFFICIENCY

Let u and v be statistics, let ψ be a (sub)parameter with variation domain Ψ and suppose that the conditional distribution of u given v depends on ω through ψ only, and is in fact parametrized by ψ. Making inference on ψ from u and from the conditional model for u given the observed value of v is an act of *separate inference*. In such a connection it is customary to speak of ψ as the *parameter of interest*. That part of ω which is complementary to ψ will be termed *incidental*. (The more commonly used, but somewhat emotional, term *nuisance* for this complementary part is avoided here.)

It may or it may not be the case that, in the given context, that part of the data x and the model $\mathfrak{P}$ which is complementary to the above-mentioned model–data basis for the separate inference can be considered as irrelevant or containing no available information with respect to ψ. If it is the case then the complementary part will be called *nonformative* with respect to ψ.

On the general principle that if something is irrelevant for a given problem then one should effect (if possible) that it does not influence the solution of the problem, one is led to conclude that it is proper to draw the inference on ψ separately provided that, as discussed above, the complementary part is nonformative. This conclusion, which may be called the ***principle of nonformation***, specializes to the *principle of ancillarity* by taking $u = x$ (the identity mapping on $\mathfrak{X}$), and to the *principle of sufficiency* by taking v constant. A detailed discussion of these two latter principles is given in the following.

Principle of Ancillarity. Suppose that, for some statistic t and some parameter ψ,

(i) The conditional model for x given the statistic t is parametrized by ψ.
(ii) The conjunction $(\mathfrak{P}_t, t)$ is nonformative with respect to ψ.

Then inference on ψ should be performed from the conjunction $(\mathfrak{P}(\cdot|t), x)$. ▶

When (i) and (ii) above are satisfied T is said to be *ancillary* with respect to ψ.

In comprehending the ancillarity principle it is essential to think of the experiment, E say, with outcome x as a *mixture experiment*, i.e. as being composed of two consecutive experiments, the first, E_t, having outcome t and the second, E^t, corresponding to the sample space $\mathfrak{X}_t = \{x: t(x) = t\}$ and the distribution family $\mathfrak{P}(\cdot|t)$. Under the conditions set out in the principle, the first experiment does not in itself give any information on the value of ψ and the second experiment yields information on the value of ψ only and thus does not provide possibility for, as it were, drawing inference within the subfamilies $\mathfrak{P}_\psi = \{P: \psi(P) = \psi\}$, $\psi \in \Psi$. One may judge, therefore, that if it is known that the experiment E^t has been performed and has led to the outcome x then the additional knowledge that the value t has been determined by the experiment E_t is irrelevant as far as inference on ψ is concerned. The direct way to ensure that this irrelevant knowledge does not influence the inference concerning ψ is to base the inference on $(\mathfrak{P}(\cdot|t), x)$, as the principle prescribes.

In many cases the precision of the inference on ψ which the (conditional) experiment E^t allows varies in a simple systematic way with the value t of the ancillary statistic, so that this value has an immediate interpretation as an index of precision. Thus t has a function analogous to that of the sample size. According to the principle of ancillarity, the appropriate precision to report is that indicated by t, not the precision relative to repetitions of the whole experiment E. These aspects were repeatedly stressed by Fisher (see, e.g., Fisher 1935, 1956). To illustrate them in the simplest possible, though somewhat unrealistic, setting, suppose an experimenter chooses between two measuring instruments by throwing a fair coin, it being known that these instruments yield normally distributed observations with mean value equal to the parameter of interest, μ say, and standard deviations, respectively, 1 and 10. Whichever instrument is selected, only one measurement will be taken. Let $t = 0$ or 1 indicate the outcome of the coin throw and let u be the measurement result; thus $x = (t, u)$. In a hypothetical, infinite sequence of independent repetitions of the whole experiment the estimate u of μ would follow the distribution with probability function

$$\frac{1}{2}\left[\varphi(u-\mu)+\frac{1}{10}\varphi\left(\frac{u-\mu}{10}\right)\right] \tag{1}$$

where φ is the density of $N(0, 1)$. However, the precision in the estimation of μ provided by the one experiment actually performed is described not by (1) but by either $\varphi(u-\mu)$ or $\varphi[(u-\mu)/10]/10$, depending on which of the two instruments was in fact used. In other words, the precision is given by the conditional model given the observed value of the ancillary statistic t.

Principle of Sufficiency. Suppose that, for some statistic t and some parameter ψ,

(i) The marginal model for t is parametrized by ψ.
(ii) The conjunction $(\mathfrak{P}(\cdot|t), x)$ is nonformative with respect to ψ.

Then inference on ψ should be performed from the conjunction $(\mathfrak{P}_t, t)$. ▶

The statistic t is called *sufficient* with respect to ψ when both (i) and (ii) of the sufficiency principle hold.

As with the ancillarity principle, it is important for understanding the primitive content of the principle of sufficiency to think of x as being obtained by a mixture, or two-stage, experiment, where here it is the knowledge about the second experiment which is irrelevant for inference on ψ.

In any given situation it is necessary to make precise what is meant by nonformativeness before it can be decided whether the principle of nonformation applies. Various mathematical definitions of nonformation are given in Section 4.3. The simplest of these is *B-nonformation* which, in particular, yields the classical concepts of ancillarity and sufficiency. In the present book these classical concepts are called *B-ancillarity* and *B-sufficiency* while the words ancillarity and sufficiency are reserved for general indicative use. (A statistic t is B-ancillary if its distribution is the same for all $P \in \mathfrak{P}$, and t is B-sufficient if the conditional distribution of x given t is the same for all $P \in \mathfrak{P}$.) For practical expository reasons the well-known theory of B-ancillary and B-sufficient statistics is dealt with separately in the next section, whereas the ancillarity and sufficiency notions which flow from the other specifications of nonformation presented in Section 4.3, i.e. *S-*, *M- and G-nonformation*, will be treated in Section 4.4. Some indications of the roles of all these concepts are however given next through discussion of some key examples.

The earliest concrete example of what is here called a B-ancillary statistic was treated by Fisher (1934) in his first discussion of fiducial inference with respect to the location and scale parameters α and β of a continuous type, one-dimensional family of distributions, the density being

$$\frac{1}{\beta}f\left(\frac{x-\alpha}{\beta}\right)$$

where f is supposed to be known. For a sample of n observations $x_1, \ldots, x_n$ the statistic

$$\left(\frac{x_1 - x_3}{x_1 - x_2}, \frac{x_1 - x_4}{x_1 - x_2}, \ldots, \frac{x_1 - x_n}{x_1 - x_2}\right),$$

which Fisher talked of as specifying the configuration or complexion of the sample, is clearly B-ancillary.

Perhaps the most commonly occurring instance of a process of separate inference is that of making a linear regression analysis from n pairs of observations $(x_1,y_1), \ldots, (x_n,y_n)$ in the cases where the xs vary randomly but are considered as given in the analysis. Here $\psi = (\alpha, \beta, \sigma^2)$ where α and β are the position and slope parameters of the regression line and σ^2 is the residual variance. The separation logic in this situation is however seldom explicated (but see Fisher 1956, §4.3, and Sverdrup 1966). Under the usual assumptions of stochastic independence and identical distribution of $(x_1, y_1), \ldots, (x_n, y_n)$ (as well as normality of the conditional distribution of y given x), the marginal model for the xs together with the observation $(x_1, \ldots, x_n)$ is indeed nonformative with respect to ψ in one of the specific senses, that of S-nonformation, given in Section 4.3, on the proviso that for every value of $(\alpha, \beta, \sigma^2)$ the class of distributions of $(x_1, \ldots, x_n)$ is the same. In other words, under the condition stipulated the statistic $(x_1, \ldots, x_n)$ is S-ancillary with respect to $(\alpha, \beta, \sigma^2)$. The condition is satisfied, in particular, if the family of distributions of a single pair (x, y) is the family of two-dimensional normal distributions.

The regression situation also gives an obvious illustration of the point made earlier about the function of ancillary statistics in providing the relevant indication of the precision in inference on the parameter of interest.

One of the very first examples adduced by Fisher to illustrate his idea of ancillarity, and which has been of decisive importance for the development of separate inference, concerns the 2×2 table

x_1	$n_1 - x_1$	n_1
x_2	$n_2 - x_2$	n_2
$x_.$	$n_. - x_.$	$n_.$

for two independent binomial variates, having probability parameters p_1 and p_2, respectively (see Fisher 1935). Suppose the object is to test the hypothesis that the odds ratio

$$\psi = \frac{p_1}{1 - p_1} \bigg/ \frac{p_2}{1 - p_2} \tag{2}$$

has a particular value, ψ_0 say, the main possibility being, of course, $\psi_0 = 1$ which corresponds to $p_1 = p_2$. In relation hereto Fisher remarks:

> Let us blot out the contents of the table, leaving only the marginal frequencies. If it be admitted that these marginal frequencies by themselves supply no information on the point at issue,...we may...recognize that we are concerned only with the relative probabilities of occurrence of the different ways in which the table can be filled in, subject to these marginal frequencies.

From a remark in Fisher's reply in the discussion to his paper it is apparent that he was considering the marginal frequencies, i.e. in effect $x.$, as ancillary not just with respect to ψ_0 but with respect to ψ.

The conditional model given $x.$ is parametrized by ψ, and $x.$ is M-ancillary with respect to ψ.

Suppose now that the data consist of a sample of n observations from a multivariate normal distribution. whose parameters are assumed to vary unrestrictedly. Inference on the matrix $\boldsymbol{\rho}$ of correlation coefficients may be performed separately, from its empirical counterpart $\boldsymbol{r}$, without loss of information. In fact, the marginal distribution of $\boldsymbol{r}$ depends on $\boldsymbol{\rho}$ only and $\boldsymbol{r}$ is obtainable by sufficient reduction in two steps, first reducing to the set of empirical means, variances, and covariances by B-sufficiency and then to $\boldsymbol{r}$ by G- (or M-) sufficiency.

Separation of a submodel for inference on a parameter of interest ψ may involve a sequence of applications of various of the ancillarity and sufficiency definitions (the latter example shows a simple case of this), and occasionally it is necessary to invoke even more general concepts of nonformation than those which have a decomposition that corresponds to such a sequence (see Section 4.7(**iv**)).

Sometimes different separation procedures, relating to one and the same interest parameter ψ and justifiable on grounds of nonformation, are applicable and it can happen that they do not lead ultimately to the same submodel. This lack of uniqueness is exemplified in Section 4.7(**vi**). (The remark on non-uniqueness in statistical conclusions made in Section 1.1 is relevant here.) Certain uniqueness results will be mentioned at the end of Sections 4.4 and 4.5.

A striking illustration of the difference it may make whether the inference is performed separately or not is furnished by the problem of estimating the standard error σ attaching to a measuring instrument, from duplicate measurements of n different items. Assuming independent normal variation of the measurements x_{ij}, $i = 1, \ldots, n$, $j = 1, 2$, the mean value of x_{ij} being ξ_i, one finds that the maximum likelihood estimate of σ^2 in this original model is $s^2/2$ where $s^2 = n^{-1}\,\Sigma(x_{ij} - \bar{x}_{i.})^2$. Thus the estimator is not consistent, tending to $\sigma^2/2$ as $n \to \infty$. In contrast, the maximum likelihood estimate from the conditional model given $\bar{x}_{*.} = (\bar{x}_{1.}, \ldots, \bar{x}_{n.})$ is the usual estimate s^2. (The mean vector $\bar{x}_{*.}$ is G- (and also M-) ancillary with respect to σ^2.)

The kind of breakdown of the direct method of maximum likelihood exhibited here is a common phenomenon in the class of cases where the number of incidental parameters tends to infinity with the number of observations. But in many important instances, primarily with exponential or partly exponential models, the situation can be remedied, in analogy with the above, by separation through conditioning. The consistency and asymptotic normality of the conditional maximum likelihood estimator in such cases have been shown by Andersen (1973). Most of his results do not presuppose nonformativeness of the conditioning statistic, so that in general a loss of information (and, hence, efficiency) may be involved. However, such a loss may well be negligible, or at least acceptable, in comparison with the advantages gained by the separation.

4.2 B-SUFFICIENCY AND B-ANCILLARITY

The main parts of the mathematical theory of the classical concepts of sufficiency and ancillarity—here called B-sufficiency and B-ancillarity—are presented in this section.

A statistic T at a statistical field $(\mathfrak{X}, \mathfrak{A}, \mathfrak{P})$ is *B-sufficient* if the members of $\mathfrak{P}$ have a common conditional probability measure given T. In other words, T is sufficient provided there exists a Markov kernel $M(\cdot\,;\cdot)$ on $\mathfrak{A} \times \mathfrak{X}$ such that for each $A \in \mathfrak{A}$ the function $M(A, \cdot)$ is measurable with respect to $\sigma(T)$, the σ-algebra generated by T, and such that

$$P(AB) = \int_B M(A, \cdot)\, dP$$

for each $P \in \mathfrak{P}$, $B \in \sigma(T)$. A *minimal B-sufficient* statistic is a B-sufficient statistic T such that if $\tilde{T}$ is any other B-sufficient statistic then $\sigma(T) \subset \sigma(\tilde{T}) \vee \tilde{\varnothing}$, where $\tilde{\varnothing}$ denotes the σ-algebra generated by those sets $A \in \mathfrak{A}$ for which $P(A) = 0$ for every $P \in \mathfrak{P}$.

If the probability measures in $\mathfrak{A}$ are mutually absolutely continuous let P_0 be an element of $\mathfrak{P}$, and generally take P_0 to be of the form $P_0 = \sum_{n=1}^{\infty} c_n P_n$ with $P_n \in \mathfrak{P}$, $c_n > 0$ $(n = 1, 2, \ldots)$ and $\sum_{n=1}^{\infty} c_n = 1$, and such that a set $A \in \mathfrak{A}$ belongs to $\tilde{\varnothing}$ if and only if $P_0(A) = 0$. (The existence of such a P_0 is well known and was established by Halmos and Savage (1949).)

In the following, when the symbol $[\mathfrak{P}]$ is put after some relation this indicates that the relation holds up to set(s) of P-measure 0 for every $P \in \mathfrak{P}$.

Theorem 4.1. *A statistic T is B-sufficient if and only if for each $P \in \mathfrak{P}$ there exists a T-measurable version of dP/dP_0.*

Proof. It causes no loss of generality to assume that the elements of $\mathfrak{P}$ are mutually absolutely continuous. (To see this, consider the family $\{\frac{1}{2}(P + P_0): P \in \mathfrak{P}\}$, which has this property.)

Suppose T is B-sufficient, let M be the Markov kernel for the common conditional distribution given T and, for a fixed $P \in \mathfrak{P}$, set

$$U(x) = \int \frac{dP}{dP_0} dM(\cdot\,; x), \qquad x \in \mathfrak{X}.$$

It is obvious that U is T-measurable. Furthermore, U is a version of dP/dP_0 because

$$U = E_0^T \frac{dP}{dP_0} \qquad [P_0]$$

and hence, for every $A \in \mathfrak{A}$,

$$\begin{aligned}
\int_A U\, dP_0 &= \int 1_A \left(E_0^T \frac{dP}{dP_0} \right) dP_0 \\
&= \int E_0^T \left(1_A E_0^T \frac{dP}{dP_0} \right) dP_0 \\
&= \int E_0^T \left(\frac{dP}{dP_0} E_0^T 1_A \right) dP_0 \\
&= \int \frac{dP}{dP_0} M(A;\cdot)\, dP_0 \\
&= P(A),
\end{aligned}$$

Conversely, let $dP'dP_0$ be T-measurable and define $M(\cdot;\cdot)$ as the Markov kernel for the conditional distribution given T under P_0. Then for every $A \in \mathfrak{A}$, $B \in \sigma(T)$

$$\begin{aligned}
\int_B M(A;\cdot) dP &= \int 1_B (E_0^T 1_A) \frac{dP}{FP_0} dP_0 \\
&= \int E_0^T \left(1_{AB} \frac{dP}{dP_0} \right) dP_0 \\
&= P(AB)
\end{aligned}$$

showing that $M(\cdot\,;\cdot)$ is also the Markov kernel for the conditional distribution under P. ▶

For the proof of the next theorem and its corollary the following four lemmas will be needed. In these lemmas the members of $\mathfrak{P}$ are presupposed to be mutually absolutely continuous, A denotes an element of $\mathfrak{A}$, and $\mathfrak{B}$ a sub-σ-algebra of $\mathfrak{A}$.

Lemma 4.1. *One has*

$$\mathfrak{B} \vee \tilde{\varnothing} = \{A : 1_A = E_0^{\mathfrak{B}} 1_A [P_0]\}.$$

Lemma 4.2. *Let Y be an integrable stochastic variable and suppose that* $\sigma(Y) \subset \mathfrak{B} \cup \tilde{\varnothing}$. Then $Y = E_0^{\mathfrak{B}}\, Y\, [P_0]$.

Proof. This may be verified by showing that $E_0^{\mathfrak{B}} Y$ is a version of $E_0^{\mathfrak{B} \vee \tilde{\varnothing}} Y$, which is straightforward to do, using Lemma 4.1. ▶

The σ-algebra $\mathfrak{B}$ is said to be *separable* if there exist sets $B_i \in \mathfrak{B}$, $i \in N$, such that $\mathfrak{B} = \sigma(1_{B_i}: i \in N)$.

Lemma 4.3. *$\mathfrak{B}$ is separable if and only if it is generated by a stochastic variable.*

Proof. The if assertion is trivial.

Suppose $\mathfrak{B} = \sigma(1_{B_i}: i \in N)$ and define Y by

$$Y = \sum_{i=1}^{\infty} 3^{-i} 1_{B_i}.$$

It is enough to prove that any set of the form $B_1 \cap B_2 \cap \ldots \cap B_n$ where B_i equals either A_i or A_i^c ($n = 1,2,\ldots; i - 1,2,\ldots, n$) is contained in $\sigma(Y)$. Set

$$\delta_i = \begin{cases} 0 & \text{if } B_i = A_i^c \\ 1 & \text{if } B_i = A_i \end{cases}$$

and

$$b = \sum_{i=1}^{n} 3^{-i} \varepsilon_i.$$

Then $B_1 \cap B_2 \cap \ldots \cap B_n = \{b \leq Y < b + 3^{-n}\}$. ▶

Lemma 4.4. *For any $\mathfrak{B}$ there exists a separable σ-algebra $\mathfrak{B}_0$ such that $\mathfrak{B}_0 \subset \mathfrak{B} \subset \mathfrak{B}_0 \vee \tilde{\varnothing}$.*

Proof. The σ-algebra $\mathfrak{A}$ is separable and hence $\mathfrak{A} = \sigma(1_A: i \in N)$ for some family $\{A_i: i \in N\}$. Define $\mathfrak{B}_0$ by

$$\mathfrak{B}_0 = \sigma(E_0^{\mathfrak{B}} 1_{A_i}: i \in N).$$

This σ-algebra is separable since it is generated by a countable set of stochastic variables, and clearly $\mathfrak{B}_0 \subset \mathfrak{B}$. Furthermore,

$$\mathfrak{A} = \{A: \text{there exists a } \mathfrak{B}_0\text{-measurable version of } E_0^{\mathfrak{B}} 1_A\},$$

from which follows that $1_B = E_0^{\mathfrak{B}_0} 1_B$ $[P_0]$ for every $B \in \mathfrak{B}$. Consequently, by Lemma 4.1, $\mathfrak{B} \subset \mathfrak{B}_0 \vee \tilde{\varnothing}$. ▶

Theorem 4.2. *A statistic T is minimal B-sufficient if and only if $\sigma\{T\} = \sigma\{dP/dP_0: P \in \mathfrak{P}\}\,[\mathfrak{P}]$.*

Proof. As in the proof of Theorem 4.1, one may assume mutual absolute continuity of the members of $\mathfrak{P}$.

Set $\mathfrak{B}_0 = \sigma\{T\}$ and $\mathfrak{B} = \sigma\{dP/dP_0: P \in \mathfrak{P}\}$.

If T is minimal B-sufficient then, by Theorem 4.1, $\mathfrak{B} \subset \mathfrak{B}_0\,[\mathfrak{P}]$. Let $\tilde{T}$ be a statistic such that $\sigma(\tilde{T}) \subset \mathfrak{B} \subset \sigma(\tilde{T}) \vee \tilde{\varnothing}$. The existence of such a statistic is apparent from Lemmas 4.3 and 4.4. Applying Lemma 4.2 one obtains

$$\frac{dP}{dP_0} = E_0^{\tilde{T}} \frac{dP}{dP_0} \qquad [P_0]$$

and this, again by Theorem 4.1, shows that $\tilde{T}$ is B-sufficient, and the minimal B-sufficiency of T then yields $\mathfrak{B}_0 \subset \mathfrak{B}\,[\mathfrak{P}]$.

On the other hand, suppose $\mathfrak{B}_0 = \mathfrak{B}[\mathfrak{P}]$. From Lemma 4.2 it follows that $E_0^{\mathfrak{B}_0}(dP/dP_0)$ is a version of dP/dP_0 for every $P \in \mathfrak{P}$, whence, using Theorem 4.1, T is B-sufficient. If $\tilde{T}$ is any B-sufficient statistic then $\mathfrak{B} \subset \sigma(\tilde{T})[\mathfrak{P}]$ which implies $\mathfrak{B}_0 \subset \sigma(\tilde{T})[\mathfrak{P}]$, as required for minimal sufficiency of T.

Corollary 4.1. *A minimal B-sufficient statistic exists.*

Proof. Invoke Theorem 4.2 and Lemmas 4.3 and 4.4. ▶

With $\mathfrak{P}$ parametrized, $\mathfrak{P} = \{\mathfrak{P}_\omega : \omega \in \Omega\}$, set

$$q(x;\omega) = \frac{dP_\omega}{dP_0}(x)$$

and let r be the mapping which maps a point $x \in \mathfrak{X}$ to the likelihood function

$$r(x) = q(x;\cdot).$$

Furthermore, let the range space $R^{\mathfrak{P}}$ of r be endowed with the product σ-algebra $\mathfrak{B}^{\mathfrak{P}}$, where $\mathfrak{B}$ is the Borel σ-algebra in R. Then r is measurable and

$$\sigma\{r\} = \sigma\{q(\cdot;\omega) : \omega \in \Omega\}. \tag{1}$$

(Here, and until the end of Example 4.2, equalities, inclusions, etc. are strict, i.e. not modulo null sets.) This proposition represents one precise interpretation of the common phrase 'the likelihood function is minimal sufficient' (cf. Theorem 4.2 and Corollary 4.1). However, rather than this interpretation, the phrase reflects the useful fact that if t is a statistic generating the same partition of $\mathfrak{X}$ as the mapping r, i.e.

$$t(x) = t(\tilde{x}) \Leftrightarrow q(x;\omega) = q(\tilde{x};\omega) \qquad \text{for every } \omega \in \Omega, \tag{2}$$

then, as a rule, t is minimal B-sufficient. That some regularity condition is needed to ensure the minimal sufficiency of such a statistic t is illustrated by:

Example 4.1. Let x_1 and x_2 be independent and normally distributed, $x_1 \sim N(0,1)$, $x_2 \sim N(\omega,1)$ with $\omega \in \Omega = R$. Then x_2 is minimal sufficient with respect to the family $\mathfrak{P} = \{P_\omega : \omega \in \Omega\}$ of joint distributions of $x = (x_1, x_2)$.

With $\mathfrak{X} = R^2$, a version of dP_ω/dP_0 is given by

$$q(x;\omega) = [1 - \delta(x_1 - \omega)]e^{\omega x_2 - \omega^2/2}$$

where δ is the function on R which is 1 at the origin and 0 otherwise. Clearly,

$$q(x;\cdot) = q(x';\cdot) \Leftrightarrow x = x'$$

which means that x generates the same partition of $\mathfrak{X}$ as does $r: x \to q(x;\cdot)$, although x is not minimal sufficient. ▶

Theorem 4.3. *Suppose t is a statistic which generates the same partition of $\mathfrak{X}$ as the mapping $r: x \to q(x;\cdot)$.*

If either $\mathfrak{P}$ is discrete or $q(x; \omega)$ is continuous in ω for each fixed x then t is minimal sufficient.

Remark. The discrete case is straightforward to verify. If, in a given case, the conditions of the theorem are not fulfilled it may well be that minimal sufficiency can be established by the method of proof below which, as should be apparent, offers scope for considerable generalizations.

Proof. First some general results on mappings and σ-algebras determined by the mappings will be stated.

For any mapping f on an arbitrary measure space $(E, \mathfrak{C})$, let $\delta(f)$ denote the partition σ-algebra determined by f, i.e. the σ-algebra of those sets $C \in \mathfrak{C}$ which are unions of elements of the partition of E generated by f. Clearly, $\delta(f) = \{C \in \mathfrak{C}: C = f^{-1}(f(C))\}$ and, if f is a measurable mapping,

$$\sigma(f) \subset \delta(f). \tag{3}$$

Under mild regularity conditions one has, in fact, that $\sigma(f) = \delta(f)$. This will be seen from the following proposition which is a special case of Theorem 3, p. 145, in Hoffmann–Jørgensen (1970).

Lemma 4.5. *Let E, F and G be Borel subsets of complete, separable metric spaces, endowed with the Borel σ-algebras. Let f and g be measurable mappings from E into F and G, respectively, and suppose that the partition of E generated by f is finer than the partition generated by g. Then there exists a measurable mapping h from F into G such that $g = h \circ f$.*

Taking g to be the indicator function of an element of $\delta(f)$ one finds that this element belongs to $\sigma(f)$. Hence one has: ▶

Corollary 4.2. *Suppose f is a measurable mapping from a Borel subset of a complete, separable metric space into a complete, separable metric space. Then $\sigma(f) = \delta(f)$.*

It follows that, always,

$$\sigma(T) = \delta(T) = \delta(r) \tag{4}$$

(and hence, in view of (1) and (3), that t is sufficient).

Now, suppose that $q(x;\cdot)$ is continuous on Ω for every $x \in \mathfrak{X}$. Let Ω_0 be a dense subset of Ω and let r_0 be the mapping on $\mathfrak{X}$ such that $r_0(x)$ is the restriction of $q(x; \cdot)$ to Ω_0. Then, by the continuity, $\sigma(r) = \sigma(r_0)$ and r_0 determines the same partition of $\mathfrak{X}$ as r (and t). Therefore, on account of (4) and Corollary 4.2,

$$\sigma(t) = \delta(r_0) = \sigma(r_0) = \sigma(r)$$

and since $\sigma(r)$ is minimal sufficient so is t. Thus Theorem 4.3 is verified. ▶

Corollary 4.3. *Suppose that t is a statistic which generates the same partition of* $\mathfrak{X}$ *as the likelihood function* $p(x;\cdot)$, *i.e.*

$$t(x) = (t\tilde{x}) \Leftrightarrow cp(x;\omega) - \tilde{c}p(\tilde{x};\omega) \qquad \text{for every } \omega \in \Omega$$

for some positive c and $\tilde{c}$ *which do not depend on* ω *but may depend on x and* $\tilde{x}$, *respectively.*

If either $\mathfrak{P}$ *is discrete or* $p(x;\omega)$ *is positive and continuous in* ω *for each fixed x then T is minimal sufficient.*

Proof. P_0 is of the form $\sum c_n P_{\omega_n}$. Taking the version of $dP_0/d\mu$ given by

$$p_0(x) = \sum c_n p(x;\omega_n)$$

and setting $q(x;\omega) = p(x;\omega)/p_0(x)$ one obtains that Theorem 4.3 applies. ▶

Example 4.2. The model function for a sample $x_1,\ldots,x_n$ from the Cauchy distribution with mode ω is

$$p(x;\omega) = \pi^{-n} \prod_{j=1}^{n} \frac{1}{1 + (x_j - \omega)^2}.$$

If $x = (x_1,\ldots,x_n)$ and $\tilde{x} = (\tilde{x}_1,\ldots,\tilde{x}_n)$ satisfy $cp(x;\cdot) = \tilde{c}p(\tilde{x};\cdot)$ then

$$c\prod_{j=1}^{n} (1 + (\tilde{x}_j - \omega)^2) = \tilde{c}\prod_{j=1}^{n} (1 + (x_j - \omega)^2).$$

Both sides of this equation are polynomials in ω and hence the equality holds for all $\omega \in R$ precisely when these two polynomials have the same roots. Since the roots are $\tilde{x}_j \pm i, j = 1,\ldots,n$, respectively $x_j \pm i, j = 1,\ldots,n$, one sees that the order statistic $(x_{(1)},\ldots,x_{(n)})$ is minimal B-sufficient. ▶

We now turn to a discussion of *B-ancillary* statistics, i.e. statistics t such that P_t does not depend on $P \in \mathfrak{P}$, and of relations between B-ancillarity and B-sufficiency. (For most of this discussion the presupposition that the family $\mathfrak{P}$ is dominated is not needed.)

Only seldom does it happen that a B-ancillary statistic t exists which is *maximal* in the sense that if $\tilde{t}$ is any other B-ancillary statistic then $\sigma(\tilde{t}) \subset \sigma(t)$ $[\mathfrak{P}]$. Another way of expressing this is that if t_1 and t_2 are B-ancillary then $t = (t_1, t_2)$ will, in general, not be B-ancillary.

A B-ancillary statistic t will be called *relatively maximal* if for any B-ancillary statistic $\tilde{t}$ with $\sigma(t) \subset \sigma(\tilde{t})$ $[\mathfrak{P}]$ one has, in fact, $\sigma(t) = \sigma(\tilde{t})$ $[\mathfrak{P}]$.†

Example 4.3. Let $(x_i, y_i), i = 1,\ldots,n$, be independent, two-dimensionally normally distributed random variables with $E(x_i) = E(y_i) = 0$, $V(x_i) = V(y_i) = 1$, $V(x_i, y_i) = \rho$ where ρ varies in $(-1, 1)$. Then $(x_1,\ldots,x_n)$ and $(y_1,\ldots,y_n)$ are both B-ancillary but together they constitute the whole sample. ▶

† In the literature on ancillarity the terms 'unique maximal' and 'maximal' are commonly used to designate what is in this book called maximal and relatively maximal, respectively.

Example 4.4. Consider two geneloci each having two allelic genes denoted, respectively, by A, a and B, b. Both A and B are assumed dominant. Suppose one has observed the phenotypes of n individuals sampled randomly among the offspring of a population consisting entirely of doubleheterozygotes of trans type (chromosomal arrangement Ab/aB), the observations being set out as in the table below.

	A–	aa	
B–	x_{11}	x_{12}	$x_{1\cdot}$
bb	x_{21}	x_{22}	$x_{2\cdot}$
	$x_{\cdot 1}$	$x_{\cdot 2}$	n

On the assumptions of random union of gametes and no selection, the corresponding table of probabilities is

	A–	aa	
B–	$\frac{2+\pi}{4}$	$\frac{1-\pi}{4}$	$\frac{3}{4}$
bb	$\frac{1-\pi}{4}$	$\frac{\pi}{4}$	$\frac{1}{4}$
	$\frac{3}{4}$	$\frac{1}{4}$	1

where the parameter π is the product of the recombination probabilities for males and females.

$x_{1\cdot}$ and $x_{\cdot 1}$ are obviously both B-ancillary, but jointly they are not B-ancillary. This is apparent for instance from the fact that the probability that both $x_{1\cdot}$ and $x_{\cdot 1}$ are 0 equals $(\pi/4)^n$. ▶

In Examples 4.3 and 4.4 the observations do not constitute a minimal B-sufficient statistic. This, however, is the case in the next example.

Example 4.5. Consider a two-by-two contingency table

$$\begin{matrix} x_{11} & x_{12} \\ x_{21} & x_{22} \end{matrix}$$

with the total fixed, and cell probabilities

$$\begin{matrix} \frac{1}{6}(1+\pi) & \frac{1}{6}(2-\pi) \\ \frac{1}{6}(1-\pi) & \frac{1}{6}(2+\pi) \end{matrix}$$

with π varying in $(-1, 1)$. Here $(x_{11}, x_{12}, x_{21}, x_{22})$ is minimal B-sufficient while $x_{1.}$ and $x_{.1}$ are each relatively maximal B-ancillary. Thus a maximal B-ancillary statistic does not exist. ▶

Examples of this kind were first given by Birnbaum (1961, 1962) and Basu (1964).

At a number of places later in the following it will be convenient to use the notion of (bounded) completeness. The family $\mathfrak{P}$ is called *(boundedly) complete* if for every (bounded) real-valued function f on $\mathfrak{X}$ which satisfies

$$\int f\, dP = 0, \qquad P \in \mathfrak{P},$$

one has

$$f = 0 \qquad [\mathfrak{P}].$$

This notion does not seem to have any significant statistical meaning, but completeness or bounded completeness implies various properties of considerable statistical interest.

The concepts of B-sufficiency and B-ancillarity are both special cases of conditional B-ancillarity. A statistic u is called *conditionally B-ancillary* given the statistic w if the members of $\mathfrak{P}$ have a common conditional distribution of u given w. B-sufficiency obtains for $u = x$, while B-ancillarity corresponds to w being a constant.

Suppose u is conditionally B-ancillary given w, let A be a u-measurable event and let $M(A; \cdot)$ be the common conditional probability of A given w. Clearly

$$\int (M(A; \cdot) - 1_A)\, dP = 0, \qquad P \in \mathfrak{P}.$$

Hence, if $\mathfrak{P}$ is boundedly complete

$$M(A; \cdot) = 1_A \qquad [\mathfrak{P}].$$

This conclusion may be paraphrased as follows.

Theorem 4.4 *If $\mathfrak{P}$ is boundedly complete then there are no nontrivial instances of conditionally B-ancillary statistics.*

As one would expect, if there are no nontrivial conditionally B-ancillary statistics then, in particular, x is minimal B-sufficient. To prove this, suppose t is B-sufficient, let A be any event and set $B = \{M(A; \cdot) = 1\}$ where $M(\cdot; \cdot)$ is the common Markov kernel. By assumption, $M(A; \cdot) = 1_A[\mathfrak{P}]$ and consequently $1_B = 1_A\ [\mathfrak{P}]$, as was to be verified.

Let t and u be statistics, assume that t is B-sufficient, and let M denote the Markov kernel for the common conditional distribution given t. In the case t and

u are independent under a $P \in \mathfrak{P}$, then for any $C \in \sigma(u)$

$$P(C) = P(C|t) = M(C;\cdot) \qquad [\mathfrak{P}].$$

This implies that u is B-ancillary.

A converse assertion, due to Basu (1955), holds for $\mathfrak{P}_t$ boundedly complete. This is stated below as a corollary to:

Theorem 4.5. *Let t, u, and w be statistics, suppose that (t, w) is B-sufficient and that u is conditionally B-ancillary given w. If $\mathfrak{P}_{(t,w)}$ is boundedly complete then t and u are conditionally independent given w.*

Proof. Let M and M_0 be the Markov kernels of the common conditional distributions of, respectively, x given (t, w) and u given w Then for every $C \in \sigma(u)$ and $P \in \mathfrak{P}$ one has

$$\int M(C;\cdot) = M_0(C;\cdot))\,dP = 0$$

whence, by bounded completeness,

$$Y(C;\cdot) = Y_0(C;\cdot) \qquad [\mathfrak{P}]$$

which means that the conditional distribution of u is the same given w as given (t, w). ▶

Corollary 4.4. *Suppose t is B-sufficient and u is B-ancillary. If $\mathfrak{P}_t$ is boundedly complete then t and u are independent.*

Illustrations of Theorem 4.5 and its corollary—the latter is known as Basu's Theorem (Basu, 1955, 1958)—will be presented in Section 8.1, after a general sufficient condition for completeness of exponential families has been established.

4.3 NONFORMATION

As mentioned in Section 4.1, a key concept of inferential separation is *nonformation*, i.e. the concept of a submodel containing no information with respect to a specified parameter function. Four mathematical definitions of nonformation, namely B-, S-, G-, and M-nonformation, are given in the present section. These specify circumstances under which, whatever the outcome of the experiment, no information on the parameter function is contained in the conjunction of the submodel and the part of the data with which the submodel is concerned. More generally, one may have situations where this conjunction is nonformative for certain experimental outcomes but not for others. It is possible to extend the definitions of B-, S-, and M-nonformation to include such cases of *pointwise nonformation*.

Let u and v be statistics and let ψ be a parameter function. Then u, v, and the submodel $\{P_\omega u(\cdot|v)\colon \omega \in \Omega\}$, consisting of the conditional distributions of u given v, may be nonformative with respect to ψ in one or more of the following senses.

B-nonformation: For every value of v the conditional distribution $P_\omega u(\cdot|v)$ does not depend on ω. ▶

S-nonformation: For every value of v the family $\{P_\omega u(\cdot|v)\colon \psi(\omega) = \psi\}$ does not depend on ψ. ▶

G-nonformation: For every value of v and ψ the family $\{P_\omega u(\cdot|v)\colon \psi(\omega) = \psi\}$ is a union of families each of which is generated by a transitive group of transformations. ▶

M-nonformation: For every value of v and ψ the family $\{P_\omega u(\cdot|v)\colon \psi(\omega) = \psi\}$ is universal.

(In the definitions of G- and M-nonformation the requirements of transitivity respectively universality refer to the set of all realizable values of u under the whole family $\mathfrak{P}u(\cdot|v)$, not just the family $\{P_\omega u(\cdot|v)\colon \psi(\omega) = \psi\}$.)

Each of these four definitions satisfies the natural requirement that nonformation with respect to ψ implies nonformation with respect to any parameter function which depends on ω through ψ only.

Obviously, B-nonformation implies S-nonformation. Moreover, G-nonformation does, as a rule, entail M-nonformation, cf. Lemma 2.1.

The above four definitions are global in that they state that irrespective of which values of u and v are realized no information on ψ can be extracted from these values plus the submodel in question. However, in general, different realizations carry different amounts of evidence and it is reasonable to ask for pointwise versions of these definitions, concerned with whether any particular observed values u and v are nonformative. G-nonformation does not seem to lend itself naturally to such an extension, but the other three notions do. It is natural to formulate the pointwise definitions in terms of the conditional probability function of u given v, and for notational convenience this function will be indicated by $p(\cdot;\omega|v)$ (rather than $pu(\cdot;\omega|v)$).

Pointwise B-nonformation: For the observed values u and v, $p(u;\omega|v)$ does not depend on ω. ▶

Pointwise S-nonformation: For the observed values u and v the family $\{p(u;\omega|v)\colon \psi(\omega) = \psi\}$ does not depend on ψ. ▶

Pointwise M-nonformation: For the observed values u and v, and for every value ψ, the family $\{p(\cdot;\omega|v)\colon \psi(\omega) = \psi\}$ has u as mode point, and the complementary family $\{p(\cdot;\omega|v)\colon \psi(\omega) \neq \psi\}$ is universal. ▶

Clearly, pointwise B-, S-, and M-nonformation for all u and v entails, respectively, B-, S-, and M-nonformation.

The reasoning behind the definition of pointwise M-nonformation, and hence behind M-nonformation, is the following. If, under the conditions specified in the definition, the observation u and the submodel $\{p(\cdot;\omega|v): \omega \in \Omega\}$ did contain available information on ψ then it would be possible on the basis of their conjunction alone to say that some value ψ_0 of ψ is less credible than some other value. But this is not warranted because

(*a*) there is perfect fit (or complete concordance) between u and the model $\{p(\cdot;\omega|v): \psi(\omega) = \psi_0\}$ (in the sense that u is a mode point of the model);

(*b*) the alternative model $\{p(\cdot;\omega|v): \psi(\omega) \neq \psi_0\}$ will fit any value u perfectly;

and because it is not scientifically reasonable to call a hypothesis (or model) which is in complete agreement with the data less plausible than an alternative hypothesis which is capable of explaining any of the possible outcomes.

M-nonformation and pointwise M-nonformation are, clearly, related in spirit to the concept of plausibility.

The main uses of the various notions of nonformation are in the processes of conditioning on ancillary statistics or margining to sufficient statistics, and many examples of these notions are contained in Sections 4.2 and 4.4 and in Chapter 10. Here, then, just two, somewhat special, examples will be given which illustrate that a statistic can be pointwise S- or M-nonformative without being globally so.

Example 4.6. Suppose an individual (or item) is subjected to two kinds of events, occurring in two independent Poisson processes with intensities λ and μ, respectively. The individual responds, e.g. dies (or the item is destroyed), at the moment both kinds of events have occurred. For each of n individuals independently, it is recorded which type of event occurred first and how long elapsed between that occurrence and the response. Let v be the number of individuals for which the first event that happened was from the Poisson process having the intensity λ. Then v follows a binomial distribution with probability parameter $\psi = \lambda/(\lambda + \mu)$. Conditionally on v, the set of recorded time intervals is S-nonformative at $v = n$ with respect to ψ (on the proviso that λ and μ vary independently, both in $(0, \infty)$). (Another way of expressing this would be to say that v is S-sufficient at $v = n$ with respect to ψ, cf. the definition of S-sufficiency in Section 4.4.) ▶

Example 4.7. Consider the 2×2 contingency table with one marginal given, as discussed in Section 4.1. If ψ is given by equation (2) of that section, then $x.$ is M-nonformative with respect to that parameter, as will be shown in Example 10.12.

Suppose, however, that

$$\psi = p_1 - p_2 .$$

Then $x_.$ is not M-nonformative with respect to ψ and it may indeed yield information on ψ. For instance, if $x_.$ is observed equal to 0 then there is reason to believe that $|\psi|$ is not very near to 1. On the other hand, for $n_1 = n_2 = n$ the realization n of $x_.$ is M-nonformative with respect to ψ. This follows from the facts that every distribution with $p_1 + p_2 = 1$ has n as mode point, which may be proved by remarking that any such distribution is symmetric around n and unimodal (by Theorem 6.6 it is, in fact, strongly unimodal), and that the subfamily of distributions of $x_.$ determined by $\psi = 0$, being a binomial family, is universal. ▶

In general, a statement that something contains no information or evidence with respect to a certain question is a *relative* statement, i.e. it is valid only on the proviso that certain other things are unknown (or not taken into consideration). Sometimes it is a far-fetched thought that such other things should become wholly or partly known, but in other connections it may be an obvious possibility. Typically, in the case where one of the definitions of (pointwise) S-nonformation, G-nonformation, or (pointwise) M-nonformation is satisfied it will nevertheless be possible to extract information on ψ from u, v, and the submodel if to these can be added further evidence on ω, available for instance from some other, independent experiment or perhaps even from another part of the same model. It is particularly important to stress this in relation to G- and M-nonformation. For these, in contrast to S-nonformation, two independent observations of (u, v), conjoined with the submodel for u given v does, as a rule, provide evidence on ψ. The difference between S-nonformation on the one hand and G- or M-nonformation on the other indicated here may be further illuminated by consideration of a situation where only one observation of (u, v) is at hand but where one gets to know that ω belongs to a subset Ω_0 of Ω such that for each value ψ there is precisely one $\omega \in \Omega$ with $\psi(\omega) = \psi$. This extra knowledge would clearly be of no help in the S-nonformation case, but would normally be relevant under G- or M-nonformation and could, for certain models, even show definitively which value of ψ is the correct one. For the often encountered cases where ω is of the form (χ, ψ) with χ and ψ variation independent, these circumstances may be briefly summarized and stressed by saying that G- or M-nonformation means no available information on ψ in the absence of knowledge on χ.

4.4 S-, G-, AND M-ANCILLARITY AND -SUFFICIENCY

The general ideas of ancillarity and sufficiency were discussed in Section 4.1. In order to obtain a fully specified concept of ancillarity or sufficiency one must give precise definition to the phrase that a submodel contains no information with respect to a specified parameter function. Various such definitions have been presented in the previous section. The most elemental of these definitions is that of B-nonformation which corresponds to B-ancillarity and B-sufficiency, i.e.

classical ancillarity and sufficiency. The properties of B-ancillary and B-sufficient statistics have, for practical reasons, been treated already in Section 4.2, and the present section is devoted mainly to illustrations of the ancillarity and sufficiency concepts derived from, respectively, S-, G-, and M-nonformation.

Let t be a statistic and ψ a subparameter, and consider the factorization

$$p(x;\omega) = p(t;\omega)p(x;\omega|t) \tag{1}$$

of the probability function for x into the marginal density for t and the conditional density given t.

Assuming that ψ parametrizes the conditional distributions, (1) may be written

$$p(x;\omega) = p(t;\omega)p(x;\psi|t). \tag{2}$$

In this case, the statistic t is ancillary with respect to ψ if $(\mathfrak{P}_t, t)$ is nonformative with respect to ψ (cf. Section 4.1).

On the other hand, assume that the marginal distributions of t are parametrized by ψ, so that (1) has the form

$$p(x;\omega) = p(t;\psi)p(x;\omega|t). \tag{3}$$

Then t is sufficient for x with respect to ψ if $(\mathfrak{P}^t, x)$ is nonformative with respect to ψ.

More generally, if t is ancillary or sufficient with respect to ψ, as above, it will also be called ancillary respectively sufficient with respect to any parameter function which depends on ψ only.

In connection with the discussion of S-ancillarity and S-sufficiency it is convenient to introduce the concept of a cut. Let t be a statistic. Each $P \in \mathfrak{P}$ may be broken into two pieces P_t and P^t, i.e. one has a mapping on $\mathfrak{P}$ into $\mathfrak{P}_t \times \mathfrak{P}^t$ given by $P \to (P_t, P^t)$. Now, t is said to be a *cut* if this mapping is actually onto $\mathfrak{P}_t \times \mathfrak{P}^t$, or, in other words, if any of the marginal distributions of t combined with any of the conditional distributions given t gives a probability measure in $\mathfrak{P}$. Clearly, if t is B-ancillary or B-sufficient then t is a cut. A cut which is neither B-ancillary nor B-sufficient is called *proper*.

With $\mathfrak{P}$ being parametrized and dominated, the fact that a statistic t is a cut means that for a suitable parametrization of $\mathfrak{P}$ one has $\omega = (\chi, \psi)$ where χ and ψ are variation-independent components of ω, and

$$p(x;\omega) = p(t;\chi)p(x;\psi|t). \tag{4}$$

In relation to t, the components χ and ψ are spoken of as a *corresponding pair of L-independent parameters* (cf. the definition of L-independence in Section 3.3).

Now, the concepts of *S-ancillarity* and *S-sufficiency* are obtained from the general stipulations of ancillarity and sufficiency by invoking the definition of S-nonformation. One sees that t is S-ancillary or S-sufficient (with respect to some parameter function) if and only if t is a cut. In that case and with χ and ψ

as indicated by (4), t is S-ancillary with respect to ψ (and any function thereof) and S-sufficient with respect to χ (and any function thereof).

The examples (3.2–3.7) of L-independence are readily seen to be also instances of cuts, and several other instances appear in Sections 10.2 and 10.3 Suffice it therefore here to give just two further illustrations.

Example 4.8. Poisson regression. In log-linear Poisson regression situations the total count is typically a cut.

In the simplest example of such a situation the observation at the value t of the regression parameter is Poisson distributed with mean value

$$e^{\alpha+\beta t},$$

and independent observations $x_1, x_2, \ldots, x_n$ are actually made at $t_1 < t_2 < \cdots < t_n$. The probability of $(x_1, \ldots, x_n)$ is

$$\exp\{e^{-n\alpha-\beta t.}\}\frac{1}{x_1!\ldots x_n!}\exp\left\{\alpha x_. + \beta\sum_{j=1}^{n} t_j x_j\right\},$$

and $x_.$ follows the Poisson distribution with parameter

$$e^{\alpha}\sum_{j=1}^{n} e^{\beta t_j}$$

while conditionally on $x_.$ the variate $(x_1, \ldots, x_n)$ has a (singular) multinomial distribution whose cell probabilities depend on β only. Thus, if α and β are variation independent and if α varies in R then $x_.$ is cut.

More generally, suppose m series of observations have been taken at $t_1 < \cdots < t_n$, and that the parameter α may be varying from series to series. The distribution of x_{ij}, the jth observation from the ith series, is then Poisson with mean value $\exp\{\alpha_i + \beta t_j\}$ and the joint distribution of the observations may be factored into the marginal distribution of $x_{..}$ which is Poisson with mean value

$$\sum_{i=1}^{m} e^{\alpha_i}\sum_{j=1}^{n} e^{\beta t_j},$$

the conditional distribution of $x_{*.}$ given $x_{..}$ which is multinomial with cell probabilities

$$\left(\sum_{i=1}^{m} e^{\alpha_i}\right)^{-1}(e^{\alpha_1}, \ldots, e^{\alpha_m}),$$

and, finally, the conditional distribution of x_{**} given $x_{*.}$, this latter depending only on β. It follows that both $x_{..}$ and $x_{*.}$ are cuts, provided that α_* and β are variation independent and that α_* has R^m as domain of variation. This was noted and used by Kalbfleisch and Sprott (1974) to split the analysis of certain dilution series data from virological experiments into three separate parts. ▶

Example 4.9. Poisson process with log-linear trend. Suppose a Poisson process with intensity function

$$\lambda(t) = e^{\alpha + \beta t}$$

has been observed in the time interval $[0, T]$. Let n be the number of events in this interval and let $t_1 < t_2 < \cdots < t_n$ be the times at which they occurred. The distribution of n is Poisson with mean value

$$e^{\alpha} \int_0^T e^{\beta t}\, dt$$

and, given n, the vector t_* has the density

$$n!\left(\frac{\beta}{e^{\beta T} - 1}\right)^n e^{\beta t.}$$

which, incidentally, is the density for the order statistic of a sample of n from the exponential distribution truncated to $[0, T]$. Consequently, for (α, β) varying in R^2, the number of events n is a cut, and hence, in particular, inference on β should be performed conditionally on n, cf. Cox and Lewis (1966). ▶

Next, a couple of general remarks concerning G-sufficiency will be made, and these will be followed by some examples of *G-ancillarity* and *G-sufficiency*, i.e. of factorizations as in (2) or (3) with $(\mathfrak{P}_t, t)$ respectively $(\mathfrak{P}^t, x)$ being G-nonformative. These examples may also be viewed as examples of *M-ancillarity* and *M-sufficiency*, cf. Section 4.3. Further illustrations of the latter two concepts will be mentioned subsequently.

Suppose there exists a group G of transformations of $\mathfrak{X}$ and a statistic t such that

(i) For each value of the interest parameter ψ the corresponding family of probability measures is generated by G, and Ω is in one-to-one correspondence with $\Psi \times G$ (in the obvious manner).
(ii) The orbits of G are the elements of the partition of $\mathfrak{X}$ induced by t.

Note that, under (i) and (ii), ψ and t are corresponding maximal invariants which implies that the distribution of t depends on ψ only. It is thus clear that (i) and (ii) are nearly sufficient to ensure G-sufficiency. Furthermore, most examples of G-sufficiency satisfy (i) and (ii). (The properties (i) and (ii) are, in essence, the conditions required by Barnard (1963a) in his definition of 'sufficiency of t with respect to ψ in the absence of knowledge of g $(\in G)$'.)

Ordinarily, the group G will be unitary, so that for each value of ψ the corresponding subfamily $\mathfrak{P}_\psi$ of $\mathfrak{P}$ is a group family.

Example 4.10. In the joint distribution of $\bar{x}$ and s^2 from a normal sample, with (ξ, σ^2) unknown, the empirical variance s^2 is G-sufficient with respect to σ^2 and, moreover, $\bar{x}$ is G-ancillary with respect to σ^2. ▶

Example 4.11. Let x_1 and x_2 be independent random variables, x_i following a gamma distribution with scale parameter β_i and known shape parameter λ_i $(i = 1, 2)$. Thus x_i has density

$$\frac{1}{\Gamma(\lambda_i)\beta_i^{\lambda_i}} x^{\lambda_i - 1} e^{-x/\beta_i} \qquad (x > 0).$$

Suppose $\chi = \beta_1/\beta_2$ is the parameter of interest and let $t = x_1/x_2$. The marginal distribution of t is the generalized F-distribution with density

$$\frac{\Gamma(\lambda_.)}{\Gamma(\lambda_1)\Gamma(\lambda_2)} \frac{1}{\chi^{\lambda_1}} \frac{t^{\lambda_1 - 1}}{(1 + t/\chi)^{\lambda_.}}.$$

Furthermore, the conditional distribution of x_2 given t is a gamma distribution with form parameter $\lambda_.$ and scale parameter

$$\left(1 + \frac{t}{\chi}\right)\beta_2.$$

Hence if the domain of variation B of (β_1, β_2) is such that χ and β_2 are variation independent with β_2 varying in $(0, \infty)$—which is the case for $B = (0, \infty)^2$ or $B = \{0 < \beta_1 \leq \beta_2\}$—then t is G-sufficient with respect to χ. ▶

Example 4.12. Consider the family of distributions of $(x_., \Sigma(x_i - \bar{x})'(x_i - \bar{x}))$ where $x_1, \ldots, x_n$ is a sample of multidimensional normal variates. The distribution of the matrix $\boldsymbol{r}$ of empirical correlations depends only on the correlation matrix $\boldsymbol{\rho}$, and $\boldsymbol{r}$ is G-sufficient for $(x_., \Sigma(x_i - \bar{x})'(x_i - \bar{x}))$ with respect to $\boldsymbol{\rho}$. (Barnard 1966.) ▶

Example 4.13. Let $v_1, \ldots, v_n$ be a sample from the k-dimensional von Mises–Fisher distribution with mean direction μ and precision χ (cf. Example 8.1). The resultant length $|v_.|$ has a distribution which depends on the precision χ only, and given $|v_.|$ the distribution of the mean direction $v_./v_.|$ is the k-dimensional von Mises–Fisher distribution having mean direction μ and precision $|v_.|\chi$. Thus, for freely varying (μ, χ), the resultant length is G-sufficient with respect to χ, in the family of distributions of the (minimal B-sufficient) statistic $v_.$. ▶

Example 4.14. A model for regression analysis of lifetime data is specified by assuming that the lifetime x of an individual with regressor covariate $z = (z_1, \ldots, z_m)$ follows a distribution with hazard function

$$\lambda_0(x) e^{\beta \cdot z}, \qquad x \in (0, \infty), \tag{5}$$

where β is an unknown parameter vector and $\lambda_0(\cdot)$ is an unknown function which is not identically 0 over any open interval. For a fixed β the family of distributions having hazard (5) is generated from the distribution with hazard $\exp\{\beta \cdot z\}$ by the group of time transformations $x \to g(x)$ where g is differentiable and strictly increasing. The connection between $g(\cdot)$ and $\lambda_0(\cdot)$ is given by

$$g(x) = \int_0^x \lambda_0(y)\,dy.$$

Suppose now that a sample of n lifetimes $x_1, \ldots, x_n$, with covariates $z_1, \ldots, z_n$ is observed. Let G be the group of transformations of $x = (x_1, \ldots, x_n)$ of the form

$$(x_1, \ldots, x_n) \rightarrow (g(x_1), \ldots, g(x_n))$$

with g as above. The rank statistic $t = ((1), \ldots, (n))$, which gives the permutation of $1, \ldots, n$ such that $x_{(1)} < x_{(2)} < \cdots < x_{(n)}$, is G-sufficient with respect to β.

The regression model discussed here was proposed by Cox (1972) who also indicated the possibility of drawing inference separately on β. That the evidence about β may be isolated through G-sufficiency was shown by Kalbfleisch and Prentice (1973) who also derived the likelihood function for β based on the marginal distribution of t. This likelihood function is

$$\exp\{\beta \cdot z\} \Big/ \prod_{i=1}^{n} \sum_{l=i}^{n} \exp\{\beta \cdot z_{(l)}\}.$$

A particular interest of the present example lies in the fact that $\lambda_0(\cdot)$, the incidental part of the specification when β is the parameter of interest, is of nonparametric character. ▶

Example 4.15. Two diallelic, autosomal loci may carry the genes G and g respectively T and t. A double heterozygotic individual is chosen at random from a population in which a proportion λ of all heterozygotes is of cis type (chromosomal arrangement GT/gt) while the proportion $1 - \lambda$ is in trans (chromosomal arrangement Gt/gT). A cross between this individual and a double recessive ggtt yields n offspring which are classified according to genotype, giving the table

GgTt	Ggtt	ggTt	ggtt.
a	b	c	d

For a recombination probability of π, this table has probability

$$\frac{n!}{a!b!c!d!}\{\lambda(1 - \pi)^{a+d}\pi^{b+c} + (1 - \lambda)\pi^{a+d}(1 - \pi)^{b+c}\}.$$

Thus

$$x = a + d$$

is B-sufficient and the probability of x is

$$p(x; \lambda, \pi) = \binom{n}{x}\{\lambda(1 - \pi)^x\pi^{n-x} + (1 - \lambda)\pi^x(1 - \pi)^{n-x}\}.$$

The statistic

$$y = \min\{x, n - x\}$$

has distribution

$$p(y; \pi) = \binom{n}{y}\{(1-\pi)^y \pi^{n-y} + \pi^y (1-\pi)^{n-y}\}$$

which does not involve λ. Furthermore, the conditional distribution of x given y is concentrated on the two points y and $n - y$, and

$$p(x; \lambda, \pi | y) = \begin{cases} \dfrac{\lambda(1-\pi)^y \pi^{n-y} + (1-\lambda)\pi^y (1-\pi)^{n-y}}{(1-\pi)^y \pi^{n-y} + \pi^y (1-\pi)^{n-y}} & \text{for } x = y \\[2ex] \dfrac{(1-\lambda)(1-\pi)^y \pi^{n-y} + \lambda \pi^y (1-\pi)^{n-y}}{(1-\pi)^y \pi^{n-y} + \pi^y (1-\pi)^{n-y}} & \text{for } x = n - y. \end{cases}$$

Note that

$$(6) \qquad p(x; \lambda, \pi | y) = p(n - x; 1 - \lambda, \pi | y).$$

Suppose λ and π are variation independent and that π varies in $(0, \frac{1}{2}]$. (In most situations of genetical interest, values of the recombination probability greater than $\frac{1}{2}$ are not a realistic possibility. Moreover, by restricting π to the interval $(0, \frac{1}{2}]$ one ensures that different values of (λ, π) yield different distributions for x.) Denote the domain of variation for λ by Λ.

If $\Lambda = (0, 1)$ then, as follows immediately from (6) and the definition of G-nonformation, the statistic y is G-sufficient for x with respect to π.

Next, consider the case $\Lambda = (0, \frac{1}{2})$. This corresponds to an experimental situation where the base population, from which the double heterozygote used for the test cross was sampled, has arisen from a pure trans population by reproduction over an unknown number of generations. (λ increases from 0 to $\frac{1}{2}$ as the number of generations increases to infinity.) Here, G-nonformation does not hold, but y is still M-sufficient with respect to π, since

$$p(x; \lambda, \pi | y) \to \tfrac{1}{2} \qquad \text{for } \lambda \uparrow \tfrac{1}{2}.$$

Finally, suppose λ is known to be very small, such as would be the case if the base population is an originally pure trans type population which has, accidentally, been contaminated with a few cis individuals. Then y is not M-sufficient, and the conditional model given y may in fact yield strong evidence concerning π. To see this, note that if $x > n/2$ one has

$$p(x; \lambda, \pi | y) \to \begin{cases} \lambda \\ \frac{1}{2} \end{cases} \qquad \text{as } \pi \to \begin{cases} 0 \\ \frac{1}{2} \end{cases}.$$

Thus, if the observed value of x is greater than $n/2$ then conditionally on y small values of π are strongly counter-indicated. ▶

A number of examples of M-ancillarity in exponential families will be exhibited in Section 10.5. Furthermore, all the instances of S-ancillary statistics mentioned in the foregoing are also M-ancillary, as is simple to see. Therefore just one other case of M-ancillarity is presented here.

Example 4.16. Let $x_1, \ldots, x_n$ be independent and identically distributed, according to the negative binomial distribution which has point probabilities

$$(1-\pi)^{\chi}\binom{\chi+i-1}{i}\pi^{i}, \qquad i = 0, 1, 2, \ldots .$$

Let the domain of variation for the parameter (χ, π) be $(0, \infty) \times (0, 1)$.

The family of conditional distributions of $x = (x_1, \ldots, x_n)$ given $x_.$ is parametrized by χ alone, and $x_.$ follows the negative binomial distribution with parameter $(n\chi, \pi)$. For any fixed χ the class of distributions of $x_.$ is universal and hence $x_.$ is M-ancillary with respect to χ. ▶

Finally, after these exemplifications of S-, G-, and M-ancillarity and -sufficiency, a few, more general, points will be taken up.

As mentioned in Section 4.1, there is in general no guarantee that different processes of separation, based on the various definitions of nonformation, will lead ultimately to the same submodel, to be used as the framework for inference on the interest parameter ψ. However, certain kinds of uniqueness hold under regularity conditions.

Suppose that t and u are statistics such that for each fixed value of ψ one has that t is B-ancillary and u is B-sufficient. Thus

$$p(x; \omega) = p(t; \psi)p(x; \omega|t) = p(u; \omega)p(x; \psi|u). \tag{7}$$

Let it moreover be assumed that ψ parametrizes both the marginal distributions for t and the conditional distributions given u. If it is known or can be proved that

(*a*) x is a one-to-one function of (t, u),

(*b*) for each fixed value of ψ the family of marginal distributions for u is boundedly complete,

then, by (*b*) and Corollary 4.1, t and u are independent and hence the marginal model for t and the conditional model given u are, in fact, the same. As is simple to see, (*a*) and (*b*) are satisfied, in particular, if t is S-sufficient, u is S-ancillary, and $\mathfrak{P}$ is boundedly complete (the uniqueness conclusion in this case was, in essence, given by Dawid (1975)). Assuming (*b*), condition (*a*) also holds if x is minimal sufficient, $\mathfrak{P}$ is discrete and t is M-sufficient. To see this, note that, since t and u are independent,

$$p(x; \omega) = p(t; \psi)p(u; \omega)p(x; \omega|t, u) = p(u; \omega)p(x; \psi|u)$$

and hence $p(x; \omega|t, u)$ depends on ω through ψ only. The assumption of M-sufficiency then entails, on account of Corollary 2.1, that the conditional distribution of x given (t, u) is uniform and since x is minimal sufficient this is possible only if (t, u) stands in one-to-one correspondence with x.

Another type of uniqueness is ensured if a maximal ancillary or minimal sufficient statistic exists (see the end of the next section).

4.5 QUASI-ANCILLARITY AND QUASI-SUFFICIENCY

Two new notions termed quasi-ancillarity and quasi-sufficiency are introduced here, mainly because of their technical usefulness. These notions have a theoretical status between the general ideas of ancillarity and sufficiency, discussed in Section 4.1, and the specific ancillarity and sufficiency concepts designated by B-, S-, G-, and M-.

The results of this section will be used in Section 10.1.

Recall, from Section 4.2, that a statistic u is said to be conditionally B-ancillary given a statistic t if the conditional distribution of u given t does not depend on $P(\in \mathfrak{P})$.

For any statistic t, let ψ_t be a parameter function which induces the same partition of $\mathfrak{P}$ as does the mapping $P \to P_t$, i.e. ψ_t parametrizes the distributions of t. Then t is called *quasi-sufficient* with respect to ψ_t (and any function thereof) if any statistic u, such that t is a function of u and such that ψ_u induces the same partition of $\mathfrak{P}$ as ψ_t, is conditionally B-ancillary given t.

Similarly, let ψ^t denote a parameter function inducing the same partition of $\mathfrak{P}$ as the mapping $P \to P^t$. The statistic t is *quasi-ancillary* with respect to ψ^t (and any function thereof) if for any statistic u, such that ψ^u and ψ^t induce the same partition of $\mathfrak{P}$ and such that u is a function of t, one has that t is conditionally B-ancillary given u.

The content of the definition of quasi-ancillarity is this. Let t and u be statistics with u being a function of t, let ψ be a parameter function and suppose the three mappings $P \to P^t$, $P \to P^u$, and ψ all induce the same partition of $\mathfrak{P}$. Then $p(x; \omega)$ factorizes as

$$(1) \qquad p(x; \omega) = p(u; \omega)p(t; \psi|u)p(x; \psi|t).$$

If the conditional distribution of t given u depends effectively on ψ then the conditional distribution of x given t cannot be said to contain all the available information on ψ, i.e. t cannot be considered ancillary with respect to ψ. Thus quasi-ancillarity is a necessary condition for ancillarity.

Example 4.17. If x_1 and x_2 are independent Poisson variates having mean values λ_1 and λ_2, and if the domain of variation of (λ_1, λ_2) is given by $\lambda_2 \leq \lambda_1$ (cf. Example 3.4) then x_2 can, obviously, not be considered as ancillary with respect to λ_1. However, as will be shown in Section 10.1, x_2 is quasi-ancillary with respect to λ_1. Quasi-ancillarity is therefore not a sufficient condition for ancillarity. ▶

The *raison d'être* for quasi-sufficiency is similar to that for quasi-ancillarity.

It will now be shown that both S-ancillarity and M-ancillarity, at least for discrete type families $\mathfrak{P}$, imply quasi-ancillarity. Let t, u, and ψ be as in the paragraph preceding Example 4.17. If t is S-ancillary then it is a cut and hence the class of distributions of t for fixed ψ does not depend on ψ. From (1) one sees that the conditional distribution of t given u cannot, therefore, depend on ψ, whence t is quasi-ancillary. Next, suppose t is M-ancillary. If there exists a factorization (1) such that $p(t;\omega) = p(u;\omega)p(t;\psi|u)$ is the density considered in checking the universality condition then, by the remark following Corollary 2.1, $p(t;\psi|u)$ does not depend on ψ (and t) which implies that t is conditionally B-ancillary given u. Thus, if every one of the possible statistics u allows a factorization of the kind mentioned, which is certainly the case if $\mathfrak{P}$ is of discrete type, then t is quasi-ancillary.

It is natural to consider the following general definitions of minimal sufficiency and maximal ancillarity. Let the symbol $\square$ stand for either B, S, G, M, or quasi. The statistic t is *minimal* $\square$*-sufficient* w.r.t. the parameter function ψ if t is $\square$-sufficient w.r.t. ψ and if for any other statistic u, which is $\square$-sufficient w.r.t. ψ, one has $\sigma(t) \subset \sigma(u) \vee \breve{\varnothing}$. (Here, as in Section 4.2, $\breve{\varnothing}$ is the class of sets which have P-measure 0 for every $P \in \mathfrak{P}$.) Also, t is *maximal* $\square$*-ancillary* w.r.t. ψ if t is $\square$-ancillary w.r.t. ψ and if for any other statistic u, which is $\square$-ancillary w.r.t. ψ, one has $\sigma(u) \subset \sigma(t) \vee \breve{\varnothing}$. Minimal B-sufficiency and maximal B-ancillarity have already been discussed in Section 4.2. As was mentioned there, maximal B-ancillary statistics only rarely exist. However, for exponential families it is possible to establish a sufficient condition for maximal quasi-ancillarity (see Theorem 10.2) and, according to this condition, most of the S- and M-ancillary statistics mentioned in the examples of Section 4.4 and in Chapter 10 are maximal.

4.6 CONDITIONAL AND UNCONDITIONAL PLAUSIBILITY FUNCTIONS

Suppose again that the probability function for x factorizes as

$$p(x;\omega) = p(t;\omega)p(x;\psi|t).$$

In this section the relation between the unconditional and the conditional plausibility functions $\Pi(\omega;x)$ and $\Pi(\psi;x|t)$, in particular between the unconditional and conditional maximum plausibility estimates of ψ, will be discussed; those estimates are denoted, respectively, by $\check{\psi}(x) - \psi(\check{\omega}(x))$ and $\check{\psi}(x|t)$. (A parallel discussion of the corresponding likelihood quantities would be possible but trivial since it would, essentially, require as an assumption that t is a cut.)

It is not presupposed in what follows that t is (M-) ancillary (but the conditions in Theorem 4.7 below come close to implying M-ancillarity).

Let x be a mode point of the family $\mathfrak{p}$ of probability functions for x. If $\check{\omega}(x)$ is non-empty and $\omega \in \check{\omega}(x)$ then

$$p(x;\omega) \geq p(\tilde{x};\omega) \qquad \text{for all } \tilde{x} \in \mathfrak{X}$$

whence, for $\psi = \psi(\omega)$ and $t = t(x)$,

$$p(x;\psi|t) \geq p(\tilde{x};\psi|t) \qquad \text{for all } \tilde{x} \text{ with } t(\tilde{x}) = t. \tag{1}$$

Therefore, if x is a mode point for $\mathfrak{p}$ then

$$\check{\psi}(x) \subset \check{\psi}(x|t). \tag{2}$$

Without the assumption of x being a mode point, (2) is not, in general, true as is simple to see by example. Note moreover that if both sides of (2) are one-point sets, which is often the case with continuous type distributions, then the unconditional and conditional estimates are simply equal (in contrast to what usually occurs for maximum likelihood estimation). In Theorems 4.6 and 4.7 conditions will be given which are sufficient to ensure equality in (2), whatever the type of distribution.

Theorem 4.6. *Let x be a mode point of $\mathfrak{p}$ and suppose that for all values ψ there exists an ω such that $\psi = \psi(\omega)$ and $t \in t(\check{x}(\omega))$ (where $t = t(x)$).*
Then

$$\check{\psi}(x) = \check{\psi}(x|t)$$

and

$$\sup_{\omega|\psi} \Pi(\omega;x) = \Pi(\psi;x|t), \qquad \psi \in \Psi. \tag{3}$$

(Since x is a mode point, $\Pi(\omega;x) = \overline{\Pi}(\omega;x)$ and $\Pi(\psi;x|t) = \overline{\Pi}(\psi;x|t)$).

Proof. It was shown above that $\check{\psi}(x) \subset \check{\psi}(x|t)$ on account of x being a mode point. Thus, suppose $\psi \in \bar{\psi}(x|t)$. By Theorem 2.1, x is also a mode point for the conditional model given t and hence (1) holds. This allows the conclusion

$$p(x;\omega) \geq p(\tilde{x};\omega) \qquad \text{for all } \tilde{x}, \omega \text{ with } t(\tilde{x}) = t \text{ and } \psi(\omega) = \psi.$$

Now, choosing ω as indicated by the second assumption of the theorem and letting $\check{x}$ denote a point in $\check{x}(\omega)$ for which $t(\check{x}) = t$, one obtains

$$p(x;\omega) \geq p(\check{x};\omega) = \sup_x p(x;\omega),$$

which shows that $\omega \in \check{\omega}(x)$, and hence $\psi \in \check{\psi}(x)$.

The plausibility function for ω may be rewritten as follows

$$\Pi(\omega;x) = \frac{p(t;\omega)p(x;\psi|t)}{\sup_x p(x;\omega)}$$

$$= \frac{p(x;\psi|t)}{\sup_{x|t} p(x;\psi|t)} \frac{p(t;\omega)\sup_{x|t} p(x;\psi|t)}{\sup_{x} p(x;\omega)}$$

$$= \Pi(\psi;x|t)\frac{\sup_{x|t} p(x;\omega)}{\sup_{x} p(x;\omega)}.$$

Thus, in order to verify (3) it must be shown that

$$\sup_{\omega|\psi} \frac{\sup_{x|t} p(x;\omega)}{\sup_{x} p(x;\omega)} = 1,$$

but this equality is implied by the assumption made in the theorem. ▶

It may be noted that if the assumption in Theorem 4.6 is fulfilled for every mode point of $\mathfrak{p}$ then ψ and $\check{t}$ are variation independent (where $\check{t}(\omega) = t(\check{x}(\omega))$. On the other hand, if ψ and $\check{t}$ are variation independent then Theorem 4.6 applies to every mode point x of $\mathfrak{p}$ such that $t \in \check{t}(\Omega)$. Further, see Theorem 4.7 below.

Let $\check{x}(\psi|t)$ denote the set of modes of the conditional distribution $p(\cdot;\psi|t)$.

Theorem 4.7. *Assume that* $\mathfrak{p}$ *is universal. Then the two conditions*

(i) $\check{\psi}(x) = \check{\psi}(x|t)$ *for all* x

(ii) $\{x{:}x \in \check{x}(\omega),\ \psi(\omega) = \psi,\ t(x) = t\} = \check{x}(\psi|t)$ *for all values of* ψ *and* t *are equivalent. Moreover,* (i) *and* (ii) *and also*

(iii) $\sup_{\omega|\psi} \Pi(\omega;x) = \Pi(\psi;x|t)$ *for all values of* ψ *and* x

are implied by either of the following two conditions

(iv) ψ *and* $\check{t}$ *are variation independent, and* $\check{t}(\Omega) = t(\mathfrak{X})$

(v) t *and* $\check{\psi}$ *are variation independent, and* $\check{\psi}(\mathfrak{X}) = \psi$*, and* (iv) *and* (v) *are equivalent.*

The four conditions (i), (ii), (iv), *and* (v) *are equivalent provided the range of* $\check{\psi}(\cdot|t)$ *is* Ψ *for all values of* t.

Proof. By the universality of $\mathfrak{p}$ one has $x \in \check{x}(\omega) \Leftrightarrow \omega \in \check{\omega}(x)$ and (cf. Corollary 2.1) $x \in \check{x}(\psi|t) \Leftrightarrow \psi \in \check{\psi}(x|t)$, from which the equivalence of (i) and (ii) as well as the equivalence of (iv) and (v) follows simply. Conditions (i) and (iii) are consequences of (iv), on account of Theorem 4.6. Finally, if the range of $\check{\psi}(\cdot|t)$ equals Ψ then (i) implies (v). ▶

If $\mathfrak{p}$ is strictly universal then the relation $\check{t}(\Omega) = t(\mathfrak{X})$ in condition (iv) is automatically true. Note also that the precondition that $\check{\psi}(\cdot|t)$ has range Ψ is very weak; it is, in particular, satisfied if $\mathfrak{X}_t = \{x; t(x) = t\}$ is finite, cf. Section 2.2.

Example 4.18. Suppose $x = (x_1, \ldots, x_m)$ where $x_1, \ldots, x_m$ are independent Poisson variates having mean values $\lambda_1, \ldots, \lambda_m$, and let $t = x_.$ and $\psi = (\lambda_1/\lambda_., \ldots, \lambda_m/\lambda_.)$. To show that ψ and $\check{t}$ are variation independent, let $x_.$ and ψ be given. For any $\mu > 0$ let $\lambda_*(\mu) = (\lambda_1(\mu), \ldots, \lambda_m(\mu)) = \mu\psi$. Then $\lambda_i(\mu)$ is a nondecreasing function of μ for all i and hence it is possible to choose a μ such that $[\lambda_1(\mu)\} + \cdots + [\lambda_m(\mu)\} = x_.$, where the symbol $[\ \}$ carries the meaning given in Section 1.1. Since, with $\lambda_*(\mu)$ as the parameter value, the set of mode points for the distribution of x is determined by $([\lambda_1(\mu)\}, \ldots, [\lambda_m(\mu)\})$ one has that $x_. \in \check{t}(\lambda_*(\mu))$. Moreover, by construction, $\psi(\lambda_*(\mu)) = \psi$. This establishes the variation independence of ψ and $\check{t}$. In fact, all the conditions of Theorem 4.7 are fulfilled.

A simple derivation of Finucan's characterization of the modes of a multinomial distribution, mentioned in Section 2.3(i) is now possible. On account of Theorem 4.7(ii), the set of modes of the multinomial distribution having trial parameter x and probability vector ψ may be expressed as

$$\check{x}(\psi|x_.) = \{([\lambda_1\}, \ldots, [\lambda_m\}) : (\lambda_1/\lambda_., \ldots, \lambda_m/\lambda_.) = \psi, [\lambda_1\} + \cdots + [\lambda_m\} = x_.\}$$

or, equivalently,

$$\check{x}(\psi|x) = \{([\mu\psi_1\}, \ldots, [\mu\psi_m\}) : \mu > 0, \quad [\mu\psi_1\} + \cdots + [\mu\psi_m\} = x_.\},$$

which is Finucan's result. ▶

Example 4.19. Let $x = x_{**}$ be an $r \times c$ contingency table of independent Poisson variates, let t be the set of marginals $(x_{.*}, x_{*.})$ and let $\psi = \psi_{**}$ be the interaction parameter whose (i,j)th element is $\psi_{ij} = \theta_{ij} - \theta_{i.} - \theta_{.j} + \theta_{..}$ where $\theta_{ij} = \ln \lambda_{ij}$ and λ_{ij} is the mean value of x_{ij}.

Again, all the conditions of Theorem 4.7, and hence of Theorem 4.6, are satisfied. For r or c equal to 2 it is possible to show this in a simple way by the kind of argument used in Example 4.18. A proof for general r and c has been established by Jensen (1976). The difficult part of the proof consists in showing that boundary points of $\check{\psi}(x)$ are also boundary points of $\check{\psi}(x|t)$, which, in view of (2), is essentially what is needed to verify condition (i) of Theorem 4.7.

Note that since, by (2),

$$\text{(4)} \qquad \check{\psi}(x_{**}) \subset \check{\psi}(x_{**}|x_{..}) \begin{array}{c} \subset \ \check{\psi}(x_{**}|x_{.*}) \ \subset \\ \\ \subset \ \check{\psi}(x_{**}|x_{*.}) \ \subset \end{array} \check{\psi}(x_{**}|x_{.*}, x_{*.})$$

and, since, as just mentioned, the two uttermost estimates are equal, all the inclusions in (4) must be equalities, i.e. the maximum plausibility estimate of the interaction is the same whether one conditions on none, some, or both marginals. The estimate is given by

$$\check{\psi} = \psi\left(\prod_{i,j} [x_{ij}, x_{ij} + 1]\right).$$

In the simplest case, $r = c = 2$, this estimate is

$$\tilde{\psi} = \left[\ln\frac{x_{11}x_{22}}{(x_{12}+1)(x_{21}+1)}, \ln\frac{(x_{11}+1)(x_{22}+1)}{x_{12}x_{21}}\right].$$

An analogue, for $r \times c$ tables, of Finucan's result is derivable as in the previous example (see Jensen 1976). This yields, in particular, a description of the mode points for the multivariate hypergeometric distribution (which occurs for $\psi = 0$).

The paper by Jensen (1976) also contains a discussion of (conditional) plausibility inference for contingency tables of arbitrary dimensions. In particular, the usefulness of (3) is illustrated in that work. ►

For strongly unimodal, exponential families of continuous type the conditions of Theorem 4.7 are usually met (see Section 9.6).

4.7 COMPLEMENTS

(i) If, for an observed x and a parameter function ψ, the likelihood function $L(\omega) = p(x;\omega)$ has the property that there exists a value $\hat{\psi}$ of ψ such that for any ω it is possible to find an $\tilde{\omega}$ with $\psi(\tilde{\omega}) = \hat{\psi}$ and $L(\tilde{\omega}) \geq L(\omega)$ then it seems reasonable to speak of $\hat{\psi}$ as a *maximum likelihood estimate of* ψ even though a maximum likelihood estimate of ω itself may not exist. Similarly for plausibility functions and other ods functions.

Example 4.20. For a single observation from the normal distribution, where $\omega = (\xi, \sigma^2)$, the likelihood function has no maximum but x is a maximum likelihood estimate of ξ in the above sense. ►

(ii) *Model control.* A rather often advocated way of controlling a proposed model $\mathfrak{P}$ consists in seeking out a statistic u, say, which is conditionally B-ancillary under the model and investigating whether it is tenable to consider u as following the exactly known, possibly conditional, distribution which this statistic has, according to the model and the conditional B-ancillarity.

Thus, for instance, a specification of a location-scale model for a sample $x_1, \ldots, x_n$ may be controlled by testing that the B-ancillarity statistic

$$\left(\frac{x_1 - x_3}{x_1 - x_2}, \ldots, \frac{x_1 - x_n}{x_1 - x_2}\right) \tag{1}$$

has the parameter-free distribution prescribed by the model.

The above procedure raises a number of interesting questions, such as to what extent the procedure is, in any given case, exhaustive and specific for the model control problem. Without attempting anything like a comprehensive discussion of those questions, a couple of points will be mentioned here.

Suppose u is B-ancillary. It is then pertinent to ask whether x together with the conditional model given u contains accessible evidence with respect to the question of validity of the model. Similarly, if $u = x$ and the considered distribution of x is that conditional on a (minimal) B-sufficient statistic t, one may inquire as to the possible controllability of the model on the basis of t alone.

A model which is not universal will in general be controllable. For the two situations just considered it is therefore essential to inquire whether the conditional model given u, respectively the marginal model for t, is universal. In many cases the answer is affirmative. The location-scale example with u given by (1) is among these.

Concerning the specificity of the model control it is of some interest to know whether the probability measures of the given model are the only ones which, under some mild general regularity conditions, assign to u the (conditional) distribution at hand, i.e. whether the distribution of u is characteristic for the model. In a number of instances that is indeed the case, cf. the examples and references given below. It should however be kept in mind that such a characterization result may say very little of how sharp a check of the model the control based on the parameter-free distribution of u does yield.

Example 4.21. If $x_1, \ldots, x_n$ are independent and identically normally distributed then

$$u = \left(\frac{x_1 - \bar{x}}{s}, \ldots, \frac{x_n - \bar{x}}{s} \right)$$

(where $\bar{x} = x_./n$ and $s^2 - \Sigma (x_i + \bar{x})^2/(n - 1)$) follows the uniform distribution on the hypersphere in R^n given by $\Sigma u_i = 0$, $\Sigma u_i^2 = 1$. The converse is true under weak regularity conditions (see Zinger and Linnik 1964). ▶

Example 4.22. Suppose $x_1, \ldots, x_n$ with $n > 2$ are independent, identically distributed, and positive random variates of continuous type and let

$$y_i = \sum_{j=1}^{i} (x_j/x_.), \qquad i = 1, 2, \ldots, n-1$$

and

$$u = -\sum_{i=1}^{n-1} 2 \ln y_i.$$

Then u has χ^2-distribution with $2(n-1)$ degrees of freedom if and only if the x_i are exponentially distributed (Csörgö and Seshadri 1970).

Example 4.23. In a 2×2 contingency table

x_1	$n_1 - x_1$	n_1
x_2	$n_2 - x_2$	n_2
$x_.$	$n_. - x_.$	$n_.$

(with x_1 and x_2 independent) the conditional distribution of x_1 given $x_.$ is hypergeometric if and only if x_1 and x_2 are both binomially distributed with a common probability parameter. This is a consequence of a theorem due to Patil and Seshadri (1964) and Menon (1966). ▶

For further results of a similar kind see Bolger and Harkness (1965), Bolshev (1965), Csörgö and Seshadri (1970), Menon (1966), Prohorov (1966), and Rasch (1974).

(iii) It is possible for the components of a partition $(\omega^{(1)},\ldots,\omega^{(m)})$ of ω to be L-independent without this independence being induced by cut(s).

Example 4.24. For a birth and death process, observed during the time interval $[0, T]$, the likelihood function is

$$\lambda^b \mu^d \mathrm{e}^{-(\lambda+\mu)z},$$

and λ and μ are L-independent (see Example 3.8). If this L-independence corresponded to a cut then the cut would be S-sufficient for either λ or μ, say λ. For fixed μ the pair (b, z) is minimal sufficient (cf., for instance, Theorem 4.3), and (b, z) would therefore have to be such a cut, which is impossible since the distribution of (b, z) depends on μ, as is apparent for instance from the formula (Puri 1968)

$$Ez = \frac{\mathrm{e}^{(\lambda-\mu)T} - 1}{\lambda - \mu} l.$$ ▶

(iv) *Ancillarity-sufficiency; combination.* The present subsection contains a brief and somewhat informal discussion of certain possibilities for extending and combining some of the more basic aspects of the considerations in Sections 4.1–4.4.

Let t and u be statistics, assume that t is a function of u, and let

$$p(x;\omega) = p(t;\omega)p(u;\omega|t)p(x;\omega|u) \qquad (2)$$

be the factorization of the probability function for x into the marginal density of t, the conditional density of u given t and the conditional density of x given u. Corresponding to this factorization the experiment yielding the observation x may be viewed as being composed of three successive experiments. The first experiment leads to observation of t. In the next experiment u is observed, the experiment being such that the distribution of u has probability function $p(\cdot;\omega|t)$. Finally, the third experiment yields x with distribution $p(\cdot;\omega|u)$.

Suppose that $p(u;\omega|t)$ depends on the interest parameter ψ only so that (2) has the form

$$p(x;\omega) = p(t;\omega)p(u;\psi|t)p(x;\omega|u).$$

One may ask then whether the information that the quantity t, which determines the second experiment, was arrived at by a random experiment with distribution $p(t; \omega)$ and that the second experiment was followed by a third with distribution $p(x; \omega|u)$, is in its totality irrelevant as regards inference on ψ.

The question will not be treated in any detail here. Suffice it to mention the rather obvious idea of contemplating combinations of B-, S-, G-, and M-ancillarity with B-, S-, G-, and M-sufficiency, thus introducing, for instance, the concept of M-B-nonformation defined as follows. The statistic (t, x) is said to be *M-B-nonformative* with respect to ψ provided: (i) for each ψ the corresponding family of probability functions for t is universal; (ii) the conditional distribution of x given u does not depend on ω. If (t, x) is M-B-nonformative then it is arguable that $(\mathfrak{P}_u(\cdot|t), u)$ contains all the information on ψ given by $(\mathfrak{P}, x)$.

The above remarks are related to the paper by Cox (1975) which discusses a generalization of the ideas of conditional and marginal likelihood.

(v) *On Birnbaum's Theorem.* Birnbaum's Theorem states that the *sufficiency axiom* (S) and the *conditionality axiom* (C) together implies the *likelihood axiom* (L) (see Birnbaum 1962, 1969). Here, (S) specifies that if t is a B-sufficient statistic then the statistical evidence on ω contained in $(\mathfrak{P}, x)$ is equivalent to that obtained in $(\mathfrak{P}_t, t)$, (C) is the analogous specification for a B-ancillary statistic, while (L) says that the evidence is entirely conveyed by the likelihood function corresponding to the observed x.

This result has caused much discussion because many statisticians have considered (C) and (S) as necessary, or at least acceptable, building blocks of a satisfactory theory of statistical inference, whereas they have found (L) unacceptable for various reasons, the most prominent being the contradictions existing between the approach to inference which flows from (L) and the classical way of performing significance tests. The prototype of such contradictions is Armitage's (1961) example, which builds on the fact that the likelihood function is independent of the stopping rule.

The attempts by Durbin (1970) (see also Birnbaum 1970, and Savage 1970) and Kalbfleisch (1975) to eschew the problem by introducing modified versions of (S) and (C) and rules for the order in which these are to be applied, although interesting, have not yielded convincing and comprehensive solutions.

As pointed out in Barndorff-Nielsen (1975), Birnbaum's result may be paraphrased as saying that if it is set up as a requirement that application of the ideas of (B-) sufficiency and (B-) ancillarity does never lead to conflicting, or non-equivalent, conclusions then these conclusions have to obey the likelihood principle. However, as mentioned in Section 1.1, such a uniqueness requirement appears unwarrantable.

(vi) *Some non-uniqueness examples.* A number of examples will be mentioned which show that different applications of various of the ancillarity, sufficiency, and nonformation concepts to one and the same statistical situation $(\mathfrak{P}, x)$ may lead to essentially different derived situations. Cases of this kind certainly present difficulties, but non-uniqueness in the conclusions of statistical investigations is not in general to be considered as extraordinary or unacceptable, cf. Section 1.1.

Example 4.25. When no maximal B-ancillary statistic exists, conditioning on different, relatively maximal, B-ancillary statistics will generally result in differing inferential statements about the parameter. ▶

Example 4.26. Durbin (1969) indicated, by an example, that the order in which the principles of B-sufficiency and -ancillarity are applied is not irrelevant. Developing this example, Dawid (1975) showed the following.

Let $x = (u, y, z)$, let y and z be conditionally independent given u, and assume that u takes the values 1 and 2, each with probability $\frac{1}{2}$, while given u the variate y has the negative binomial distribution

$$\binom{y+u-1}{y}\phi^u(1-\phi)^y,$$

and the variate z follows the binomial distribution

$$\binom{n}{z}\psi^z(1-\psi)^{n-z}$$

for $u = 1$, and the negative binomial distribution

$$\binom{z+k-1}{z}(1-\psi)^k\psi^z$$

for $u = 2$. Here n and k are known integers with $k \leq n$ and the parameter $\omega = (\phi, \psi)$ has the domain $(0, 1)^2$.

Obviously, (u, y) is S-ancillary with respect to ψ. On the other hand, the statistic

$$t = \begin{cases} (u, z) & \text{for } z \neq n-k \\ (1, n-k) & \text{for } z = n-k \end{cases}$$

is (minimally) S-sufficient with respect to ψ. The conditional situation is not the same as the marginal, and they cannot be reconciled by further application of S-ancillarity or -sufficiency. (It may, however, be noted that when the observed value of z is different from $n - k$, reconciliation is achievable by applying the concept of pointwise B-nonformation to the marginal situation, thereby recovering the conditional distribution of z given u as the appropriate model for inference on ψ.) ▶

Example 4.27. One of the means of investigating the reading ability of children is to let them read one or several short texts aloud, recording for each child and text the time taken to carry through the reading. Let t_{ij} denote the time recorded for child i, $i = 1, \ldots, l$, and text j, $j = 1, \ldots, m$, let n_j be the number of words in text j

and suppose that within each of the texts the words are fairly homogeneous. For such situations Rasch (1960) proposed the model which specifies that the t_{ij} are independent with t_{ij} being distributed as a sum of n_j independent exponential variates having parameter of the form $\delta_i \varepsilon_j$.

Set $x_{ij} = \ln t_{ij}, \alpha_i = \ln \delta_i, \beta_j = \ln \varepsilon_j$ and

$$x_{i.} = \sum_j w_j x_{ij}, x_{.j} = \sum_i v_i x_{ij}, x_{..} = \sum v_i w_j x_{ij}$$

$$\alpha_. = \sum v_i \alpha_i, \beta_. = \sum w_j \beta_j$$

with the v_i and w_j being arbitrarily selected, positive weights which satisfy

$$\sum v_i = 1, \qquad \sum w_j = 1.$$

It was observed by Rasch that the distribution of $x_{..}$ depends on $\alpha_. + \beta_.$ only, while the distributions of the vectors $(x_{1.} - x_{..}, \ldots, x_{l.} - x_{..})$ and $(x_{.1} - x_{..}, \ldots, x_{.m} - x_{..})$ depend on $(\alpha_1 - \alpha_., \ldots, \alpha_l - \alpha_.)$ and $(\beta_1 - \beta_., \ldots, \beta_m - \beta_.)$, respectively; further, the distribution of the matrix $[x_{ij} - x_{i.} - x_{.j} + x_{..}]$ is independent of all of the parameters.

The model belongs to the general class of additive models

$$x_{ij} = \alpha_i + \beta_j + u_{ij} \tag{3}$$

for which the matrix $[u_{ij}]$ of error variables has a known distribution. (In the particular case of the reading speed model, the u_{ij} are independent and $\exp(u_{ij})$ follows a gamma distribution with shape parameter n_j and scale parameter 1.)

It is immediately obvious from (3) that Rasch's observation holds for any of the models in this class. (It may also be noted that for each of these models the family of distributions of x_{**} is a group family.)

Moreover, for any such model, the statistic $(x_{1.} - x_{..}, \ldots, x_{l.} - x_{..})$ is G-sufficient with respect to $(\alpha_1 - \alpha_., \ldots, \alpha_l - \alpha_.)$. Similarly for $x_{..}$ together with $\alpha_. + \beta_.$, and for $(x_{.1} - x_{..}, \ldots, x_{.m} - x_{..})$ together with $(\beta_1 - \beta_., \ldots, \beta_m - \beta_.)$.

Since $[x_{ij} - x_{i.} - x_{.j} + x_{..}]$ is B-ancillary, inference concerning the parameters should according to the ancillarity principle be drawn in the conditional model given $[x_{ij} - x_{i.} - x_{.j} + x_{..}]$. This conditional model does (clearly) also belong to the general class considered here. Thus it is arguable that a proper distribution for inference on $(\alpha_1 - \alpha_., \ldots, \alpha_l - \alpha_.)$, say, is that of $(x_{1.} - x_{..}, \ldots, x_{l.} - x_{..})$ given $[x_{ij} - x_{i.} - x_{.j} + x_{..}]$. But this distribution is different from the marginal distribution of $(x_{1.} - x_{..}, \ldots, x_{1.} - x_{..})$ unless $(x_{1.} - x_{..}, \ldots, x_{1.} - x_{..})$ and $[x_{ij} - x_{i.} - x_{.j} + x_{..}]$ are independent; if the u_{ij} are independent then the two distributions are equal only if the u_{ij} are normally distributed, as follows from the Skitovič—Darmois theorem (see e.g. Linnik 1964) which states that independence of two linear combinations, with nonzero coefficients, of a set of independent random variates implies normality of these latter variates.

As a basis for inference on $(\alpha_1 - \alpha_., \ldots, \alpha_l - \alpha_.)$ the conditional distribution of

$(x_{1.} - x_{..}, \ldots, x_{1.} - x_{..})$ given $[x_{ij} - x_{i.} - x_{.j} + x_{..}]$ may be found preferable to the marginal, on the ground that the control of the model based on the ancillary statistic $[x_{ij} - x_{i.} - x_{.j} + x_{..}]$ appears as completely untangled as possible from inference on $(\alpha_1 - \alpha_., \ldots, \alpha_I - \alpha_.)$ when the conditional distribution is employed. ▶

4.8 NOTES

The ideas of sufficiency and ancillarity were introduced, respectively, in Fisher (1920) and Fisher (1934, 1935). A precursor of the ancillarity idea had however been mentioned in the last section of Fisher (1925); this earlier idea is commented on in Efron (1975). It may also be added that the conditional, exact test for the 2×2 contingency table was proposed in the fifth edition of Fisher's *Statistical Methods for Research Workers* (1934). Fisher's discussion of sufficiency in the papers Fisher (1920, 1921, 1925) has often been taken as being concerned with what has here been called B-sufficiency, but it seems that he may have had a more general concept in mind, rather like: a statistic t is said to be sufficient with respect to a parameter of interest ψ if for any statistic u such that the joint distribution of t and u depends on ψ only it holds that the conditional distribution of u given t is independent of ψ.

In the mid-thirties, separate inference was also advocated by Bartlett (1936, 1937). The first of these papers concerns conditional estimation, the second conditional testing. In the latter paper Bartlett presented: (i) the derivation of the conditional likelihood ratio test for the identity of the variances in k independent, normal samples; (ii) the viewpoint that control of whether a sample is normally distributed ought to be carried out in the, parameter-free, conditional distribution given $(\bar{x}, s^2)$, and similarly for the exponential distribution; (iii) the viewpoint that test for the identity of k Poisson distributions, when one observation is available from each distribution, in principle ought to be carried out in the conditional distribution given the sum of the observations, and similarly for the binomial distribution; (iv) a discussion of the conditional likelihood ratio test for independence in the general two-dimensional contingency table.

The cornerstone of the Neyman–Pearson test theory is the paper by Neyman and Pearson (1933) in which test power is proposed as the central criterion for evaluation of the quality of tests. In that work the concept of a similar test is also introduced and it is shown that the requirement of similarity implies, in a wide class of cases, that the test is composed of conditional tests with the same level as the test itself. However, although this result is much used for the construction of tests in the Neyman–Pearson approach (see, for instance, Lehmann 1959), it has never been part of that approach to perform the tests separately, in the conditional model. (Conditional inference given a B-ancillary statistic has been criticized by Welch (1939) on the ground that it may yield confidence intervals which if viewed unconditionally are inefficient. Apart from the fact that this kind of comparison is begging the question, Welch did not, in his argument, allow the conditional procedure its full flexibility and accordingly his conclusion does not hold, cf. Barnard (1976).)

A highly clarifying discussion of the meaning of ancillarity, in which the idea of mixture experiments was brought to the fore, was given by Cox (1958).

The work by G. Rasch on what he has called measurement models and specific objectivity should also be mentioned as a very considerable impetus in the field of inferential separation (see Rasch 1960, 1961, 1968).

The abstract theory of B-sufficiency built up primarily by Halmos and Savage (1949) and Bahadur (1954) is more general than is needed for the main part of statistical inference purposes. Moreover, it has turned out that, on this general level, the proposed definitions of sufficiency and minimal sufficiency do not possess certain basic properties which are, essentially, always met in applications. For instance, a minimal sufficient σ-algebra and a minimal sufficient statistic, in the sense of Bahadur (1954), may both (Pitcher 1957) or each one separately (Landers and Rogge 1972b) fail to exist. (A minimal sufficient σ-algebra always exists provided the family of probability measures is dominated, but even under this assumption it can happen that a minimal sufficient statistic, in the Bahadur sense, is not available, as demonstrated in the latter paper.) Moreover, a common regular conditional probability measure given a sufficient σ-algebra need not exist, even if the original σ-algebra is separable and each member of the family of probability measures admits a regular conditional probability measure given the sufficient σ-algebra, cf. Landers and Rogge (1972a). These problems, which from the statistical viewpoint are fictitious, do not occur within the framework for B-sufficiency discussed in Section 4.2, due to the following facts: (i) the sample space is Euclidean (it would have been sufficient to assume that the basic σ-algebra was separable); (ii) only sufficiency of statistics, and not sub-σ-algebras, is considered, and statistics are defined as taking values in a Euclidean space; (iii) $\mathfrak{P}$ is dominated by a σ-finite measure; (iv) a B-sufficient statistic is defined as one for which there exists a common regular conditional probability measure given the statistic (rather than in terms of conditional expectations, as in the Halmos–Savage–Bahadur approach). In essence, Theorem 4.1 is due to Halmos and Savage (1949) while Theorem 4.2 and Corollary 4.1 were given by Bahadur (1954). Theorem 4.3 and Corollary 4.3 have been presented in Barndorff-Nielsen, Hoffmann-Jørgensen, and Pedersen (1976).

Basu (1959) has discussed the classes of ancillary statistics and events from an abstract viewpoint. In concrete cases it is often an intricate mathematical problem to determine these families. An outstanding classical example is that of finding similar tests for the Behrens–Fisher problem, see Linnik (1968). (The problem of finding the class of similar (nonrandomized) tests of a given hypotheses is the same as the problem of finding the class of B-ancillary events for a certain sub-family $\mathfrak{P}_0$ of $\mathfrak{P}$.)

The term nonformation and the definitions of (pointwise) B-, S-, G-, and M-nonformation were introduced in Barndorff-Nielsen (1976c).

The notions of S-sufficiency and S-ancillarity are due respectively to Fraser (1956) and to Sverdrup (1966) and Sandved (1967). Some properties of S-ancillarity were discussed in Sandved (1972). The definition of G-sufficiency is virtually equivalent to Barnard's (1963a) definition of a concept of sufficiency (cf. Section 4.4), and M-ancillarity and M-sufficiency were proposed in Barndorff-Nielsen (1973a,c). A critique of certain other proposals for definition of ancillarity or sufficiency may be found in Barndorff-Nielsen (1973a).

The material on quasi-ancillarity and quasi-sufficiency, presented in Section 4.6, is new, while the core of the material in Section 4.5 was presented in Barndorff-Nielsen (1976b) (see also Jensen 1976).

The population to which a proposition is referred and by which its probability is determined was termed the *reference set* by R. A. Fisher. Conditional inference involves a change of reference set. An overview of Fisher's ideas on the role of the reference set in statistical, particularly fiducial, inference is available in Pedersen (1976).

PART II

Convex Analysis, Unimodality, and Laplace Transforms

A concise account is given of those parts of the subjects of convexity, unimodality, and Laplace transforms which will be invoked, for statistical purposes, in Part III. In particular, conjugate convex functions and certain convex duality properties are discussed.

CHAPTER 5

Convex Analysis

5.1 CONVEX SETS

The convex hull of a set M in R^k will be denoted by conv M. The relative interior of a convex set C is denoted by ri C, the relative boundary of C by rbd C, and dim C will stand for the dimension of C (i.e. the dimension of the affine hull of C).

Theorem 5.1. *For any convex set C in R^k,* $\text{cl}(\text{ri}\, C) = \text{cl}\, C$ *and* $\text{ri}(\text{cl}\, C) = \text{ri}\, C$.

For a proof see Rockafellar (1970), p. 46. ▶

Let M_1 and M_2 be arbitrary subsets of R^k. A hyperplane H is said to *separate* M_1 and M_2 if M_1 is contained in one of the two closed halfspaces determined by H and M_2 is contained in the other closed halfspace. H separates M_1 and M_2 *properly* if M_1 and M_2 are not both contained in H. It is said to separate M_1 and M_2 *strongly* provided it separates M_1 and M_2 and provided M_1 and M_2 are both at a positive distance from H.

Theorem 5.2. *Let C_1 and C_2 be convex sets in R^k. In order that there exists a hyperplane separating C_1 and C_2 properly it is necessary and sufficient that* ri C_1 *and* ri C_2 *have no point in common.*

For a proof see Rockafellar (1970), p. 97. ▶

Theorem 5.3. *Let M be a subset of R^k. If $x \in \text{conv}\, M$ then x can be written as a convex combination of $k + 1$ points of M. Moreover, if $x \in \text{int}\,(\text{conv}\, M)$ then there exists a natural number $m \leq 2k$ and points $x_1, x_2, \ldots, x_m$ in M such that $x \in \text{int conv}\, \{x_1, \ldots, x_m\}$.*

The first assertion is, of course, Carathéodory's theorem. For a proof of the second assertion see Valentine (1964), p. 41. ▶

Let x be a point of a convex set C. A vector x^* is said to be *normal* to C at x provided

$$(z - x) \cdot x^* \leq 0 \qquad \text{for all } z \in C.$$

The set of vectors which are normal to C at x form a convex cone $K(x)$ called the *normal cone* to C at x. Clearly, $K(x) = \{0\}$ for $x \in \text{int}\, C$, while $K(x)$ contains a halfline for every $x \in C \setminus \text{int}\, C$. Defining the normal cone to C at a point $x \notin C$ by $K(x) = \emptyset$ one has then established a mapping K on all of R^k, the *normal cone mapping*.

Any halfline or nonzero vector in R^k determines a direction of R^k. Formally, a *direction* of R^k is defined as an equivalence class of closed halflines in R^k, two closed halflines being equivalent if they are translates of each other. A convex subset C of R^k is said to *recede in the direction D* if C includes all the halflines in the direction D which start at points of C. The set consisting of the zero vector and of all the vectors which determine the directions of recession of C is a convex cone; it is called the *recession cone* of C and is denoted by $0^+ C$ (cf. Rockafellar 1970). If C is closed or open and if it contains some closed halfline in the direction D then it actually recedes in the direction D. Moreover, for any convex set C one has

$$0^+(\text{ri}\, C) = 0^+(\text{cl}\, C) \supset 0^+ C. \tag{1}$$

The recession cone of a closed convex set is closed and if K is a closed convex cone then

$$0^+ K = K. \tag{2}$$

Let C be a convex set in R^k and let x^* be a nonzero vector in R^k. C is said to be *bounded in the direction determined by* x^* provided there exists an $\alpha \in R$ such that $x \cdot x^* \leq \alpha$ for every $x \in C$. The *barrier cone* of C, denoted by bar C, is the set consisting of the zero vector and of the vectors in the direction of which C is bounded. Any such cone is convex. For any convex set C

$$\text{bar}\, C = \text{bar}\,(\text{ri}\, C) = \text{bar}\,(\text{cl}\, C). \tag{3}$$

The *polar* of a convex cone K is the convex cone K^0 defined by $K^0 = \{x^* : x^* \leq 0\}$. The polar K^0 is closed and

$$K^{00} = \text{cl}\, K. \tag{4}$$

Theorem 5.4. *The polar of the barrier cone of a closed convex set C is the recession cone of C; in symbols*

$$(\text{bar}\, C)^0 = 0^+ C. \tag{5}$$

For a proof see Rockafellar (1970), p. 123. ▶

Theorem 5.5. *A closed convex set C in R^k is bounded if and only if its recession cone $0^+ C$ consists of the zero vector alone.*

For a proof see Rockafellar (1970) p. 64. ▶

If C is a convex set, the set $(-0^+ C) \cap 0^+ C$ is called the *lineality space* of C and is denoted by line C. The set lin C is the largest subspace contained in $0^+ C$. It follows from (1) that

$$\text{lin}\,(\text{cl}\, C) = \text{lin}\,(\text{ri}\, C) \supset \text{lin}\, C. \tag{6}$$

One has (cf. Rockafellar 1970, p. 65)

$$C = \text{lin}\, C + (C \cap (\text{lin}\, C)^{\perp}). \tag{7}$$

Theorem 5.6. *Suppose K is a closed convex cone. Then*

$$(\text{lin}\, K)^{\perp} = \text{aff}\, K^{0}. \tag{8}$$

This result is a special case of Theorem 14.6 in Rockafellar (1970). ▶

Theorem 5.7. *Let C be a non-empty convex set. Then*

$$(\text{lin}\,(\text{cl}\, C))^{\perp} = (\text{lin}\,(\text{ri}\, C))^{\perp} = \text{aff}(\text{bar}\,(\text{cl}\, C)) = \text{aff}(\text{bar}\,(\text{ri}\, C)).$$

Proof. By (6) and (3) it suffices to show that if C is closed then

$$(\text{lin}\, C)^{\perp} = \text{aff}(\text{bar}\, C).$$

We have

$$\text{aff}(\text{bar}\, C) = \text{aff}(\text{cl}(\text{bar}\, C))$$

and (cf. (4) and Theorem 5.4)

$$(0^{+} C)^{0} = \text{cl}\,(\text{bar}\, C).$$

Hence, applying Theorem 5.6, we find

$$\begin{aligned} \text{aff}(\text{bar}\, C) &= \text{aff}(0^{+} C)^{0} \\ &= (\text{lin}\, 0^{+} C)^{\perp}. \end{aligned}$$

It thus remains to prove that

$$\text{lin}\, C = \text{lin}\, 0^{+} C. \tag{9}$$

Since C is closed, $0^{+}C$ is a closed convex cone and hence $0^{+}(0^{+}C) = 0^{+}C$, cf. (2). The equality (9) now follows from the definition of lineality space. ▶

A *face* of a convex set C is a convex subset F of C such that any (closed) line segment in C with a relative interior point in F has both endpoints in F. Obviously, $\varnothing$ and C are both faces of C; any other face is called *proper*.

A point of C is an *extreme point* if it cannot be expressed as a convex combination of two other points in C which are both different from the former. The extreme points of C coincide with those faces of C which consist of precisely one point.

The collection of the relative interiors of the (non-empty) faces of a convex set C constitutes a partition of C; in other words, every point in C belongs to precisely one of those relative interiors. (cf. Rockafellar 1970, p. 164).

Theorem 5.8. *Let* $C = \text{conv}\, M$, *where M is a subset of* R^k, *and let F be a non-empty face of C.*

Then $F = \text{conv}\,(M \cap F)$.

For a proof see Rockafellar (1970) p. 165. ▶

A convex set C is a *polytope* if it is the convex hull of finitely many points in R^k. Any proper face F of a polytope C is of the form $F = H \cap C$ where H is a supporting hyperplane of C.

The affine mappings considered in the remainder of the present section are all on R^k into R^k.

Lemma 5.1. *If a is an affine mapping and C a convex set in R^k then $a(C)$ is convex and* $\text{ri}\, a(C) = a(\text{ri}\, C)$, $\text{cl}\, a(C) \supset a(\text{cl}\, C)$.

For a proof see Rockafellar (1970), p. 48. ▶

The next three lemmas are simple to prove.

Lemma 5.2. *Any affine mapping a is of the form $a = p \circ a_0 + b$ where a_0 is a regular linear mapping, p is a projection and b a translation.*

Lemma 5.3. *If a is a regular affine mapping and M an arbitrary set in R^k then* $a(\text{conv } M) = \text{conv}\, a(M)$, $a(\text{cl conv } M) = \text{cl conv}\, a(M)$ *and* $a(\text{ri conv } M) = \text{ri conv}\, a(M)$.

Lemma 5.4. *If p is a projection and M an arbitrary set in R^k then* $p(\text{conv } M) = \text{conv}\, p(M)$, $\text{cl}\, p(\text{cl conv } M) = \text{cl conv}\, p(M)$ *and* $p(\text{ri conv } M) = \text{ri conv}\, p(M)$.

On combining Lemmas 5.1–5.4 one obtains

Theorem 5.9. *If a is an affine mapping and M an arbitrary subset of R^k then* $a(\text{conv } M) = \text{conv}\, a(M)$, $\text{cl}\, a(\text{cl conv } M) = \text{cl conv}\, a(M)$ *and* $a(\text{ri conv } M) = \text{ri conv}\, a(M)$.

5.2 CONVEX FUNCTIONS

Let f be a function defined on a subset D of R^k and with values in $[-\infty, \infty]$. The set

$$\{(x, \eta) : x \in D, \eta \in R, f(x) \leq \eta\}$$

is called the *epigraph* of f and is denoted by epi f. The function f is *convex* if f does not take the value $-\infty$ and if epi f is non-empty and convex as a subset of R^{k+1}. (Rockafellar (1970) uses a slightly wider definition of convex function in that he allows $-\infty$ as function value and also he does not require epi f to be non-empty. The definition used here coincides with Rockafellar's definition of a proper convex function.) This definition is equivalent to the requirement that D is convex and that f is finite for at least one $x \in D$ and satisfies

$$f((1-\lambda)x + \lambda y) \leq (1-\lambda)f(x) + \lambda f(y)$$

whenever $x \in D$, $y \in D$ and $0 \leq \lambda \leq 1$. As testified by Rockafellar's writings there are great technical advantages in allowing $+\infty$ as a value in the definition of convex function. Note that a convex function f can always be extended to a convex function on all of R^k by setting $f(x) = +\infty$ for $x \notin D$.

The *effective domain* of a convex function f is the set $\{x: f(x) < \infty\}$, which is convex. This set will be denoted by $\operatorname{dom} f$. Thus, if f is a convex function on R^k then domain $f = R^k$ while $\operatorname{dom} f$ is, in general, a genuine subset of R^k.

It is simple to prove the following two theorems:

Theorem 5.10. *If f is a convex function on R^k and $\mathbf{A}$ is a $k \times m$ matrix then the function $\tilde{f}$ defined on R^m by*

$$\tilde{f}(y) = \inf\{f(x): x\mathbf{A} = y\}$$

is convex, provided it nowhere takes the value $-\infty$. (Even if $-\infty$ is a value of $\tilde{f}$, this function is still convex in the extended sense that epi f *is convex.)*

Theorem 5.11. *Let f be a convex function on R^k and let M be a subset of R^k. Then*

$$\sup\{f(x): x \in \operatorname{conv} M\} = \sup\{f(x): x \in M\}$$

and the first supremum is attained only when the second (more restricted) supremum is attained.

A function g on R^k into $[-\infty, \infty)$ is *concave* provided $-g$ is convex.

Example 5.1. Let g be a concave, everywhere finite function on R^k, and let h be a non-increasing convex function on R.

Then $f = h \circ g$ is convex because

$$\begin{aligned} f((1-\lambda)x + \lambda y) &\leq h((1-\lambda)g(x) + \lambda g(y)) \\ &\leq (1-\lambda)f(x) + \lambda f(y) \end{aligned}$$

for every $x, y \in \operatorname{dom} f$ and $\lambda \in [0, 1]$. ▶

A function f on R^k into $(-\infty, \infty]$ is said to be *quasiconvex* if for every $\alpha \in R$ the level set $\{x: f(x) \leq \alpha\}$ is convex. (It is obvious that any convex function is quasiconvex.)

Theorem 5.12. *A function f on R^k into $(-\infty, \infty]$ is convex if and only if for every $x^* \in R^k$ the function*

$$f(x) - x^* \cdot x, \qquad x \in R^k$$

is quasiconvex.

Proof. The only if assertion is trivial.

Suppose f is not convex. Then there exist $x_0, x_1 \in \operatorname{dom} f$ and $\lambda \in (0, 1)$ such that, letting $\alpha_0 = f(x_0)$ and $\alpha_1 = f(x_1)$,

$$f((1-\lambda)x_0 + \lambda x_1) > (1-\lambda)\alpha_0 + \lambda\alpha_1.$$

Determine $x^* \in R^k$ and $c_0 \in R$ so that

$$x^* \cdot (x_1 - x_0) = \alpha_1 - \alpha_0$$

$$x^* \cdot x_0 + c_0 = \alpha_0.$$

Then the graph of the affine function

$$x^* \cdot x + c_0, \qquad x \in R^k$$

contains the points (x_0, α_0) and (x_1, α_1), and hence x_0 and x_1 belong to the level set

$$\left\{x : f(x) - x^* \cdot x \le c_0\right\}.$$

But $(1 - \lambda)x_0 + \lambda x_1$ is not an element of this set and thus

$$f(x) - x \cdot x^*, \qquad x \in R^k$$

is not quasiconvex. ▶

Recall that a function f on a set $D(\subset R^k)$ into $(-\infty, \infty]$ is, by definition, lower semi-continuous at a point $x \in D$ if

$$f(x) = \liminf_{y \to x} f(y).$$

If f is any convex function on R^k then the closure (in R^{k+1}) of its epigraph is the epigraph of a certain function which is called the *closure* of f and is denoted by $\mathrm{cl}\, f$. $\mathrm{cl}\, f$ is a convex and lower semi-continuous and $\mathrm{cl}\, f = f$ if and only if f is lower semi-continuous. f is said to be *closed* if $f = \mathrm{cl}\, f$. In any case $\mathrm{cl}\, f$ agrees with f except perhaps at relative boundary points of $\mathrm{dom}\, f$. Moreover, we have:

Theorem 5.13. *Let f be a convex function on R^k. Then for every $x \in \mathrm{ri}(\mathrm{dom}\, f)$ and every $y \in R^k$*

$$(\mathrm{cl}\, f)(y) = \lim_{\lambda \uparrow 1} f((1 - \lambda)x + \lambda y).$$

Corollary 5.1. *For a closed convex function f on R^k one has*

$$f(y) = \lim_{\lambda \uparrow 1} f((1 - \lambda)x + \lambda y)$$

for every $x \in \mathrm{dom}\, f$ and every y.

For proofs of Theorem 5.13 and its corollary see Rockafellar (1970), p. 57. ▶

Theorem 5.14. *Let f be a closed convex function on R^k.*

The non-empty sets among the level sets $\{x : f(x) \le \alpha\}$, $\alpha \in R$, are closed and convex and they all have the same recession cone.

For a proof see Rockafellar (1970), p. 58 and p. 70. ▶

A convex function f on R^k is *polyhedral* if its epigraph is the intersection of a finite collection of closed halfspaces of R^{k+1}. Such a function is closed, and it attains its infimum provided it is bounded below (cf. Rockafellar 1970, p. 268).

The *recession function* of a convex function f is the convex function whose epigraph is $0^+(\text{epi} f)$. It is denoted by $f0^+$.

Theorem 5.15. *Let f be a convex function, and let y be a vector. If one has*

$$\liminf_{\lambda \to \infty} f(x + \lambda y) < \infty$$

for a given x, then x actually has the property that $f(x + \lambda y)$ is a non-increasing function of λ, $-\infty < \lambda < \infty$. This property holds for every x if and only if $(f0^+)(y) \leq 0$. When f is closed, the property holds for every x if it holds for even one $x \in \text{dom} f$.

For a proof see Rockafellar (1970), p. 68. ▶

Example 5.2. For any open subset C of R^k there exist closed convex functions f such that $\text{dom} f = C$ and

$$\liminf_{\lambda \to \infty} f(x + \lambda y) = \infty \tag{1}$$

for every x and every $y \neq 0$. This will be shown here by exhibiting a concrete example of such a function.

Nothing is lost by assuming $0 \in C$. Moreover, if $C = R^k$ then the function $x \cdot x$ is of the desired kind, so suppose $C \neq R^k$.

Let d be the function on R^k such that $d(x)$ is the distance of x to the boundary of C provided $x \in C$, while $d(x)$ is zero otherwise. Then d is continuous and concave. The latter assertion may be proved as follows. Let $x_0, x_1 \in C$, let S_i be the sphere with centre x_i and radius $d(x_i)$, $i = 1, 2$, and set $M = \text{bd conv}(S_0 \cup S_1)$. For any $\lambda \in (0, 1)$ one finds

$$\begin{aligned} d((1 - \lambda)x_0 + \lambda x_1) &\geq \inf_{z \in M} \|(1 - \lambda)x_0 + \lambda x_1 - z\| \\ &= (1 - \lambda)d(x_0) + \lambda d(x_1). \end{aligned}$$

Now define f by

$$f(x) = x \cdot x + 1/d(x), \qquad x \in R^k.$$

This function is closed convex and it clearly satisfies (1) for $x = 0$ and hence, by Theorem 5.15, for all x. ▶

Let f be an arbitrary function on R^k into $(-\infty, \infty]$ (which is not identically $+\infty$). The *convex hull of f* is the function $\text{conv} f$ on R^k defined by

$$(\text{conv} f)(x) = \inf\{\eta : (x, \eta) \in \text{conv epi} f\}. \tag{2}$$

This function is convex, provided it does not take the value $-\infty$, and it is the greatest convex function $\leq f$. Set $S = \text{dom}\, f$ and let $\mathcal{S}$ consist of the points $(x, f(x))$, $x \in S$ and the direction of $(0, 1)$ (where $0 \in R^k$, $1 \in R$). It is simple to see that

$$\text{(3)} \qquad \text{conv epi} f = \text{conv}\, \mathcal{S}$$

where $\text{conv}\, \mathcal{S}$ is defined by $\text{conv}\, \mathcal{S} = \text{conv}\, S + \{\lambda(0, 1) : 0 \leq \lambda < \infty\}$.

Theorem 5.16. *Let f be an arbitrary function on R^k into $(-\infty, \infty]$.*
One has

$$\text{(4)} \qquad (\text{conv} f)(x) = \inf\left\{\sum_{i=1}^{k+1} \lambda_i f(x_i) : \sum_{i=1}^{k+1} \lambda_i x_i = x\right\}$$

where the infimum is taken over all expressions of x as a convex combination of $k + 1$ points in R^k.

For a proof see Rockafellar (1970), p. 157. ▶

Theorem 5.17. *Let f be a function on R^k into $(-\infty, \infty]$, set $S = \text{dom} f$ and let $\mathcal{S}$ consist of the points $(x, f(x))$, $x \in S$, and the direction (0.1). Suppose S is a finite set.*
Then $\text{conv} f$ is a closed convex function and $\text{epi}(\text{conv} f) = \text{conv}\, \mathcal{S}$.

Proof. $\text{conv}\, \mathcal{S}$ is a closed set and

$$(\text{conv} f)(x) = \inf\{\eta : (x, \eta) \in \text{conv}\, \mathcal{S}\}$$

cf. (2) and (3). The theorem now follows at once. ▶

Let $\{f_\omega : \omega \in \Omega\}$ be an arbitrary collection of closed convex functions on R^k and set

$$f = \sup\{f_\omega : \omega \in \Omega\}$$

the supremum being taken pointwise. Then f is also a closed convex function, unless it is identically $+\infty$.

5.3 CONJUGATE CONVEX FUNCTIONS

The conjugate of a convex function f on R^k is the function f^* on R^k defined by

$$f^*(x^*) = \sup(x \cdot x^* - f(x)), \qquad x^* \in R^k.$$

Clearly, for any convex function f on R^k

$$\text{(1)} \qquad x \cdot x^* \leq f(x) + f^*(x^*), \qquad x \in R^k, x^* \in R^k.$$

This inequality is called *Fenchel's inequality.*

Theorem 5.18. *Let f be a convex function on R^k. The conjugate is then a closed convex function on R^k. Moreover, $(\text{cl} f)^* = f^*$ and $f^{**} = \text{cl} f$. Thus, in particular, $f = f^{**}$ if and only if f is closed.* ▶

For a proof see Rockafellar (1970), p. 104.

More generally than the first assertion of Theorem 5.18 one has that for any f on R^k into $(-\infty, \infty]$ the function f^* defined on R^k by

$$f^*(x^*) = \sup\{x \cdot x^* - f(x)\}$$

is closed convex (unless it is identically $+\infty$) because it is the supremum of closed convex (in fact, affine) functions. Whether f is convex or not, f^* will be called the *conjugate* of f. Using Theorem 5.16 it is simple to see that

$$f^* = (\operatorname{conv} f)^*. \tag{2}$$

Examples of conjugate pairs of closed convex functions are:

Example 5.3. If

$$f(x) = \frac{1}{p}|x|^p, \qquad x \in R,$$

where $1 < p < \infty$, then

$$f^*(x^*) = \frac{1}{q}|x^*|^q, \qquad x^* \in R$$

with $1/p + 1/q = 1$. ▶

Example 5.4. Suppose f is a positive semi-definite quadratic form on R^k,

$$f(x) = \tfrac{1}{2}x\mathbf{c}x', \qquad x \in R^k$$

$\mathbf{c}$ being a symmetric, positive semi-definite $k \times k$ matrix.

If $\mathbf{c}$ is non-singular then

$$f^*(x^*) = \tfrac{1}{2}x^*\mathbf{c}^{-1}x^{*\prime}, \qquad x^* \in R^k.$$

Thus, in particular, for the function $f_0(x) = \frac{1}{2}x \cdot x$, $x \in R^k$, we have $f_0 = f_0^*$. This function is the only function on R^k which is equal to its own conjugate. In fact, any such function satisfies

$$x \cdot x \leq f(x) + f^*(x) = 2f(x)$$

and hence $f \geq f_0$ which in turn implies $f^* \leq f_0^*$. Since $f_0 = f_0^*$ and, by assumption, $f = f^*$ we must have $f = f_0$.

In general, for $\mathbf{c}$ arbitrary positive semi-definite

$$f^*(x^*) = \begin{cases} \frac{1}{2}x^*\tilde{\mathbf{c}}x^{*\prime} & x^* \in L \\ +\infty & x^* \notin L \end{cases}$$

where L is the orthogonal complement of the subspace $\{x : x\mathbf{c} = 0\}$ and where $\tilde{\mathbf{c}}$ is the unique symmetric positive semi-definite $k \times k$ matrix satisfying $\tilde{\mathbf{c}}\mathbf{c} = \mathbf{c}\tilde{\mathbf{c}} = \mathbf{p}$ with $\mathbf{p}$ the matrix of the orthogonal projection of R^k onto L. ▶

Example 5.5. Let f be defined on R^k by

$$f(x) = \begin{cases} x_1 \ln x_1 + \cdots + x_k \ln x_k & \text{for } x \in \Delta \\ +\infty & \text{for } x \notin \Delta \end{cases}$$

where

$$\Delta = \{x : x = (x_1, \ldots, x_k), x_1 \geqslant 0, \ldots, x_k \leq 0, x_1 + \cdots + x_k = 1\}$$

and where $0 \ln 0$ is interpreted as 0. Then f is closed convex and the conjugate of f is given by

$$f^*(x^*) = \ln(e x_1^* + \cdots e x_k^*), \qquad x^* \in R^k.$$

For a proof see Rockafellar (1970), pp. 148–149. ▶

The relationship between the effective domains of a convex function f and its conjugate depends heavily on the behaviour of f and consequently very little can be said in general about this relationship. There is however one simple and useful result in this area, noted by Fenchel (1953), p. 93.

Theorem 5.19. *Let f be any convex function on R^k. Then the barrier cone of* dom f *is contained in the recession cone of* dom f^*; in symbols

$$\text{bar}(\text{dom} f) \subset 0^+(\text{dom} f^*).$$

Proof. Suppose $y \in \text{bar}(\text{dom} f)$, $y \neq 0$ and $x^* \in \text{dom} f^*$. We have to show that $\{x^* + \lambda y : \lambda \geq 0\} \subset \text{dom} f^*$. Now, both $(y, 0)$ and $(x^*, -1)$ belong to the barrier cone of epi f and since barrier cones are convex

$$(x^*, -1) + \lambda(y, 0) = (x^* + \lambda y, -1) \in \text{bar}(\text{epi} f), \qquad \lambda \geq 0$$

which implies $x^* + \lambda y \in \text{dom} f^*$ for every $\lambda \geq 0$. ▶

That the inclusion in Theorem 5.19 may be strict can be seen e.g. by taking $f(x) = \frac{1}{2} x \cdot x$, $x \in R^k$. In this instance $\text{bar}(\text{dom} f) = \{0\}$ while $0^+(\text{dom} f^*) = R^k$.

Theorem 5.20. *Let f be a closed convex function on R^k.*

In order that $\{x : f(x) \leq \alpha\}$ be a bounded (as well as closed, convex) set in R^k for all $\alpha \in R$ it is necessary and sufficient that $0 \in \text{int}(\text{dom} f^*)$. *In this case the infimum of f is attained.*

For a proof see Rockafellar (1970), p. 123 and p. 265. ▶

The *indicator function* of a convex set C in R^k is defined by

$$\delta(x|C) = \begin{cases} 0 & \text{for } x \in C \\ \infty & \text{for } x \notin C. \end{cases}$$

The function $\delta(\cdot|C)$ is obviously convex, and it is closed if and only if C is closed. Its conjugate $\delta^*(\cdot|C)$ is equal to the *support function* of C, i.e.

$$\delta^*(x^*|C) = \sup\{x \cdot x^* : x \in C\}, \qquad x^* \in R^k.$$

Theorem 5.21. *Let f be a convex function. The support function of* dom f *is then the recession function f^*0^+ of f^*. If f is closed, the support function of* dom f^* *is the recession function $f0^+$ of f.*

For a proof see Rockafellar (1970), p. 116. ▶

Let $f_1, \ldots, f_m$ be convex functions on R^k. The *infimal convolution* of $f_1, \ldots, f_m$ is defined on R^k by

$$f_1 \square f_2 \square \cdots \square f_m(x) = \inf\{f_1(x_1) + f_2(x_2) + \cdots + f_m(x_m) : x_1 + x_2 + \cdots + x_m = x\}$$

and is again a convex function, provided it does not take the value $-\infty$ for any $x \in R^k$. The operation of infimal convolution is, essentially, dual to that of addition of convex functions. In fact

$$(f_1 \square \cdots \square f_m)^* = f_1^* + \cdots + f_m^*$$

and if the sets $\operatorname{ri}(\operatorname{dom} f_i)$, $i = 1, \ldots, m$, have a point in common then

$$(f_1 + \cdots + f_m)^* = f_1^* \square \cdots \square f_m^*, \tag{3}$$

cf. Rockafellar (1970), p. 145.

Consider a convex function f on R^{l+m} and denote its argument by (u, v) where $u \in R^l$, $v \in R^m$. The function

$$h(u, v^*) = \sup_v \{v \cdot v^* - f(u, v)\}, \qquad u \in R^l, v^* \in R^m$$

is called a *partial conjugate* of f. For each fixed u it is either a closed convex function of v^* or identically $-\infty$. And for each fixed v^* it is, on account of Theorem 5.10, either concave or, if it takes the value $+\infty$, concave in the extended sense indicated by that theorem. (h is a so-called saddle function, cf. Rockafellar (1970), p. 349.) The set of points (u, v^*) for which $h(u, v^*)$ is finite is denoted dom h.

Set

$$D = \{u : (u, v) \in \operatorname{dom} f \text{ for some } v\}$$
$$E = \{v^* : (u^*, v^*) \in \operatorname{dom} f^* \text{ for some } u^*\}.$$

It will now be shown that

$$\operatorname{ri} D \times E \subset \operatorname{dom} h \subset D \times \operatorname{cl} E \tag{4}$$

provided f is closed.

Note first that the functions $f_u(\cdot) = f(u, \cdot)$, $u \in D$, are closed convex functions on R^m which all have the same recession function. Hence, by Theorem 5.21, the effective domains of the conjugates $(f_u)^*$ of these functions must be equal, modulo relative boundary points. Moreover, for $u \in \text{ri}\, D$ one has

$$\text{dom}\,(f_u)^* = E. \tag{5}$$

To see this, observe that $v^* \in \text{dom}\,(f_u)^*$ if and only if $(u, v^*) \in \text{dom}\,(f + \delta(\cdot|u \times R^m))^*$ and that

$$\begin{aligned}(f + \delta(\cdot|u \times R^m))^*(u, v^*) &= (f^* \,\square\, \delta^*(\cdot|u \times R^m))(u, v^*)\\ &= \inf_{u^*} \{f^*(u^*, v^*) - u \cdot u^*\} + u \cdot u.\end{aligned}$$

It is now simple to verify (4).

In the case where D and E are both open then

$$\text{dom}\, h = D \times E. \tag{6}$$

5.4 DIFFERENTIAL THEORY

Let f be a convex function on R^k and let x^* be a vector in R^k. Then x^* is said to be a *subgradient* of f at a point x if

$$f(z) \geq f(x) + (z - x) \cdot x^* \qquad \forall\, z \in R^k. \tag{1}$$

The set of all subgradients of f at x is called the *subdifferential* of f *at* x and is denoted by $\partial f(x)$. The (possibly) multivalued mapping $\partial f : x \to \partial f(x)$ is the *subdifferential* of f and f is said to be *subdifferentiable at* x provided $\partial f(x) \neq \emptyset$. Obviously

$$f^*(x^*) = x \cdot x^* - f(x) \Leftrightarrow x^* \in \partial f(x). \tag{2}$$

In other words: for each $x^* \in R^k$ the concave function $l(\cdot; x^*)$ on R^k defined by $l(x; x^*) = x . x^* - f(x)$, $x \in R^k$, attains its supremum at x if and only if $x^* \in \partial f(x)$. If f is closed then the latter condition is equivalent to $x \in \partial f^*(x^*)$. Consequently, the mappings ∂f and ∂f^* are each other's inverse provided f is closed.

The domain of ∂f, i.e. the set $\{x : \partial f(x) \neq \emptyset\}$, will be denoted by $\text{dom}\, \partial f$.

Example 5.6. Consider the indicator function $\delta(\cdot|C)$ of a convex set C in R^k. It is obvious from the defining relation (1) that $\partial\delta(x|C) = \emptyset$ if $x \notin C$, while for $x \in C$ the vector x^* is a subgradient of $\delta(\cdot|C)$ at x if and only if $0 \geq (z - x) \cdot x^*$ for every $z \in C$, i.e. x^* is a normal to C at x. Thus

$$\partial\delta(\cdot|C) = K$$

where K denotes the normal cone mapping on C. ▶

Theorem 5.22. *Let f be a convex function on R^k. Then*

$$\text{ri}\,(\text{dom}\, f) \subset \text{dom}\, \partial f \subset \text{dom}\, f.$$

A point x in the relative boundary of $\operatorname{dom} f$ *is not contained in* $\operatorname{dom} \partial f$ *if and only if*

$$\frac{f(x + \lambda(x_0 - x)) - f(x)}{\lambda} \to -\infty \qquad \text{as } \lambda \downarrow 0$$

for one—and hence every—point $x_0 \in \operatorname{ri}(\operatorname{dom} f)$.

The range of ∂f *is contained in* $\operatorname{dom} f^*$.

The proof of this theorem is easy to deduce by means of Theorems 23.4, 23.2, and 23.3 on pp. 216–217 in Rockafellar (1970). ▶

Theorem 5.23. *Let f be a function on R^k into $(-\infty, \infty]$ for which* $\operatorname{dom} f$ *is finite. Set* $\tilde{f} = \operatorname{conv} f$.

Then $\operatorname{dom} \partial \tilde{f} = \operatorname{dom} \tilde{f}$.

Proof. Set $S = \operatorname{dom} f$ and let S' consist of the points $(x, f(x))$, $x \in S$, and the direction $(0, 1)$ (where $0 \in R^k$, $1 \in R$). From Theorem 5.17 we have $\operatorname{conv} S' = \operatorname{epi} \tilde{f}$.

Consider a point $x \in \operatorname{dom} \tilde{f}$ and set

$$M = \{(x, \eta) : \eta \leq \tilde{f}(x)\}.$$

Then $\operatorname{conv} S' \cap \operatorname{ri} M = \emptyset$ and hence, by Rockafellar (1970) Theorem 20.2, there exists a hyperplane separating $\operatorname{conv} S'$ and M properly and not containing M. Such a hyperplane is obviously a nonvertical supporting hyperplane to $\operatorname{epi} \tilde{f}$ at $(x, \tilde{f}(x))$ and consequently $\partial \tilde{f}(x) \neq \emptyset$.

Theorem 5.24. *Let $f_1, \ldots, f_m$ be convex functions on R^k and let $f = f_1 + \cdots + f_m$. Then*

$$\partial f(x) \supset \partial f_1(x) + \cdots + \partial f_m(x) \qquad \forall\, x \in R^k.$$

If the convex sets $\operatorname{ri}(\operatorname{dom} f_i)$, *$i = 1, 2, \ldots, m$, have a point in common, then actually*

$$\partial f(x) = \partial f_1(x) + \cdots + \partial f_m(x) \qquad \forall\, x \in R^k.$$

For a proof see Rockafellar (1970), p. 223. ▶

Corollary 5.2. *Suppose $f_1, \ldots, f_m$ are closed convex functions on R^k such that the sets* $\operatorname{ri}(\operatorname{dom} f_i^*)$, *$i = 1, \ldots, m$, have a point in common.*

If $x_1, \ldots, x_m$ are points in R^k for which

$$\partial f_1(x_1) \cap \ldots \cap \partial f_m(x_m) \neq \emptyset$$

then

$$(f_1 \,\square \cdots \square\, f_m)(x_.) = f_1(x_1) + \cdots + f_m(x_m)$$

(where $x_. = x_1 + \cdots + x_m$).

Proof. Let $x^* \in \partial f_1(x_1) \cap \ldots \cap \partial f_m(x_m)$, then

$$x_. \in \partial f_1^*(x^*) + \cdots + \partial f_m^*(x^*) = \partial(f_1^* + \cdots + f_m^*)(x^*)$$

and hence, by (3) of Section 5.3 and (2) above,

$$\begin{aligned}(f_1 \square \cdots \square f_m)(x_.) &= (f_1^* + \cdots + f_m^*)^*(x_.) \\ &= x_. \cdot x^* - (f_1^* + \cdots + f_m^*)(x^*) \\ &= f_1(x_1) + \cdots + f_m(x_m).\end{aligned}$$ ▶

The following theorem describes the relationship between the concepts of subgradient and gradient of a convex function.

Theorem 5.25. *Let f be a convex function on R^k and let x be a point in R^k. Then $\partial f(x)$ contains exactly one element if and only if f is finite and differentiable at x (which implies $x \in \operatorname{int}(\operatorname{dom} f)$). In this case the element is $Df(x)$.*

For a proof see Rockafellar (1970), p. 242. ▶

The entire subdifferential mapping ∂f can, in fact, be constructed from the gradient mapping Df when f is a closed convex function with $\operatorname{int}(\operatorname{dom} f) \neq \emptyset$. Specifically:

Theorem 5.26. *Let f be a closed convex function with* $\operatorname{int}(\operatorname{dom} f) \neq \emptyset$. *Then*

$$\partial f(x) = \operatorname{cl}(\operatorname{conv} M(x)) + K(x) \qquad \forall x \in R^k$$

where $M(x)$ is the set of all limits of sequences of the form $Df(x_1), Df(x_2), \ldots$ such that f is differentiable at x_i and x_i tends to x and where $K(x)$ is the normal cone to $\operatorname{dom} f$ *at x.*

For a proof see Rockafellar (1970), p. 246. ▶

Consider a convex function f on R^k for which $\operatorname{int}(\operatorname{dom} f) \neq \emptyset$ and f is differentiable throughout $\operatorname{int}(\operatorname{dom} f)$. Such a function will be said to be *steep at x*, where x is a boundary point of $\operatorname{dom} f$, if

$$|Df(x_i)| \to \infty$$

whenever $x_1, x_2, \ldots$ is a sequence of points in $\operatorname{int}(\operatorname{dom} f)$ converging to x. Furthermore, f will be called *steep* if it is steep at all boundary points of $\operatorname{dom} f$.

Theorem 5.27. *Let f be a convex function on R^k such that* $\operatorname{int}(\operatorname{dom} f) \neq \emptyset$ *and f is differentiable throughout* $\operatorname{int}(\operatorname{dom} f)$.

Then f is steep if and only if

$$\frac{d}{d\lambda} f(x + \lambda(z - x)) \downarrow -\infty \qquad \text{as } \lambda \downarrow 0$$

for any $z \in \operatorname{int}(\operatorname{dom} f)$ and any boundary point x of $\operatorname{dom} f$.

For a proof see Rockafellar (1970), p. 252. ▶

Corollary 5.3. *Let f be a closed convex function on R^k such that* dom f *is open and f is differentiable* (*on* dom f).

Then f is steep.

Proof. Use Corollary 5.1.

A convex function f on R^k is *essentially smooth* if $\text{int}(\text{dom} f) \neq \varnothing$, f is differentiable on $\text{int}(\text{dom} f)$ and steep.

Theorem 5.28. *Let f be a closed convex function on R^k. Then ∂f is a single-valued mapping (i.e. for every x, $\partial f(x)$ contains at most one element) if and only if f is essentially smooth. In this case, ∂f reduces to the gradient mapping Df, i.e. $\partial f(x)$ consists of the vector $Df(x)$ alone when $x \in \text{int}(\text{dom} f)$ while $\partial f(x) = \varnothing$ when $x \notin \text{int}(\text{dom} f)$.*

For a proof see Rockafellar (1970), p. 252. ▶

A convex function f on R^k will be called *essentially strictly convex* if f is strictly convex on every convex subset of dom ∂f.

Theorem 5.29. *Let f be a closed convex function. Then f is essentially strictly convex if and only if $\partial f(x_1) \cap \partial f(x_2) = \varnothing$ whenever $x_1 \neq x_2$.*

For a proof see Rockafellar (1970), p. 254. ▶

Theorem 5.30. *A closed convex function on R^k is essentially strictly convex if and only if its conjugate is essentially smooth.*

For a proof see Rockafellar (1970), p. 254. ▶

The concept of conjugacy for convex functions is closely related to the classical concept of Legendre transformation. We shall now describe this correspondence. Let f be a differentiable real-valued function defined on an open subset U of $\mathbf{R}^k$. The *Legendre transform* of the pair (U, f) is defined to be the pair (V, g) where $V = Df(U)$ is the range of the gradient mapping Df and g is the function on V given by the formula

$$g(x^*) = x \cdot x^* - f(x)$$

where x satisfies

$$x^* = Df(x).$$

(The gradient mapping Df does not have to be one-to-one in order for g to be well defined, i.e. single-valued. For this it suffices that

$$x_1 \cdot x^* - f(x_1) = x_2 \cdot x^* - f(x_2),$$

whenever $Df(x_1) = Df(x_2) = x^*$. Then the value of $g(x^*)$ can be obtained unambiguously from the formula by replacing $(Df)^{-1}(x^*)$ by any of the vectors it contains.)

The classical areas of application of the Legendre transformation lie within the theory of differential equations (cf. Kamke 1930, 1974) and the calculus of variations (cf. Courant and Hilbert 1953, pp. 231–242).

If U and f are both convex, f can be extended (in unique manner) to a closed convex function on all of R^k with U as the interior of its effective domain. There is then the following relation between the Legendre transform of (U,f) and the conjugate of the extended f.

Theorem 5.31. *Let f be any convex function on R^k such that the set $U = \text{int}(\text{dom}\, f)$ is non-empty and f is differentiable on U. The Legendre transform (V, g) of (U,f) is then well-defined. Moreover, $V(= Df(U))$ is a subset of $\text{dom}\, f^*$ and g is the restriction of f^* to V.*

For a proof see Rockafellar (1970), p. 256. ▶

On adding the assumption that f is essentially smooth one obtains the following sharpening of Theorem 5.31:

Theorem 5.32. *Let f be any essentially smooth closed convex function on R^k and let $U = \text{int}(\text{dom}\, f)$ Then the Legendre transform (V, g) of (U,f) is well-defined. One has $V = \text{dom}\, f^*$ so that V is almost convex in the sense that $\text{ri}(\text{dom}\, f^*) \subset V \subset \text{dom}\, f^*$. Furthermore g is the restriction of f^* to V and g is strictly convex on every convex subset of V.*

For a proof see Rockafellar (1970), p. 257. ▶

The condition of steepness in the definition of essential smoothness is needed for the conclusion (in Theorem 5.32) that V is almost convex. Suppose, for example, that f is the closed convex function on $\mathbf{R}^2$ determined by

$$f(x) = \frac{x_1^2}{4x_2} \qquad \text{for } x_1 \in R, x_2 > 0.$$

The steepness condition fails at the origin $(0,0)$ and V is a parabola

$$V = \{x^* : x_2^* = -(x_1^*)^2\}.$$

This example is due to Rockafellar.

A pair (C,f) is said to be of *Legendre type* if C is an open convex set and f is a strictly convex and differentiable function on C such that

$$|Df(x_i)| \to \infty$$

whenever $x_1, x_2, \ldots$ is a sequence of points in C converging to a boundary point of C.

Clearly, if a convex function f on R^k is essentially strictly convex and essentially smooth then $(\text{int}(\text{dom}\, f), f)$ is of Legendre type.

Theorem 5.33. *Let f be a closed convex function. Let $C = \text{int}(\text{dom}\, f)$ and $C^* = \text{int}(\text{dom}\, f^*)$. Then (C,f) is a convex function of Legendre type if and only if*

(C^*,f^*) is a convex function of Legendre type. In this case (C^*,f^*) is the Legendre transform of (C,f) and vice versa. The gradient mapping Df is then one-to-one from the open convex set C onto the open convex set C^*, continuous in both directions, and $Df^* = (Df)^{-1}$.

For a proof see Rockafellar (1970), p. 258. ▶

Let f be a closed convex function on R^{l+m}, recall the notation used at the end of Section 5.3 in relation to partial conjugation, and set

$$M = \{(u, \partial f_u(v)) : (u, v) \in \operatorname{dom} \partial f\}.$$

Suppose $u \in \operatorname{ri} D$. Then a vector v^* belongs to $\partial f_u(v)$ if and only if (u^*, v^*) belongs to $\partial(f + \delta(\cdot|u \times R^m))(u, v)$ for every $u^* \in R^l$, as follows directly from the definition of subgradient. But, by Theorem 5.24

$$\partial(f + \delta(\cdot|u \times R^m)) = \partial f + \partial\delta(\cdot|u \times R^m)$$

and hence, for every $v \in R^m$,

$$\partial(f + \delta(\cdot|u \times R^m))(u, v) = \partial f(u, v) + R^l \times \{0\}.$$

Consequently, for $u \in \operatorname{ri} D$, $\partial f_u(v)$ equals the projection of $\partial f(u, v)$ onto $\{0\} \times R^m$ (interpreted as a subset of R^m).

From this and the proof of (4) in Section 5.3 it is simple to see that

$$\operatorname{ri} D \times \operatorname{ri} E \subset M \subset D \times \operatorname{cl} E$$

and that if f is of Legendre type then

$$M = \operatorname{int} D \times \operatorname{int} E. \tag{3}$$

Theorem 5.34. *Let f be a closed convex function with* $\operatorname{dom} f$ *open and suppose f is strictly convex and differentiable on* $\operatorname{dom} f$*. For $x \in \operatorname{dom} f$, denote $Df(x)$ by $d(= d(x))$, and let $x = (x^{(1)}, x^{(2)})$ and $d = (d^{(1)}, d^{(2)})$ be similar partitions of x and d.*

Then the mapping defined on $\operatorname{dom} f$ *by $x \to (x^{(1)}, d^{(1)})$ is a homeomorphism, and $x^{(1)}$ and $d^{(2)}$ are variation independent.*

Proof. Using Corollary 5.2 one finds that f is of Legendre type, and the theorem then follows from Theorem 5.33 and formula (3).

The idea of the proof of this theorem is due to Rockafellar. ▶

5.5 COMPLEMENTS

(i) A convex set is called polyhedral if it is the intersection of a finite collection of closed halfspaces.

Theorem 5.35. *Let f be a closed convex function on R^k, and let K be a non-empty* *closed convex cone in* R^k. *Let* K^* *be the negative of the polar of* K, *i.e.*

$$K^* = \{x^* : x \cdot x^* \geq 0 \qquad \forall\ x \in K\}.$$

One has

$$\inf\{f(x) : x \in K\} = -\inf\{f^*(x^*) : x^* \in K^*\}$$

if either of the following conditions hold:

(*a*) $\operatorname{ri}(\operatorname{dom} f) \cap \operatorname{ri} K \neq \varnothing$;

(*b*) $\operatorname{ri}(\operatorname{dom} f^*) \cap \operatorname{ri} K^* \neq \varnothing$.

Under (*a*), *the infimum of* f^* *over* K^* *is attained, while under* (*b*) *the infimum of* f *over* K *is attained.*

If K *is polyhedral,* ri K *and* ri K^* *can be replaced by* K *and* K^* *in* (*a*) *and* (*b*). *In general,* x *and* x^* *satisfy*

$$f(x) = \inf_K f = -\inf_{K^*} f^* = -f^*(x^*)$$

if and only if

$$x^* \in \partial f(x), x \in K, x^* \in K^*, x \cdot x^* = 0.$$

For a proof see Rockafellar (1970), p. 335. ▶

(**ii**) *Convex support.* Let π denote a probability measure on R^k. A point $x \in R^k$ is said to be a *point of support* for π provided every neighbourhood of x has positive π-measure. It follows immediately from Lindelöf's covering theorem that every Borel set with positive π-measure contains a point of support. Let $S = S_\pi$ denote the set of support points for π. S is called the *support* of π. It is a closed set and may be characterized as the smallest closed set having π-measure 1, i.e. the intersection of all closed sets with measure 1. The *convex support* $C = C_\pi$ of π is the closed convex hull of S. It is the smallest closed convex set with measure 1. (A more precise name for C would be 'closed convex support' but since the convex hull of S does not, in itself, play any prominent part in what follows, the shorter 'convex support' is adopted here.) Finally, aff S is called the *affine support* of π.

Let h denote a continuous mapping on R^k into R^k. Then $h(S_\pi) \subset S_{\pi h}$ and $h(S_\pi)$ is dense in $S_{\pi h}$. If h is a homeomorphism onto R^k then $h(S_\pi) = S_{\pi h}$.

Theorem 5.36. *If* a *is an affine mapping on* R^k *into* R^k *then* $\operatorname{cl} a(C_\pi) = C_{\pi a}$ *and* $a(\operatorname{ri} C_\pi) = \operatorname{ri} C_{\pi a}$.

Proof. If M_0 and M are arbitrary subsets of R^k with $M_0 \subset M$ and M_0 dense in M then $\operatorname{cl}\operatorname{conv} M_0 = \operatorname{cl}\operatorname{conv} M$. Hence, by Theorem 5.9,

$$\operatorname{cl} a(C_\pi) = \operatorname{cl} a(\operatorname{cl}\operatorname{conv} S_\pi) = \operatorname{cl}\operatorname{conv} a(S_\pi) = \operatorname{cl}\operatorname{conv} S_{\pi a} = C_{\pi a}$$

and

$$a(\text{ri } C_\pi) = a(\text{ri conv } S_\pi) = \text{ri conv } a(S_\pi) = \text{ri cl conv } a(S_\pi) = \text{ri } C_{\pi a}.$$ ▶

Let X denote the identity mapping on R^k. If the mean value of X with respect to π exists then $E_\pi X \in C$.

Suppose that $X_1, X_2, \ldots, X_n, \ldots$ is a sequence of independent k-dimensional random vectors with common marginal distribution π. Let S_n be the support of the distribution π_n of

$$\bar{X}_n = \frac{X_1 + \cdots + X_n}{n}.$$

Then

$$S_n = \text{cl}\{\bar{x} : \bar{x} = (x_1 + \cdots + x_n)/n, x_i \in S \text{ for } i = 1, \ldots, n\} \tag{1}$$

and the convex support of π_n is equal to C for all n, i.e.

$$\text{cl conv } S_n = C. \tag{2}$$

Moreover S_n is asymptotically dense in $\text{conv}\, S$ in the sense that to every $x_0 \in \text{conv}\, S$ and every neighbourhood U of x_0 there exists an n_0 such that

$$S_n \cap U \neq \emptyset \qquad \forall\, n \geq n_0.$$

(iii) *Jensen's inequality.* Let π be a probability measure on R^k with convex support C and suppose that its mean value

$$\mu = \int x\, d\pi$$

exists. Furthermore, let f be a convex function on R^k such that $C \subset \text{dom} f$ and $\mu \in \text{dom}\, \partial f$. Then Jensen's inequality

$$f(\mu) \leq \int f\, d\pi$$

is valid.

This may be seen by taking a $\mu^* \in \partial f(\mu)$, noting that

$$f(x) \geq f(\mu) + (x - \mu) \cdot \mu^* \qquad \forall\, x \in R^k,$$

and integrating with respect to π.

Suppose equality holds in Jensen's inequality. Then on a convex set having probability one f coincides with an affine function. And if f is strictly convex on C then $\pi\{\mu\} = 1$.

CHAPTER 6

Log-concavity and Unimodality

6.1 LOG-CONCAVITY

A function g on R^k and into $[0, \infty)$ is called logarithmically concave of *log-concave* if $\ln g$ is concave. Otherwise expressed, g is log-concave if $f = -\ln g$ is convex.

Theorems 6.1 and 6.2 below (which, except for a sharpening of Theorem 6.2, have previously been presented in Barndorff-Nielsen (1973b)) give various criteria for summability of log-concave functions. These are formulated in terms of f rather than g.

Theorem 6.1. *Let f be a closed convex function on R^k with* $\operatorname{int}\operatorname{dom} f \neq \emptyset$. *The following four conditions are equivalent*

(i)
$$\int e^{-f}\, d\lambda < \infty$$

(*where λ denotes Lebesgue measure on R^k*).

(ii) *There exists an $x \in \operatorname{dom} f$ such that*

$$\liminf_{\lambda \to \infty} f(x + \lambda y) = \infty$$

for every $y \in R^k$ with $y \neq 0$.

(iii) *There exist two scalars λ and α such that $\lambda > 0$ and*

$$f(x) \geq \lambda |x| + \alpha, \qquad x \in R^k.$$

(iv)
$$0 \in \operatorname{int}\operatorname{dom} f^*.$$

Proof. By Theorem 5.15, condition (ii) above is equivalent to the condition

(ii)′
$$(f0^+)(y) > 0, \qquad y \in R^k, y \neq 0.$$

Condition (ii)′, in turn, is equivalent to (iv) on account of Theorem 5.21.

Next it will be shown that (iv) implies (iii). f^* *is* continuous on $\operatorname{int}\operatorname{dom} f^*$ and hence, if $0 \in \operatorname{int}\operatorname{dom} f^*$, there exists a $\lambda > 0$ and an α such that

$$f^*(\lambda e(x)) \leq \alpha, \qquad x \neq 0$$

where $x \in R^k$ and $e(x)$ denotes the unit vector in the direction determined by x. From the definition of f^* it follows that

$$\lambda e(x) \cdot x - f(x) \leq f^*(\lambda e(x)), \qquad x \neq 0.$$

Consequently

$$f(x) \geq \lambda |x| - \alpha, \qquad x \neq 0$$

and thus (iii) holds.

It is simple to demonstrate that if (iii) is valid then so is (i).

Thus it only remains to prove that (i) entails (ii). Suppose (ii) is not fulfilled, then there exist an $x \in \operatorname{dom} f$ and a $y \neq 0$ for which

$$\liminf_{\lambda \to \infty} f(x + \lambda y) < \infty$$

and this implies, by Theorem 5.15, that $f(x' + \lambda y)$ is a non-increasing function of λ, $-\infty < \lambda < \infty$, for every $x' \in R^k$. Clearly, then

$$\int e^{-f} \, d\lambda = \infty. \qquad ▶$$

Theorem 6.2. *Let S be a subset of Z^k and let f be a closed convex function on R^k with* $\operatorname{int} \operatorname{dom} f \neq \varnothing$.

If

$$\int e^{-f} \, d\lambda < \infty \tag{1}$$

then

$$\sum_{i \in S} e^{-f(i)} < \infty.$$

The converse assertion holds provided $\varnothing \neq Z^k \cap \operatorname{int} \operatorname{dom} f \subset S$.

Proof. Set

$$M_\nu = \{ i : i \in S, \nu - 1 \leq |i| < \nu \}, \qquad \nu = 1, 2, \ldots .$$

Trivially, the number of elements in M_ν does not exceed $(2\nu + 1)^k$. Using this and the equivalence of (1) to condition (iii) in Theorem 6.1 one finds

$$\begin{aligned}\sum_{i \in S} e^{-f(i)} &= \sum_{\nu=1}^{\infty} \sum_{i \in M_\nu} e^{-f(i)} \\ &\leq \sum_{\nu=1}^{\infty} (2\nu + 1)^k e^{-\alpha - \lambda(\nu - 1)} \\ &< \infty.\end{aligned}$$

Suppose the conditions for the converse assertion are satisfied and assume that

$$\int e^{-f} d\lambda = \infty.$$

Then (cf. the end of the proof of Theorem 6.1) there exists a $y \neq 0$ such that $f(x + \lambda y)$ is non-increasing as a function of λ for every $x \in R^k$. Let i_0 be a point in $Z^k \cap \operatorname{int} \operatorname{dom} f$, let U be an open set such that $i_0 \in U$ and

$$\sigma = \sup\{f(x): x \in U\} < \infty,$$

and set

$$M = Z^k \cap \{x + \lambda y: x \in U, \lambda \geq 0\}.$$

Then $M \subset S$ and $f(i) \leq \sigma$ for $i \in M$. Moreover, as will be shown below, M contains infinitely many points. Consequently

$$\sum_{i \in S} e^{-f(i)} \geq \sum_{i \in M} e^{-\sigma} = \infty.$$

In proving that M is an infinite set, it causes no loss of generality to assume that $i_0 = 0$.Furthermore, it suffices to show that the distance between the two sets Z^k and $\{ny: n = 1, 2, \ldots\}$ is zero. If the coordinates $y_1, \ldots, y_k$ of y are such that they do not satisfy any equality

$$m_1 y_1 + \cdots + m_k y_k = m_0 \tag{2}$$

where $m_0, m_1, \ldots, m_k$ are integers, not all 0, then this follows from the well-known theorem of Kronecker (see e.g. Hardy and Wright (1960) p. 382 which states that the sequence ny modulo 1 ($n = 1, 2, \ldots$) is everywhere dense.) The general case, of possible dependence between $y_1, \ldots, y_k$ of the form (2), is now simply dealt with. ▶

A probability measure π on R^k is said to be *log-concave* if

$$\pi((1 - \lambda)C_0 + \lambda C_1) \geq \pi(C_0)^{1-\lambda} \pi(C_1)^{\lambda}$$

for all convex subsets C_0 and C_1 of R^k, and all $\lambda \in (0, 1)$.

It follows simply from this definition that if π is log-concave and if a is an affine mapping on R^k and into R^l then πa is also log-concave.

Theorem 6.3. *Let π be a probability measure on R^k of continuous type.*
Then π is log-concave if and only if $d\pi/d\lambda$ is log-concave.

This result is due to Prékopa (1971), who proved the if assertion, and to Borell (1975) (see also Prékopa 1973). ▶

6.2 UNIMODALITY OF CONTINUOUS-TYPE DISTRIBUTIONS

Let π be the probability measure of a continuous type distribution on R^k and let S and p denote, respectively, the support and the density of π. Set

$$\varphi = -\ln p.$$

The distribution is said to be *unimodal* if φ is quasiconvex and *strongly unimodal* if φ is convex. Clearly, these specifications are equivalent to p being quasiconcave, respectively log-concave. It may also be noted that in the one-dimensional case, π is unimodal if and only if there exists a real number m such that p is non-decreasing on $(-\infty, m]$ and non-increasing on $[m, \infty)$.

Unimodality is a rather weak property which allows only a very limited set of useful conclusions to be drawn. In particular, it does not hold that the marginal distributions of a unimodal distribution are unimodal (counterexamples are easily constructed), nor that the convolutions of unimodal distributions are again unimodal (for a one-dimensional counterexample, see Appendix II, written by K. L. Chung, in Gnedenko and Kolmogorov (1954)).

The closure, $\mathrm{cl}\, f$, of a convex function f on R^k agrees with f except, possibly, at relative boundary points of $\mathrm{dom}\, f$. Hence, when π is strongly unimodal, nothing is lost by assuming φ closed.

Suppose the distribution is strongly unimodal with φ closed. Since

$$\int e^{-\varphi}\, d\lambda = 1$$

one finds from Theorem 6.1 that $0 \in \mathrm{int\, dom}\, \varphi^*$ and hence, by Theorem 5.20, the non-empty sets among the level sets $\{x : p(x) \geq \alpha\}$, $\alpha \in R$, are all closed, bounded, and convex, and the set

$$\{x : p(x) = \sup_z p(z)\}$$

of mode points of p is non-empty.

On account of Theorem 6.3 and the remark preceding that result one has that if π is strongly unimodal and if a is an affine mapping on R^k and into R^l such that $1 \leq l \leq k$ and a has rank l then πa is strongly unimodal. In particular:

Theorem 6.4. *Marginal distributions of strongly unimodal (continuous type) distributions are again strongly unimodal.*

Corollary 6.1. *Convolutions of strongly unimodal (continuous type) distributions are again strongly unimodal.*

Proof. Let π_1 and π_2 be strongly unimodal probability measures on R^k, with densities p_1 and p_2. The product measure $\pi = \pi_1 \times \pi_2$ has density p given by $p(x) = p_1(x_1)p_2(x_2)$ where $x = (x_1, x_2) \in R^{2k}$. It is thus obvious that π is strongly unimodal, and since strong unimodality is preserved under regular affine transformations, the result follows. ▶

Corollary 6.1 was proved for $k = 1$ by Ibragimov (1956) and generally by Davidovič, Korenbljum, and Hacet (1969). In fact, it follows from Ibragimov (1956) that, in the one-dimensional case, one has the stronger result.

Theorem 6.5. *A one-dimensional, continuous-type distribution is strongly unimodal if and only if its convolution with any unimodal, continuous-type distribution is again unimodal.* ▶

(Actually, Ibragimov introduced the concept of strong unimodality by the latter property.) Theorem 6.5 does not generalize to higher dimensions; in fact, an example, due to T. W. Anderson and given in Sherman (1955), shows that the only if assertion of Theorem 6.5 is not true for two-dimensional distributions.

Example 6.1. Multivariate normal distribution. For the $N_r(\xi, \Sigma)$-distribution the function φ is, except for additive constants, given by

$$\tfrac{1}{2}x\Sigma^{-1}x' - \xi\Sigma^{-1}x'.$$

Thus $N_r(\xi, \Sigma)$ is strongly unimodal. ▶

Example 6.2. Gamma distribution. Up to additive constants the φ function of the gamma-distribution with shape parameter λ and scale parameter β is

$$-(\lambda - 1)\ln x + x/\beta.$$

The distribution is unimodal for all $(\lambda, \beta) \in (0, \infty)^2$ and strongly unimodal if $\lambda \geq 1$. ▶

Example 6.3. A one-dimensional distribution with density

$$a(\alpha, \beta)\, e^{-\alpha\varphi_0(x) + \beta x}$$

where α and β are parameters, $a(\alpha, \beta)$ is a norming constant, and φ_0 is a convex function, is obviously strongly unimodal provided $\alpha \geq 0$. Distributions of this form are: normal ($\varphi_0(x) = x^2$), gamma ($\varphi_0(x) = -\ln x$), Laplace ($\varphi_0(x) = |x|$), generalized inverse Gaussian with power parameter 1 ($\varphi_0(x) = 1/x$) and hyperbolic ($\varphi_0(x) = \sqrt{(1 + x^2)}$) (see the next example). ▶

Example 6.4. Multivariate hyperbolic distribution. The r-dimensional hyperbolic distribution has probability function

$$a(\alpha, \beta, \delta, \Sigma)\exp[-\alpha\sqrt{\{\delta^2 + (x - \xi)\Sigma^{-1}(x - \xi)'\}} + \beta \cdot (x - \xi)] \tag{1}$$

where $\alpha > 0$, $\delta > 0$, $\xi \in R^r$, $\beta \in R^r$, and Σ, a positive definite $r \times r$ matrix, are parameters and, for $\chi = \sqrt{(\alpha^2 - \beta\Sigma\beta')}$,

$$a(\alpha, \beta, \delta, \Delta) = \frac{1}{(2\pi)^{(n-1)/2}\sqrt{|\Sigma|}} \frac{\chi^{n+1}/(2\alpha)}{(\delta\chi)^{(n+1)/2}K_{(n+1)/2}(\delta\chi)},$$

$K_{(n+1)/2}(\cdot)$ denoting the modified Bessel function of the third kind and with index $(n+1)/2$. (The hyperbolic distributions were introduced in Barndorff–Nielsen (1976).) As a function of x, the logarithm of (1) determines a hyperboloid, and hence the distribution is strongly unimodal. ▶

Example 6.5. Dirichlet distribution. The density of the Dirichlet distribution with parameters $\lambda_1 > 0, \ldots, \lambda_{k+1} > 0$ is

$$\frac{\Gamma(\lambda_.)}{\Gamma(\lambda_1)\cdots\Gamma(\lambda_{k+1})} x_1^{\lambda_1 - 1} \ldots x_k^{\lambda_k - 1}(1 - x_.)^{\lambda_{k+1} - 1}$$

for $x_1 > 0, \ldots, x_k > 0$ and $x_. < 1$, and 0 otherwise. It is strongly unimodal provided $\lambda_i \geq 1$, $i = 1, \ldots, k+1$. ▶

Example 6.6. Wishart distribution. The $\binom{r+1}{2}$-dimensional Wishart-distribution with f degrees of freedom and mean value $f\Sigma$ is the distribution of a random variable of the form $x = y_1'y_1 + \cdots + y_f'y_f$ where $y_1, \ldots, y_f$ are independent and $N_r(0, \Sigma)$-distributed. It will be denoted by $W_r(f, \Sigma)$ and it is of continuous type for $f > r$, in which case its density with respect to Lebesgue measure on $R^{\binom{r+1}{2}}$ is given by

$$w_r(f)|\Sigma|^{-f/2}|x|^{(f-r-1)/2} \exp\{-\tfrac{1}{2}\operatorname{tr}(\Sigma^{-1}x)\}$$

for x (viewed as an $r \times r$ matrix) positive definite. Here

$$w_r(f)^{-1} = 2^{fr/2}\pi^{r(r-1)/4} \prod_{i=1}^{r} \Gamma\left(\frac{f+1-i}{2}\right).$$

The distribution is strongly unimodal because $-\ln|x|$ (interpreted as $+\infty$ for x not positive definite) is a convex function on $R^{\binom{r+1}{2}}$, as will be verified in Example 7.2. ▶

6.3 UNIMODALITY OF DISCRETE-TYPE DISTRIBUTIONS

Consider a discrete type probability measure π on R^k and let p be the probability function of π (i.e. $p(x) = \pi\{x\}$ for all $x \in R^k$). Set ▶

$$\varphi = -\ln p,$$

denote the support of π by S and suppose $S \subset Z^k$.

π is said to be *unimodal*, respectively *strongly unimodal*, provided S equals the intersection of Z^k and some convex set, and there exists a quasiconvex, respectively convex, function on R^k which coincides with φ on S. If S is of the form specified in these definitions—i.e. if π is c-discrete—then, as is simple to see, π is strongly unimodal if and only if φ and convϕ coincide on S. For $k = 1$, unimodality is equivalent to the sequence $p(i)$, $i \in Z$, being first non-decreasing and then non-increasing, and strong unimodality is equivalent to S being a set of integers and

$$p(i)^2 \geq p(i-1)p(i+1), \qquad i \in Z. \tag{1}$$

The analogue of Theorem 6.4 does not hold for discrete-type distributions, not even if only marginalizations, which accord with the assumed lattice structure of the supports, are considered.

Example 6.7. The two-dimensional distribution with support $\{0, 1, 2\}^2$ and point probabilities $p(i, j)$ as given in Table 1 is strongly unimodal.

Table 1. Table of $cp(i, j)$ where $c = 13 + 2\sqrt{2}$

$i \backslash j$:	0	1	2
0	2	$\sqrt{2}$	1
1	4	2	1
2	2	$\sqrt{2}$	1

However, if $x = (x_1, x_2)$ is a random variable following this distribution then the distribution of x_2 is not strongly unimodal. ▶

Also, Corollary 6.1 does not, in general, extend to the discrete case.

Example 6.8. Let π_1 and π_2 be the probability measures on R^2 having support $\{0, 1\}^2$ and point probabilities $\pi_1(0, 0) = \pi_1(1, 1) = 4/10$, $\pi_1(0, 1) = \pi_1(1, 0) = 1/10$, respectively $\pi_2(0, 0) = \pi_2(1, 1) = 1/10$, $\pi_2(0, 1) = \pi_2(1, 0) = 4/10$. The convolution $\pi = \pi_1 * \pi_2$ has support $\{0, 1, 2\}^2$, and $\pi(1, 1) = 16/100$ and $\pi(0, 1) = \pi(2, 1) = 17/100$ so that $\pi(1, 1)^2 < \pi(0, 1)\ \pi(2, 1)$. Thus π is not strongly unimodal even though both π_1 and π_2 have this property. ▶

In three dimensions it may even happen that the n-fold ($n \geq 3$) convolution of a strongly unimodal distribution with itself is not strongly unimodal (cf. the remark after Example 10.16).

Some useful, partial extensions of Corollary 6.1 are however available. Firstly, in one dimension the result carries over. (This result is contained already in a paper by Fekete (1912).) What is more, in complete analogy with Theorem 6.5 one has:

Theorem 6.6. *A one-dimensional distribution, whose support is a subset of Z, is strongly unimodal if and only if its convolution with any unimodal distribution, whose support is a subset of Z, is again unimodal.*

This result was given by Keilson and Gerber (1971). ▶

One immediate consequence of Theorem 6.6, which is used at several places in the present treatise, is that convolutions of binomial distributions (with, possibly, differing probability parameters) are strongly unimodal.

A number of sufficient conditions for two- or three-dimensional, discrete distributions to be strongly unimodal have been derived by Pedersen (1975a, b). In particular, some of these conditions imply that convolutions of trinomial distributions (with, possibly, different probability parameters) are strongly unimodal. Furthermore, Pedersen used the conditions to prove the strong unimodality of the distributions of various marginals of certain two- and three-dimensional contingency tables, cf. also Examples 10.12 and 10.13.

Example 6.9. The Poisson distribution satisfies (1) and is hence strongly unimodal. ▶

Example 6.10. The negative binomial distribution is unimodal for all values of the shape parameter χ, and, by (1), strongly unimodal if $\chi \geq 1$. ▶

Example 6.11. The multinomial distribution has point probabilities

$$p(i) = \binom{n}{i} \pi_1^{i_1} \cdots \pi_k^{i_k} (1 - \pi_.)^{n - i_1}.$$

Clearly, in order to show that this distribution is strongly unimodal it suffices to establish the existence of a convex function on R^k which coincides with $-\ln\binom{n}{i}$ for $i \in Z^k$, $i_1 \geq 0, \ldots, i_k \geq 0$ and $i_. \leq n$. Now,

$$\exp\left\{-\ln \binom{n}{i}\right\} = \frac{\Gamma(i_1 + 1) \cdots \Gamma(i_k + 1)\Gamma(n - i_. + 1)}{\Gamma(n + 1)}$$

$$= c(i)$$

where c is the Laplace transform of the continuous type probability measure on R^k having density

$$(n + k)^{(k)}(1 + e^{y_1} + \cdots + e^{y_k})^{-n-k-1} e^{y_.} \tag{2}$$

at $y = (y_1, \ldots, y_k) \in R^k$. Since Laplace transforms are log-convex (cf. Section 7.1), the result follows. ▶

Example 6.12. Negative multinomial distribution. By an argument similar to that given in Example 6.11 it may be shown that the k-dimensional negative multinomial distribution with shape parameter $\chi > k$ is strongly unimodal, the density corresponding to (2) being given by

$$(\chi - 1)^{(k)}(1 - e^{y_1} - \cdots - e^{y_k})^{\chi - k - 1} e^{y_.}$$

for $e^{y_1} + \cdots + e^{y_k} < 1$, and 0 otherwise. ▶

Example 6.13. The multivariate hypergeometric distribution is strongly unimodal. A simple proof of this will be given in Example 9.18. ▶

6.4 COMPLEMENTS

(i) *Hölder's inequality and log-convexity.* One of the possible ways of expressing Hölder's inequality is to say that if f and g are (Borel) measurable functions on R^k into $[-\infty, \infty)$, μ is a σ-finite measure on R^k, and $\lambda \in [0, 1]$ then

$$\int e^{\lambda f + (1-\lambda) g} \, d\mu \leq \left(\int e^f \, d\mu\right)^{\lambda} \left(\int e^g \, d\mu\right)^{1-\lambda} \tag{1}$$

(as follows from the elementary inequality

$$a^{\lambda}b^{1-\lambda} \leq \lambda a + (1-\lambda)b, \qquad a \geq 0, b \geq 0,$$

by setting $a = \exp f / \int \exp f \, d\mu$ and $b = \exp g / \int \exp g \, d\mu$). In other words, the mapping $f \to \int \exp f \, d\mu$ is log-convex.

Provided the integrals are finite and $0 < \lambda < 1$, equality holds in (1) if and only if f and g are equal up to an additive constant.

It follows from Hölder's inequality that if $f_1, \ldots, f_n$ are log-convex with $\operatorname{dom} \ln f_1 \cap \ldots \cap \operatorname{dom} \ln f_n \neq \varnothing$ and if $\lambda_1 > 0, \ldots, \lambda_n > 0$ then $\lambda_1 f_1 + \cdots + \lambda_n f_n$ is log-convex. To see this, it suffices to consider the case $n = 2$, $\lambda_1 = \lambda_2 = 1$. For $f = f_1 + f_2$, $0 \leq \lambda \leq 1$ and x and y in $\operatorname{dom} \ln f_1 \cap \operatorname{dom} \ln f_2$ one has by the log-convexity of f_1 and f_2

$$f(\lambda x + (1-\lambda)y) \leq e^{\lambda \ln f_1(x) + (1-\lambda) \ln f_1(y)} + e^{\lambda \ln f_2(x) + (1-\lambda) \ln f_2(y)}.$$

Viewing the sum of the two exponential terms as an integral and applying (1) one obtains the result.

(ii) Let π be a continuous type probability measure on R and let n be an integer with $n > 1$.

The product measure $\pi^{(n)}$ is unimodal if and only if π is strongly unimodal; and then $\pi^{(n)}$ is strongly unimodal.

The only nontrivial part of this proposition is the only if assertion, which was established, via the concept of *Schur concavity*, by Marshall and Olkin (1974).

(iii) Let F be the distribution function of a one-dimensional distribution and suppose (for simplicity) that F is differentiable. Khintchin showed that the distribution is unimodal with mode at 0 if and only if

$$G(x) = F(x) - xF'(x)$$

is a distribution function (see Gnedenko and Kolmogorov 1954, p. 157).

It may be noted that $-G$ is, in essence, a Legendre transform of F.

(iv) Any mixture of a Poisson distribution with a continuous type, unimodal mixing distribution is unimodal (Holgate 1970). The crux in Holgate's proof consists in showing that if the density f of the mixing distribution is differentiable with $\lambda_0 > 0$ as a mode point then the point probabilities

$$p_n = \int_0^{\infty} e^{-\lambda} \frac{\lambda^n}{n!} f(\lambda) \, d\lambda$$

of the mixture satisfy

$$(n+1)\, \Delta p_n \leq \lambda_0 \, \Delta p_{n-1}, \qquad n = 1, 2, \ldots,$$

where $\Delta p_n = p_{n+1} - p_n$.

According to the result of Khintchine mentioned under (iii)

$$G(\lambda) = F(\lambda) - (\lambda - \lambda_0)f(\lambda)$$

is a distribution function, and a simple calculation shows that

$$\lambda_0 \Delta p_{n-1} - (n + 1)\Delta p_n = \int_0^\infty e^{-\lambda} \frac{\lambda^n}{n!} dG(\lambda)$$

which is obviously non-negative.

CHAPTER 7

Laplace Transforms

7.1 THE LAPLACE TRANSFORM

Let μ be a positive, σ-finite measure on R^k such that the function c on R^k and into $(0, \infty]$ defined by

$$c(\theta) = \int e^{\theta . x} \, d\mu, \qquad \theta \in R^k$$

is not identically $+\infty$. Then c is called the *Laplace transform* of μ, and the *effective domain* of c is the set

$$\Theta = \{\theta : c(\theta) < \infty\},$$

which will be denoted by $\operatorname{dom} c$.

Note that if $\theta_0 \in \Theta$ then the function

$$c_{\theta_0}(\theta) = c(\theta + \theta_0)/c(\theta_0), \qquad \theta \in R^k \tag{1}$$

is also a Laplace transform, namely of the probability measure on R^k given by

$$\frac{d\pi}{d\mu}(x) = c(\theta_0)^{-1} e^{\theta_0 . x}. \tag{2}$$

Let S and C denote the support and the convex support of μ (i.e. C is the closed convex hull of S). Clearly, any probability measure π of the form (2) has the same support as μ.

Set

$$\kappa = \ln c.$$

Theorem 7.1. *The logarithm κ of the Laplace transform c is a closed convex function on R^k and κ is strictly convex on* $\operatorname{dom} \kappa$ *provided μ is not concentrated on an affine subspace of R^k.*

Proof. In view of (1) it causes no loss of generality to suppose that μ is a probability measure. The statements on convexity then follow from Hölder's inequality, including the criterion for equality, see Section 6.2(i). In proving that κ is closed it is convenient to apply Theorem 5.13 by means of which the problem is reducible to the, easily handled, one-dimensional case. ▶

Example 7.1. It was mentioned in Example 5.5 that the function on R which takes the value

$$\theta \ln \theta$$

for $\theta \geq 0$ and $+\infty$ otherwise is closed convex. In fact, it is the logarithm of the Laplace transform of one of the so-called *extreme stable* distributions, namely the distribution with characteristic function

$$e^{-\frac{1}{2}\pi|t| + it \log |t|}.$$

The distribution is of continuous type and has density of the form

$$f(x) = \pi^{-1} \int_0^{\pi} e^{-\alpha(v) e^x - x} \alpha(v)\, dv, \qquad x \in R, \tag{3}$$

where

$$\alpha(v) = (v/\sin v)\, e^{-v \cot v}$$

(cf. Eaton, Morris, and Rubin (1971) and references therein). ▶

Example 7.2. The fact that the density of the r-dimensional normal distribution with mean 0 and precision $\mathbf{\Delta}$ integrates to 1 may be written

$$|\mathbf{\Delta}|^{-\frac{1}{2}} = (2\pi)^{-r/2} \int_{R^r} e^{-\frac{1}{2}x\mathbf{\Delta}x'}\, d\lambda.$$

The integral on the right hand side converges for a $\mathbf{\Delta} \in R^{\binom{r+1}{2}}$ if and only if $\mathbf{\Delta}$ (viewed as a symmetric $r \times r$ matrix) is positive definite. Hence the function on $R^{\binom{r+1}{2}}$ defined by $|\mathbf{\Delta}|^{-\frac{1}{2}}$ for $\mathbf{\Delta}$ positive definite and $+\infty$ otherwise is a Laplace transform (of the measure $(2\pi)^{-r/2}\lambda t$ where $t(x) = -(x_1^2/2, \ldots, x_r^2/2, x_1x_2, x_1x_3, \ldots, x_{r-1}x_r)$).

As a consequence one finds from Theorem 7.1 that $\ln |\mathbf{\Delta}|$, interpreted as $-\infty$ if $\mathbf{\Delta}$ is not positive definite, is a closed concave function on $R^{\binom{r+1}{2}}$. ▶

Although the logarithm κ of a Laplace transform is closed, i.e. lower semicontinuous, it need not be continuous on Θ $(= \operatorname{dom} \kappa)$.

Example 7.3. Let μ be the absolutely continuous measure on R^2 whose density with respect to Lebesgue measure is

$$f(x) = \frac{1}{2\sqrt{\pi}}(1 + x_1^2)^{-3/2} e^{-x_1^2 - x_2^2/(4(1+x_1^2))}, \quad x = (x_1, x_2) \in R^2.$$

For any $\theta = (\theta_1, \theta_2) \in R^2$ one has

$$\int_{-\infty}^{+\infty} e^{\theta \cdot x} f(x)\, dx_2 = (1 + x_1^2)^{-1} e^{\theta_2^2 + \theta_1 x_1 - (1 - \theta_2^2) x_1^2}$$

whence

$$\Theta = \{\theta : |\theta_2| < 1 \quad \text{or} \quad \theta_1 = 0, \theta_2 = \pm 1\},$$

and, for $|\theta_2| < 1$

$$(4) \quad c(\theta) = e^{\theta_2^2 + \theta_1^2 \{4(1 - \theta_2^2)\}} \int_{-\infty}^{+\infty} (1 + x_1^2)^{-1} \exp\{-(1 - \theta_2^2)(x_1 - \theta_1 / [2(1 - \theta_2^2)]^2\}$$

Letting θ tend to $(0, 1)$ along the curve $(\varepsilon, \sqrt{(1 - \varepsilon^3)})$, $0 < \varepsilon < 1$, one finds that $c(\theta)$ tends to ∞ because the factor in front of the integration sign behaves as $\exp\{1/4\varepsilon\}$ for $\varepsilon \downarrow 0$ while the integral is of the order of magnitude $\varepsilon^{5/2}$. Thus c is not continuous at $(0, 1)$. ▶

Henceforth in this section only transforms of probability measures π will be considered. For such measures κ is called the *cumulant* transform. Let $x = (x_1, \ldots, x_k)$ be a random variable with distribution π. Then

$$c(\theta) = E_\pi e^{\theta . x}.$$

It is simple to see that for any vector $e \in R^k$ of unit length one has

$$(5) \qquad \lim_{\lambda \to \infty} e^{-\lambda d} c(\lambda e) = \begin{cases} 0 & \text{for } d > \delta^*(e|C) \\ \pi\{e \cdot x = \delta^*(e|C)\} & \text{for } d = \delta^*(e|C). \\ \infty & \text{for } d < \delta^*(e|C). \end{cases}$$

The *Fourier–Laplace transform* of π is the complex-valued function $\tilde{c}$ defined on the set

$$\tilde{\Theta} = \{\zeta = \theta + i\eta : \theta \in \Theta, \eta \in R^k\}$$

by

$$\tilde{c}(\zeta) = \int e^{\zeta . x}\, d\pi.$$

Theorem 7.2. *The Fourier–Laplace transform $\tilde{c}$ is an analytic function on* int $\tilde{\Theta}$, *and derivatives of $\tilde{c}$ at points in* int $\tilde{\Theta}$ *can be computed by differentiation under the integration sign.*

Proof. Let e_j be the jth unit vector, $e_j = (0, \ldots, 0, 1, 0, \ldots, 0)$, let $\zeta \in \tilde{\Theta}$ and let h be a complex number $\neq 0$. Then

$$(6) \qquad \frac{c(\zeta + he_j) - c(\zeta)}{h} = \int \frac{e^{hx_j} - 1}{h} e^{\zeta \cdot x}\, d\pi$$
$$\to \int x_j e^{\zeta \cdot x}\, d\pi \qquad \text{as } h \to 0$$

where the dominated convergence theorem has been used in conjunction with the inequality

$$(7) \qquad \left| \frac{e^{\alpha h} - 1}{h} \right| < \frac{e^{\alpha\delta} + e^{-\alpha\delta}}{\delta} \qquad \text{for } |h| \leq \delta$$

which holds for any $\alpha \in R$ and $\delta > 0$, since

$$\left| \frac{e^{\alpha h} - 1}{h} \right| \leq \sum_{\nu=1}^{\infty} \frac{|\alpha|^\nu |\delta|^{\nu-1}}{\nu!} = \frac{e^{|\alpha|\delta} - 1}{\delta} < \frac{e^{|\alpha|\delta}}{\delta}.$$

Thus, on int $\tilde{\Theta}$, $\tilde{c}$ is analytic separately in each of the coordinates $\zeta_1, \ldots, \zeta_k$ of ζ and this, by a theorem of Hartog (see e.g. Bochner and Martin 1948, p. 140) implies that $\tilde{c}$ is analytic as a function of ζ. The above also shows that first order differentiations may be performed under the integration sign, and the result for general order is provable by induction. ▶

Corollary 7.1. *The Laplace transform c is infinitely often differentiable at every point $\theta_0 \in \text{int}\, \Theta$ and the derivatives may be computed by differentiation under the integration sign. Furthermore, if $0 \in \text{int}\, \Theta$ then c can be expanded in a power series around zero,*

$$c(\theta) = \sum \frac{\theta_1^{i_1}}{i_1!} \cdots \frac{\theta_k^{i_k}}{i_k!} \mathring{\mu}_{i_1 \ldots i_k}, \qquad |\theta| < \delta$$

for some $\delta > 0$, and here $\mathring{\mu}_i$, $i = (i_1, \ldots, i_k)$, is the *i*th order moment of x_1

$$\mathring{\mu}_{i_1 \ldots i_k} = Ex_1^{i_1} \cdots x_k^{i_k}.$$

Corollary 7.2. *If $0 \in \Theta$ then κ can be expanded in a power series around zero,*

$$\kappa(\theta) = \sum \frac{\theta_1^{i_1}}{i_1!} \cdots \frac{\theta_k^{i_k}}{i_k!} \kappa_{i_1 \ldots i_k}, \qquad |\theta| < \delta$$

for some $\delta > 0$, and here κ_i, $i = (i_1, \ldots, i_k)$, is the ith order cumulant of x.

These two corollaries may also be proved without invoking Theorem 7.2, and hence—implicitly—Hartog's theorem, essentially by using the properties expressed in (6) and (7). Furthermore, it was shown, in the proof of Theorem 7.2, that c is analytic separately in each coordinate of ζ, a fact which together with the uniqueness theorem for characteristic functions (or Fourier transforms; see e.g. Kawata 1972, p. 326–327) implies:

Theorem 7.3. *(Uniqueness). Let c_1 and c_2 be the Laplace transforms of two probability measures π_1 and π_2 on R^k. If there exists an open set $M \subset R^k$ such that $c_1(\theta) = c_2(\theta) < \infty$ for $\theta \in M$ then $\pi_1 = \pi_2$.*

Proof. It causes no loss of generality to assume $0 \in M$. Let $\tilde{c}_i$ denote the Fourier–Laplace transform of π_i, $i = 1, 2$. Analytic continuation in one co-ordinate of ζ at a time shows that $\tilde{c}_1(i\eta) = \tilde{c}_2(i\eta)$ for $\eta \in R^k$, i.e. π_1 and π_2 have the same characteristic function. ▶

7.2 COMPLEMENTS

(i) As in Sections 6.2 and 6.3, let $\varphi = -\ln p$ where p denotes the probability function of a distribution on R^k of either continuous or discrete type. For a number of the standard statistical distributions that are strongly unimodal the function φ does coincide, on the support S of the distribution, with a function which is not only convex but, in fact, equal to the logarithm of a Laplace transform (cf. Theorem 7.1). That the Wishart distribution has this property was shown by Examples 6.6 and 7.2. Other instances are: the normal distribution on R^k, the gamma distribution with shape parameter $\lambda > 1$, the Poisson distribution, the multinomial distribution, and the negative multinomial distribution with shape parameter $\chi > 1$ (as concerns the latter two distributions see Examples 6.11 and 6.12).

(ii) The following result was given in Barndorff-Nielsen (1970), but the present proof, which uses convex duality, is simpler.

Theorem 7.4. *To any open convex subset Θ of R^k with $0 \in \Theta$ there exists a probability measure π on R^k such that Θ is the domain of the Laplace transform of π and such that the affine support of π is equal to R^k.*

Proof. Let φ^* denote a closed convex function on R^k with the properties that $\operatorname{dom} \varphi^* = \Theta$ and

$$\liminf_{\lambda \to \infty} \varphi^*(\lambda y) = \infty$$

for every $y \in R^k$ with $y \neq 0$. The existence of such a function was shown in Example 5.2. Set $\varphi = \varphi^{**}$. By Theorem 6.1, $0 \in \operatorname{int} \operatorname{dom} \varphi$ and thus $\operatorname{int} \operatorname{dom} \varphi \neq \varnothing$. Hence the same theorem can be applied to φ and, since $0 \in \Theta = \operatorname{int} \operatorname{dom} \varphi^*$, one obtains

$$\int e^{-\varphi} \, d\lambda < \infty.$$

For any constant d one has $(\varphi^* + d)^* = \varphi - d$ and thus, for suitable choice of φ^*,

$$\int e^{-\varphi}\, d\lambda = 1.$$

Now, let π be the absolutely continuous probability measure on R^k whose density is $\exp(-\varphi)$. Clearly, the affine support of π is R^k, and the formula

$$(\varphi(\cdot) - \theta \cdot \cdot)^* = \varphi^*(\theta + \cdot)$$

together with a straightforward, further application of Theorem 6.1, shows that the Laplace transform of π has Θ as domain. ▶

PART III

Exponential Families†

The theory of exponential families is developed from the beginning. After the introductory theory has been established, a systematic study is made of duality relations and the principal lods functions in exponential families. Finally, the character of ancillary and sufficient statistics, under exponential models, is investigated.

† Throughout Part III, $\mathfrak{P}$ stands for an exponential family.

CHAPTER 8

Introductory Theory of Exponential Families†

A major part of the more elementary properties of exponential families are discussed. However, questions concerning duality and lods functions (log-probability functions, log-likelihood functions, log-plausibility functions) for these families are deferred to the next chapter.

8.1 FIRST PROPERTIES

The family of probability measures $\mathfrak{P}$ is said to be an *exponential family* provided there exists a σ-finite measure μ on $\mathfrak{X}$, a positive integer k, real-valued functions $a, \alpha_1, \ldots, \alpha_k$ on $\mathfrak{P}$ and real-valued measurable functions $b, t_1, \ldots, t_k$ on $\mathfrak{X}$ such that $\mathfrak{P}$ is dominated by μ, $b \geq 0$, and for every $P \in \mathfrak{P}$

$$\frac{dP}{d\mu}(x) = a(P)b(x)\,e^{\alpha(P)\cdot t(x)} \tag{1}$$

where $\alpha = (\alpha_1, \ldots, \alpha_k)$, $t = (t_1, \ldots, t_k)$. In this case (1) is called an *exponential representation* of the densities of $\mathfrak{P}$ with respect to μ. The probability measures in an exponential family $\mathfrak{P}$ are mutually absolutely continuous. Hence they all have the same support. Let P_0 be an arbitrary element in $\mathfrak{P}$ then, by (1),

$$\frac{dP}{dP_0} = \frac{a(P)}{a(P_0)}\,e^{(\alpha(P)-\alpha(P_0))\cdot t} \tag{2}$$

for all $P \in \mathfrak{P}$.

Formula (2) in conjunction with Theorem 4.2 shows that if $\alpha_1, \ldots, \alpha_k$ are affinely independent then t is minimal sufficient.

When $\mathfrak{P}$ is exponential then to any σ-finite measure μ dominating $\mathfrak{P}$ there exists a representation of the form (1). For each dominating μ let $k(\mu)$ denote the

† This chapter is largely selfcontained.

smallest integer such that the densities of the probability measures in $\mathfrak{P}$ with respect to μ are representable as in (1). Factually, $k(\mu)$ is an integer independent of μ; this integer is called the *order* of $\mathfrak{P}$ and is denoted by ord $\mathfrak{P}$. Any representation (1) with $k = \text{ord}\,\mathfrak{P}$ is said to be *minimal.*

Suppose $(\mathfrak{X}, \mathfrak{A}, \mathfrak{P})$ is a statistical field such that the elements of $\mathfrak{P}$ are mutually absolutely continuous and let P_0 be an arbitrary element of $\mathfrak{P}$. It is simple to see that $\mathfrak{P}$ is exponential of order k if and only if $\dim V = k + 1$ where V denotes the linear space of functions on $\mathfrak{X}$ generated by 1 $(= 1_{\mathfrak{X}})$ and $\ln dP/dP_0$, $P \in \mathfrak{P}$.

Note that if $\mathfrak{P}$ is an exponential family having representation (1) and if

$$\frac{dP}{d\mu}(x) = \tilde{a}(P)b(x)e^{\tilde{\alpha}(P)\cdot\tilde{t}(x)}, \tag{3}$$

with $\tilde{\alpha}$ and $\tilde{t}$ of dimension $\tilde{k}$, is another representation of $\mathfrak{P}$ then

$$\Delta\alpha \cdot \Delta t = \Delta\tilde{\alpha} \cdot \Delta\tilde{t} \tag{4}$$

where

$$\Delta\alpha = \alpha(P) - \alpha(P_0), \qquad \Delta\tilde{\alpha} = \tilde{\alpha}(P) - \tilde{\alpha}(P_0)$$

$$\Delta t = t(x) - t(x_0), \qquad \Delta\tilde{T} = \tilde{t}(x) - \tilde{t}(x_0)$$

and $P_0, P \in \mathfrak{P}$, x_0, $x \in \mathfrak{X}$.

Lemma 8.1. *Let (1) and (3) be two representations of an exponential family $\mathfrak{P}$ and suppose (1) is minimal.*

Then $\tilde{k} \geq k$ and there exist two constant $\tilde{k} \times k$ matrices $\mathbf{A}$ and $\bar{\mathbf{A}}$, both of rank k, and two constant $1 \times k$ vectors B and $\bar{B}$ such that

$$\tilde{t}\mathbf{A} + B = t \tag{5}$$

$$\tilde{\alpha}\bar{\mathbf{A}} + \bar{B} = \alpha \tag{6}$$

and

$$\mathbf{A}'\bar{\mathbf{A}} = \mathbf{I}_k. \tag{7}$$

If $\tilde{t}_1, \ldots, \tilde{t}_{\tilde{k}}$ are affinely independent then

$$\tilde{\alpha} = \alpha\mathbf{A}' + \bar{D},$$

and if $\tilde{\alpha}_1, \ldots, \tilde{\alpha}_{\tilde{k}}$ are affinely independent then

$$\tilde{t} = t\bar{\mathbf{A}}' + D,$$

where D and $\bar{D}$ denote constant vectors.

Proof. Since (1) is assumed minimal, both $t_1, \ldots, t_k$ and $\alpha_1, \ldots, \alpha_k$ are affinely independent. From this fact and formula (4) the lemma follows in a simple way.

It may also be noted that (7) implies that $\bar{\mathbf{A}}\mathbf{A}'$ is idempotent. Thus $\bar{\mathbf{A}}\mathbf{A}'$ and $\mathbf{A}\bar{\mathbf{A}}'$ are projections, onto $R^k\mathbf{A}'$ and $R^k\bar{\mathbf{A}}'$, respectively.

As another consequence of Lemma 1 one has:

Corollary 8.1. *The representation (1) is minimal if and only if both of the following conditions are satisfied:*

(i) $\alpha_1, \ldots, \alpha_k$ *are affinely independent.*

(ii) $t_1, \ldots, t_k$ *are affinely independent.*

Consider the representation (1) and let $\Theta = \alpha(\mathfrak{P})$. The mapping $P \to \alpha(P)$, $P \in \mathfrak{P}$, is one-to-one. Thus $\{P_\theta : \theta \in \Theta\}$, where $P_\theta = \alpha^{-1}(\theta)$, is a parametrization of $\mathfrak{P}$. Such a parametrization is called *canonical*, and *minimal canonical* if (1) is minimal.

The statistics t occurring in the various possible representations (1) are called *canonical statistics*; t is said to be *minimal canonical* if it occurs in a minimal representation.

Where the identity mapping x on $\mathfrak{X}$ is a minimal canonical statistic, $\mathfrak{P}$ will be designated as *linear*.

An indexed family of probability measures $\mathfrak{P} = \{P_\omega ; \omega \in \Omega\}$ with probability functions of the form

$$p(x;\omega) = a(\omega) b(x) e^{\alpha(\omega) \cdot t(x)},$$

where $a(\omega) > 0$, $b(x) \geq 0$, $\alpha(\omega) \in R^k$ and $t(x) \in R^k$, is clearly exponential. If $t_1, \ldots, t_k$ are affinely independent then

$$\operatorname{ord} \mathfrak{P} = \dim(\operatorname{aff} \alpha(\Omega)),$$

and, furthermore, the mapping $\omega \to P_\omega$ is one-to-one if and only if $\omega \to \alpha(\omega)$ is one-to-one.

Example 8.1. von Mises–Fisher distributions. Let $\mathfrak{X} = S_d$, the unit sphere in R^d, and let $\mathfrak{P} = \{P_{(\mu,\chi)} : (\mu, \chi) \in S_d \times [0, \infty)\}$ be the family of *von Mises–Fisher* distributions on S_d given by

$$\frac{dP_{(\mu,\chi)}}{dP_0}(v) = a(\chi) e^{\chi \mu \cdot v} \tag{8}$$

where P_0 is the uniform distribution on S_d while μ and χ are parameters, which vary independently, $\mu \in S_d$, $\chi \in [0, \infty)$. These parameters are called the *mean direction* and the *precision*, respectively. The norming constant $a(\chi)$ depends on χ only and it may be expressed as

$$a(\chi) = \chi^{\frac{1}{2}d-1} / \{(2\pi)^{\frac{1}{2}d} I_{\frac{1}{2}d-1}(\chi)\}$$

with $I_v(\cdot)$ denoting the modified Bessel function of the first kind and order v. Thus, for $d = 2$, i.e. for the von Mises distribution,

$$a(\chi) = I_0(\chi)^{-1},$$

and for $d = 3$, i.e. for the Fisher distribution,

$$a(\chi) = \frac{\chi}{\sinh \chi}.$$

(For a comprehensive account of the history and theory of the von Mises–Fisher distributions see Mardia (1972, 1975).)

$\mathfrak{P}$ is exponential of order d and $\theta = \chi\mu$, which varies in $\Theta = R^k$, is a minimal canonical parameter. The mapping $(\mu, \chi) \to P_{(\mu,\chi)}$ is not a parametrization since $P_{(\mu,0)} = P_0$ for every μ. ▶

Henceforth, $\{P_\theta : \theta \in \Theta\}$ will denote a canonical parametrization of $\mathfrak{P}$ and, with

$$\frac{dP_\theta}{d\mu}(x) = a(\theta)b(x)\,e^{\theta \cdot t(x)} \tag{9}$$

being the exponential representation considered, S will stand for the (common) support of $P_\theta t$, $\theta \in \Theta$, and C for the convex support, i.e. $C = \text{cl conv}\, S$. Moreover, c will denote the function defined on R^k by

$$c(\zeta) = \int b(x)\,e^{\zeta \cdot t(x)}\, d\mu.$$

Thus $a(\theta) = c(\theta)^{-1}$ for $\theta \in \Theta$. Finally, set $\kappa = \ln c$.

Theorem 8.1. *For any $\theta \in \Theta$, the Laplace transform of $P_\theta t$* is

$$c(\cdot + \theta)/c(\theta). \tag{10}$$

If $\theta \in \text{int}\,\Theta$ then the statistic t has moments of all orders with respect to P_θ and

$$D^i\kappa(\theta) = \kappa_i(\theta) \tag{11}$$

where D is the differential operator, $i = (i_1, \ldots, i_k)$ is a vector of non-negative integers and $\kappa_i(\theta)$ denotes the ith *order cumulant of t under P_θ.*

Proof. Straightforward, using Theorem 7.2 and Corollary 7.2. ▶

In particular, one has by (11)

$$E_\theta t = \frac{\partial \kappa}{\partial \theta} \tag{12}$$

$$V_\theta t = \frac{\partial^2 \kappa}{\partial \theta\, \partial \theta'} \tag{13}$$

where $V_\theta t$ is non-singular.

In the majority of cases to be considered the representation (9) will be chosen so that $0 \in \Theta$ and $a(0) = 1$. Then c and κ are, respectively, the Laplace and cumulant transform of $P_0 t$. Furthermore,

$$\frac{dP_\theta}{dP_0} = a(\theta)\,e^{\theta\cdot t}. \tag{14}$$

Any exponential representation of this latter type is called a *standard representation*, and *minimal standard* provided (14) is minimal.

Example 8.2. Let $\mathfrak{X} = R$ and let $\mathfrak{P}$ be of order 1 and linear. Thus $\mathfrak{P}$ has exponential representation

$$a(\theta)b(x)\,e^{\theta x}. \tag{15}$$

Suppose that $\operatorname{int}\Theta \neq \varnothing$.

If for each member of $\mathfrak{P}$ the mean equals the variance then $\mathfrak{P}$ is a subset of the family of Poisson distributions (Kosambi 1949, Bildikar and Patil 1968; some extensions of the result may be found in the latter paper). In proving this it causes no loss of generality to assume $0 \in \operatorname{int}\Theta$ and $a(0) = 1$. Then, by assumption and formulas (12) and (13),

$$\kappa'(\theta) = \kappa''(\theta), \qquad \theta \in \operatorname{int}\Theta,$$

and $\kappa(0) = 0$. Solution of this equation leads to the result.

For another exemplification, let $x_1, \ldots, x_n, \ldots$ be independent observations following the distribution (15), set

$$\bar{x} = \frac{1}{n}\sum_{i=1}^{n} x_i, \qquad s^2 = \frac{1}{n-1}\sum_{i=1}^{n} (x_i - \bar{x})^2$$

and assume that for each $\theta \in \operatorname{int}\Theta$ the mean value of x_i is positive. Again, suppose $0 \in \operatorname{int}\Theta$ and $a(0) = 1$. Asymptotically, for $\theta \in \operatorname{int}\Theta$ and $n \to \infty$, $\bar{x}$ and the dispersion index $s^2/\bar{x}$ have a joint normal distribution whose correlation is 0 if and only if

$$\kappa'(\theta)\kappa'''(\theta) - \kappa''(\theta)^2 = 0. \tag{16}$$

Thus, for the Poisson family the statistic $s^2/\bar{x}$, which is often used to test the specification of Poisson distribution, and the statistic $\bar{x}$ are asymptotically independent. Solution of (16) shows that if $\mathfrak{P}$ has this independence property then it is Poissonian in the sense that there exists a positive constant d such that

$$P_\theta\{x_i = dr\} = e^{-\lambda}\frac{\lambda^r}{r!}, \qquad r = 0, 1, \ldots,$$

where $\lambda = \lambda_0 \exp\{\theta d\}$, λ_0 being a constant. ▶

Let $\mathfrak{X}$ be an arbitrary Euclidean sample space, and let P_0 be a probability measure and $t = (t_1, \ldots, t_k)$ a statistic on $\mathfrak{X}$. By the *exponential family generated by*

P_0 and t is meant the family $\{P_\theta : \theta \in \Theta\}$ where $\Theta = \operatorname{dom} c$, c being the Laplace transform of $P_0 t$, and where

$$\frac{dP_\theta}{dP_0} = a(\theta)\,\mathrm{e}^{\theta \cdot t}$$

with $a(\theta) = c(\theta)^{-1}$.

Consider an arbitrary exponential family $\mathfrak{P}$, suppose t is a minimal canonical statistic for $\mathfrak{P}$ and let P_0 be an element of $\mathfrak{P}$. Then the family $\tilde{\mathfrak{P}}$ generated by P_0 and t does not depend on P_0 and t, and furthermore $\mathfrak{P} \subset \tilde{\mathfrak{P}}$. Hence it is reasonable to speak of $\tilde{\mathfrak{P}}$ as the *family generated by* $\mathfrak{P}$. Note that $\operatorname{ord} \mathfrak{P} = \operatorname{ord} \tilde{\mathfrak{P}}$. If $\mathfrak{P} = \tilde{\mathfrak{P}}$ then $\mathfrak{P}$ is called *full*. When $\mathfrak{P}$ is given by an exponential representation

$$\frac{dP_\theta}{d\mu}(x) = a(\theta) b(x)\,\mathrm{e}^{\theta \cdot t(x)}, \qquad \theta \in \Theta, \tag{17}$$

then $\mathfrak{P}$ is full if

$$\Theta = \left\{\theta : \int \mathrm{e}^{\theta \cdot t(x)} b(x)\, d\mu < \infty\right\}.$$

The converse assertion holds provided (17) is minimal. Thus, for a full exponential family any of the possible minimal canonical parameter domains is convex (cf. Theorem 7.1).

$\mathfrak{P}$ is said to be *regular* if it is full and if for some (and hence for every) minimal canonical parametrization $\{P_\theta : \theta \in \Theta\}$ the set Θ is open. The reason for distinguishing this concept by the name regular will become apparent later on. Most of the standard families of distributions are regular. Obviously, any full family $\mathfrak{P}$ having finite support is regular since $\Theta = R^k$.

Example 8.3. Multivariate normal family. The family $\mathfrak{N}_r$ of r-dimensional normal distributions is the full exponential family on R^r generated by the standard normal distribution $N_r(0, \mathbf{I})$ with density function

$$(2\pi)^{-r/2}\,\mathrm{e}^{-\frac{1}{2} x \cdot x}$$

and by the statistic

$$t(x) = (x, x'x).$$

The derivative of $P_{(\xi, \Sigma)}$ with respect to $P_{(0, \mathbf{I})}$ may be written

$$|\Delta|^{\frac{1}{2}}\,\mathrm{e}^{-\frac{1}{2}\xi\Delta\xi'}\,\mathrm{e}^{\xi\Delta x' - \frac{1}{2} x(\Delta - \mathbf{I})x'}. \tag{18}$$

A canonical parameter is therefore given by

$$\theta = (\xi\Delta, \Delta) \tag{19}$$

and this relation together with the openness of Γ_r, the set of positive definite $r \times r$

matrices considered as a subset of $R^{\binom{r+1}{2}}$, shows that Θ is open. Hence $\mathfrak{N}_r$ is regular. ▶

It is an immediate corollary of Theorem 7.4 that to any open convex subset Θ of R^k there exists a regular exponential family $\mathfrak{P}$ of order k such that Θ is the parameter domain for a minimal canonical parametrization of $\mathfrak{P}$.

Suppose $\mathfrak{P}$ is full and that (9) is a minimal representation of $\mathfrak{P}$. If the convex function κ is steep then $\mathfrak{P}$ too is said to be *steep*. By Theorem 5.27, κ is steep if and only if

$$(20) \qquad (\tilde{\theta} - \theta) \cdot D\kappa(\lambda\theta + (1 - \lambda)\tilde{\theta}) \to \infty \qquad \text{as } \lambda \downarrow 0$$

for every $\theta \in \operatorname{int} \Theta$ and every $\tilde{\theta} \in \operatorname{bd} \Theta$. Clearly, κ is steep either for all or for none of the minimal representations of $\mathfrak{P}$, so that steepness is an intrinsic possible property of $\mathfrak{P}$.

On account of Corollary 5.3 and Theorem 7.1 one has:

Theorem 8.2. *If $\mathfrak{P}$ is regular then it is steep.* ▶

An instance of a steep but non-regular family $\mathfrak{P}$ is provided by:

Example 8.4. Inverse Gaussian family. Let $N^-(\mu, \lambda)$ indicate the continuous type distribution on R with density

$$(21) \qquad (\lambda/2\pi)^{\frac{1}{2}} x^{-\frac{3}{2}} e^{-\lambda(x-\mu)^2/2\mu^2 x}, \qquad x > 0.$$

Here μ and λ are parameters varying independently, both in $(0, \infty)$. The mean value of (21) is μ, and λ is a measure of precision. The distribution $N^-(\mu, \lambda)$ was coined the *inverse Gaussian* by Tweedie. The class of inverse Gaussian distributions will be denoted by $\mathfrak{N}^-$.

Rewriting (21) in the form

$$(22) \qquad (2\pi)^{-\frac{1}{2}} x^{-\frac{3}{2}} \lambda^{\frac{1}{2}} e^{\sqrt{(\alpha\lambda)}} e^{-\alpha x/2 - \lambda/2x},$$

where $\alpha = \lambda/\mu^2$, one sees that $\mathfrak{N}^-$ is not full, the full family generated by $\mathfrak{N}^-$ being obtained by allowing α to take on also the value 0. For $\alpha = 0$, (22) becomes

$$(\lambda/2\pi)^{\frac{1}{2}} x^{-\frac{3}{2}} e^{-\lambda/2x}$$

which is the density of the stable distribution with characteristic exponent $\frac{1}{2}$ and scale parameter λ^{-1} (see e.g. Feller 1966). Let $\bar{\mathfrak{N}}^-$ denote the full family.

It is apparent from (6) that the cumulant transform of the distribution with $(\alpha, \lambda) = (0, 1)$ is given by

$$\kappa(\alpha, \lambda) = \tfrac{1}{2} \ln \lambda + \sqrt{(\alpha\lambda)}, \qquad (\alpha, \lambda) \in [0, \infty) \times (0, \infty).$$

Thus κ is steep, but $\bar{\mathfrak{N}}^-$ is not regular. ▶

Let Θ be a minimal canonical parameter domain for $\mathfrak{P}$. The family $\mathfrak{P}$ is called *open*, *convex*, or *connected* if Θ has the respective property; $\mathfrak{P}$ has *open kernel* if the interior of Θ is non-empty. These possible properties of $\mathfrak{P}$ are all intrinsic, i.e. they do not depend on the particular minimal canonical parameter domain considered.

An exponential family $\mathfrak{P}$ is a *power series family* if it is linear and if the support S for the canonical statistic x (the identity mapping on $\mathfrak{X}$) is a subset of N_0^k. In this case the point probabilities of $\mathfrak{P}$ may be written

$$b(x)\lambda^x/g(\lambda) \tag{23}$$

where

$$\lambda = (\lambda_1, \ldots, \lambda_k) = (e^{\theta_1}, \ldots, e^{\theta_k})$$

$$\lambda^x = \lambda_1^{x_1} \cdots \lambda_k^{x_k}$$

and

$$g(\lambda) = c(\theta),$$

g being the generating function for the coefficients $b(x)$, $x \in S$.

Example 8.5. Sum-symmetric power series families. Let the representation (23) be minimal and suppose $\mathfrak{P}$ has open kernel. Then the power series family is said to be *sum-symmetric* provided g depends on λ through $\lambda_{.} = \lambda_1 + \cdots + \lambda_k$ only. This property is independent of which of the possible representations (23) is considered.

The class of sum-symmetric power series families was introduced by Patil (1968) and its properties have been studied in that paper and by Joshi and Patil (1970, 1971). (See Section 10.3.)

Examples of such families are the multinomial family, the multivariate Poisson family, the negative multinomial family, and the multivariate logarithmic family, the latter having point probabilities

$$\frac{(x_{.} - 1)!}{x_1! \cdots x_k!} \pi^x / \{-\ln(1 - \pi_{.})\}$$

where $x \in N_0^k \backslash \{0\}$ and $\pi \in \Pi$ with Π given by equation (2) of Section 3.3. Each of these four families is regular. ▶

Lemma 8.2. *Let t be a minimal canonical statistic for $\mathfrak{P}$.*

If $\mathfrak{P}$ has open kernel then $\mathfrak{P}_t$ is complete.

Proof. Suppose f is a t measurable function on $\mathfrak{X}$ into R and that

$$\int f(x)a(\theta)b(x)\, e^{\theta \cdot t}\, d\mu = 0, \qquad \theta \in \Theta. \tag{24}$$

Define the measures μ^+ and μ^- by

$$d\mu^\pm = f^\pm(x)b(x)\,d\mu,$$

where

$$f^\pm(x) = \max\{\pm f(x), 0\},$$

and note that (24) implies

$$\int e^{\theta\cdot t}\,d\mu^+ = \int e^{\theta\cdot t}\,d\mu^-.$$

The lemma now follows from the uniqueness theorem for Laplace transforms (Theorem 7.3). ▶

Lemma 8.2 in conjunction with Theorem 4.5 or Corollary 4.4 yields, in a simple manner, a number of important results on stochastic independence.

Example 8.6. An elementary illustration of Theorem 4.5 is provided by the $r \times c$ contingency table x_{**} of independent Poisson variates and with no interaction, i.e. the so-called multiplicative Poisson model. The well-known conditional independence of the two marginals $x_{.*}$ and $x_{*.}$ given the total $x_{..}$ is obtained from the theorem by fixing, for instance, the row parameters. ▶

Example 8.7. The independence of $\bar{x}$ and s^2 for a sample $x_1, \ldots, x_n$ from $N(\xi, \sigma^2)$ is seen by fixing σ^2 and using Corollary 4.4. It also follows immediately from that corollary that $\bar{x}$ and s^2 are independent of the B-ancillary statistic

$$u = \left(\frac{x_1 - \bar{x}}{s}, \ldots, \frac{x_n - \bar{x}}{s}\right),$$

and of the empirical measures of skewness and kurtosis

$$g_1 = m_3/m_2^{3/2} \quad \text{and} \quad g_2 = m_4/m_2^2 - 3$$

where $m_r = \Sigma(x_i - \bar{x})^2/n$, $(r = 2, 3, 4)$. ▶

Example 8.8. If $x_{1*}, \ldots, x_{n*}$ is a sample of n from the r-dimensional normal distribution $N_r(\xi, \mathbf{\Sigma})$ and if the variance $\mathbf{\Sigma}$ is a diagonal matrix then the empirical correlation matrix $\mathbf{r}$ is independent of

$$\bar{x}_{.1}, \ldots, \bar{x}_{.r}, \Sigma(x_{i1} - \bar{x}_{.1})^2/(n-1), \ldots, \Sigma(x_{ir} - \bar{x}_{.r})^2/(n-1),$$

the set of estimates of the unspecified parameters. Again, this may be seen from Corollary 4.4, on noting that the distribution of $\mathbf{r}$ does not depend on ξ and the diagonal matrix $\mathbf{\Sigma}$. ▶

See, moreover, Example 9.31.

Theorem 8.3. *Suppose* $\mathfrak{P}$ *is full and* $\{P_\theta: \theta \in \Theta\}$ *is minimal. Endow* Θ *with the usual Euclidean topology and* $\mathfrak{P}$ *with the weak topology. Then*

(i) *The mapping* $\varphi: \theta \to P_\theta$ *is continuous on* $\operatorname{int}\Theta$.

(ii) *If there exists a minimal canonical statistic which is continuous then* φ^{-1} *is continuous on* $\mathfrak{P}$.

Proof. Let

$$\frac{dP_\theta}{d\mu}(x) = a(\theta)b(x)\,e^{\theta\cdot t(x)}$$

be the minimal representation corresponding to $\{P_\theta: \theta \in \Theta\}$.

If $\theta_n \to \theta$ and $\theta_n, \theta \in \operatorname{int}\Theta$ then

$$\frac{dP_{\theta_n}}{d\mu}(x) \to \frac{dP_\theta}{d\mu}(x), \qquad x \in \mathfrak{X},$$

and this implies, by the so-called Scheffé Lemma, that $P_{\theta_n} \to P_\theta$ (weakly).

Now, suppose t is continuous. To prove continuity of φ^{-1} it must be shown that if $P_{\theta_n} \to P_\theta$ then $\theta_n \to \theta$. Let $C_0(R^k)$ denote the space of continuous functions on R^k with compact support. From $P_{\theta_n} \to P_\theta$ and the continuity of t it follows that $P_{\theta_n}t \to P_\theta t$, i.e.

$$\int f(t)\,dP_{\theta_n}t \to \int f(t)\,dP_\theta t, \qquad f \in C_0(R^k).$$

For shortness, let $\pi = P_\theta t$. Now,

$$\int f(t)\,dP_{\theta_n}t = \int f(t)\frac{a(\theta_n)}{a(\theta)}\,e^{(\theta_n - \theta)\cdot t}\,d\pi$$

and thus it suffices to prove that if $\theta = 0 \in \Theta$ and if for some sequence a_n

$$a_n \int f(t)\,e^{\theta_n\cdot t}\,d\pi \to \int f(t)\,d\pi, \qquad f \in C_0(R^k), \tag{25}$$

then $\theta_n \to 0$.

On account of the compactness of the surface of the unit sphere in R^k, one can, without loss of generality, assume

$$\frac{\theta_n}{|\theta_n|} \to e, \qquad \text{where } |e| = 1.$$

Let

$$F(d) = \pi\{t \in R^k : t\cdot e \le d\}, \qquad d \in R.$$

F is a distribution function with at least two points of increase, d_1 and $d_2 > d_1$. In fact, if F had only one point of increase d, then

$$1 = \pi\{t\cdot e = d\} = P_0\{t\cdot e = d\}$$

in contradiction to the affine independence of $t_1, \ldots, t_k$. Choose a $\delta \in (0, (d_2 - d_1)/5)$ and let $J_1 = \{t: d_1 - \delta < t \cdot e < d_1 + \delta\}$. Furthermore, let f be a non-negative function in $C_0(R^k)$ which satisfies

$$f(t) = 0, \qquad t \notin J_1$$

and

$$\int_{J_1} f(t)\, d\pi > 0.$$

Such f certainly exists. Let S_f denote the (compact) support of f. For n sufficiently large one has

$$\theta_n \cdot t = |\theta_n| \left[\left(\frac{\theta_n}{|\theta_n|} - e \right) \cdot t + e \cdot t \right] \leq |\theta_n|(d_1 + 2\delta), \qquad t \in S_f$$

and hence

$$a_n \int_{R^k} f(t)\, e^{\theta_n \cdot t}\, d\pi \leq a_n \int_{S_f} f(t)\, e^{|\theta_n|(d_1 + 2\delta)}\, d\pi = a_n\, e^{|\theta_n|(d_1 + 2\delta)} \int_{R_k} f(t)\, d\pi.$$

By letting $n \to \infty$ and employing (25), one obtains

(26) $$1 \leq \liminf a_n\, e^{|\theta_n|(d_1 + 2\delta)}.$$

In a similar way it may be shown that

(27) $$1 \geq \limsup a_n\, e^{|\theta_n|(d_2 - 2\delta)}.$$

(26), (27) and the inequality $d_1 + 2\delta < d_2 - 2\delta$ together imply $\theta_n \to 0$. ▶

In cases where $\mathfrak{P}$ is full and $\{P_\theta: \theta \in \Theta\}$ is minimal, τ will denote the mapping defined on int Θ by

$$\tau(\theta) = E_\theta t$$

and $\mathfrak{T}$ will stand for $\tau(\text{int}\, \Theta)$. The mapping τ is a one-to-one, both ways continuously differentiable mapping between the two open, connected sets int Θ and $\mathfrak{T}$; moreover, τ is strictly increasing in the sense that

(28) $$(\theta - \tilde{\theta})(\tau(\theta) - \tau(\tilde{\theta})) > 0$$

for every pair of points $\theta, \tilde{\theta}$ in int Θ. It follows, in particular, that if Θ is open (i.e. $\mathfrak{P}$ is regular) then $\mathfrak{P}$ may be parametrized by the mapping $\tau \to P_\theta$ (where $\tau = \tau(\theta)$). Such a parametrization is called a *mean value parametrization.*

Besides canonical and mean value parametrizations of regular exponential families, also parametrizations which are, so to speak, a mixture of the two have interest. Consider a partition $(\theta^{(1)}, \theta^{(2)})$ of θ and the similar partition $(\tau^{(1)}, \tau^{(2)})$ of τ. Observing that $\tau = D\kappa$ and applying Theorem 5.34 to κ one finds that the mapping

$$\theta \to (\tau^{(1)}, \theta^{(2)}) \tag{29}$$

is a homeomorphism. Thus $(\tau^{(1)}, \theta^{(2)})$ furnishes a parametrization of $\mathfrak{P}$ which is called a *mixed* parametrization. Theorem 5.34 moreover yields:

Theorem 8.4. *If $\mathfrak{P}$ is regular then for any mixed parametrization of $\mathfrak{P}$ with partition $(\tau^{(1)}, \theta^{(2)})$ the components $\tau^{(1)}$ and $\theta^{(2)}$ are variation independent.* ▶

Example 8.9. For the family $\mathfrak{N}_r$ of r-dimensional normal distributions the pair $(\xi, \mathbf{\Delta})$, where $\mathbf{\Delta} = \mathbf{\Sigma}^{-1}$, is a mixed parameter.

Another mixed parametrization of $\mathfrak{N}_r$ is determined as follows.

Suppose x is normally distributed with mean ξ and variance $\mathbf{\Sigma}$, let $(x^{(1)}, x^{(2)})$ and $(\xi^{(1)}, \xi^{(2)})$ be similar partitions of x and ξ, and let

$$\mathbf{\Sigma} = \begin{Bmatrix} \mathbf{\Sigma}_{11} & \mathbf{\Sigma}_{12} \\ \mathbf{\Sigma}_{21} & \mathbf{\Sigma}_{22} \end{Bmatrix} \quad \text{and} \quad \mathbf{\Delta} = \begin{Bmatrix} \mathbf{\Delta}_{11} & \mathbf{\Delta}_{12} \\ \mathbf{\Delta}_{21} & \mathbf{\Delta}_{22} \end{Bmatrix}$$

be the corresponding partitions of the variance and the precision. Then

$$\tau^{(1)} = (\xi^{(1)}, \mathbf{\Sigma}_{11}), \qquad \theta^{(2)} = (\xi^{(1)}\mathbf{\Delta}_{12} + \xi^{(2)}\mathbf{\Delta}_{22}, \mathbf{\Delta}_{12}, \mathbf{\Delta}_{22})$$

constitute a mixed parameter. The variables $\tau^{(1)}$ and $\theta^{(2)}$ are, respectively, the parameters of the marginal (normal) distribution of $x^{(1)}$ and of the conditional (normal) distribution of $x^{(2)}$ given $x^{(1)}$ (cf. formulas (1) and (2) of Section 1.1, and Example 3.7). ▶

Example 8.10. Genotype distributions. Let there be given a random sample of size n from an infinite genetical, diploid population. Suppose that a certain locus carries m allelic genes $A_1, \ldots, A_m$ and let $p_{ij} > 0$ denote the proportion of individuals in the population having genotype $A_i A_j$. The probability that the sample contains x_{ij} individuals of genotype $A_i A_j (1 \le i \le j \le m)$ is

$$\frac{n!}{\prod_{i \le j} x_{ij}!} \prod_{i \le j} p_{ij}^{x_{ij}}. \tag{30}$$

The population gene frequency of A_i will be denoted by p_i. Note that

$$p_i = p_{ii} + \tfrac{1}{2}(p_{1i} + \cdots + p_{i-1i} + p_{ii+1} + \cdots + + p_{im})$$

and

$$p_1 + \cdots + p_m = 1.$$

If the population has been created through random union of gametes and if no selection has taken place then

$$\begin{aligned} p_{ii} &= p_i^2, & 1 \le i \le m \\ p_{ij} &= 2p_i p_j, & 1 \le i \le j \le m \end{aligned} \tag{31}$$

Formula (31) defines the hypothesis of *Hardy–Weinberg distribution* of the observed genotype numbers x_{ij}.

Let $\mathfrak{P}$ be the exponential family given by (30). $\mathfrak{P}$ is regular and a convenient canonical (non-minimal) parametrization of $\mathfrak{P}$ is determined by

$$\theta = (\theta_1, \dots, \theta_m, \theta_{12}, \theta_{13}, \dots, \theta_{m-1m})$$

where

$$\theta_i = \frac{1}{2} \ln p_{ii}, \qquad 1 \le i \le m$$

$$\theta_{ij} = \frac{1}{2} \ln \frac{p_{ij}^2}{4p_{ii}p_{jj}}, \qquad 1 \le i < j \le m.$$

To this parametrization corresponds the canonical statistic

$$t = (t_1, \dots, t_m, t_{12}, t_{13}, \dots, t_{m-1m})$$

where

$$t_i = 2x_{ii} + x_{1i} + \cdots + x_{i-1i} + x_{ii+1} + \cdots + x_{im}, \qquad 1 \le i \le m$$

$$t_{ij} = x_{ij}, \qquad 1 \le i < j \le m.$$

Thus t_i is the number of A_i-genes and t_{ij} is the number of $A_i A_j$-heterozygotes in the sample. There is exactly one linear constraint between 1, $t_1, \dots, t_m$, $t_{12}, t_{13}, \dots, t_{m-1m}$, which is satisfied with probability 1, namely

$$t_1 + \cdots + t_m = 2n.$$

The hypothesis of Hardy–Weinberg distribution is equivalent to

$$\theta_{ij} = 0, \qquad 1 \le i < j \le m,$$

and the parameters θ_{ij} express possible deviations from the hypothesis in a useful way: θ_{ij} is positive if there is excess of $A_i A_j$-heterozygotes in the population and it is negative in the adverse instance. Moreover, if the population was initially in Hardy–Weinberg distribution but has been subjected to selection, the fitness component of genotype $A_i A_j$ being w_{ij}, then

$$\theta_{ij} = \frac{1}{2} \ln \frac{w_{ij}^2}{w_{ii} w_{jj}}.$$

The pair

$$(p_*, \theta^{(2)}),$$

where

$$p_* = (p_1, \dots, p_m)$$

$$\theta^{(2)} = (\theta_{12}, \theta_{13}, \dots, \theta_{m-1m}),$$

furnishes an interesting example of a mixed parametrization, and the variation independence of p_* and $\theta^{(2)}$ (cf. Theorem 8.4) seems worth noting. ▶

Mixed parametrizations are of particular interest in connection with the notions of L-independence and cuts in exponential families, cf. Sections 9.2, 10.2, and 10.3.

Throughout what follows, $(\theta^{(1)}, \theta^{(2)})$, $(t^{(1)}, t^{(2)})$, and $(\tau^{(1)}, \tau^{(2)})$ will be similar partitions of θ, t, and τ, and the common dimension of $\theta^{(i)}$, $t^{(i)}$, and $\tau^{(i)}$ is denoted by $k^{(i)}$, $i = 1, 2$.

The next lemma, which will be particularly useful in Chapter 10, gives information on the relations between the various possible such partitions.

Lemma 8.3. *Let* $t = (t^{(1)}, t^{(2)})$ *and* $\tilde{t} = (\tilde{t}^{(1)}, \tilde{t}^{(2)})$ *be minimal canonical statistics for* $\mathfrak{P}$ *and denote the dimensions of* $t^{(1)}$ *and* $\tilde{t}^{(1)}$ *by* $k^{(1)}$ *and* $\tilde{k}^{(1)}$. *Furthermore, let* $\theta = (\theta^{(1)}, \theta^{(2)})$ *and* $\tilde{\theta} = (\tilde{\theta}^{(1)}, \tilde{\theta}^{(2)})$ *be the corresponding minimal canonical parameters and let*

$$\tilde{t} = t\mathbf{A} + B, \qquad \tilde{\theta} = \theta\overline{\mathbf{A}} + \overline{B},$$

be the affine connections whose existence was established in Lemma 8.1. Here $\mathbf{A}' = \overline{\mathbf{A}}^{-1}$. *Partition* $\overline{\mathbf{A}}$

$$\overline{\mathbf{A}} = \begin{Bmatrix} \overline{\mathbf{A}}_{11} & \overline{\mathbf{A}}_{12} \\ \overline{\mathbf{A}}_{21} & \overline{\mathbf{A}}_{22} \end{Bmatrix}$$

such that $\overline{\mathbf{A}}_{11}$ *is a* $k^{(1)} \times \tilde{k}^{(1)}$ *matrix and assume that*

(*a*) *for some value of* $\theta^{(2)}$

$$\dim \operatorname{aff} \Theta_{\theta^{(2)}} = k^{(1)}$$

and

(*b*) $\tilde{\theta}^{(2)}$ *depends on* θ *only through* $\theta^{(2)}$.

Then

(*i*) $$\overline{\mathbf{A}}_{12} = 0$$

(*ii*) $$t^{(1)} = \tilde{t}^{(1)}\overline{\mathbf{A}}'_{11} - B^{(1)}\overline{\mathbf{A}}'_{11}$$

(*iii*) $$\tilde{\theta}^{(2)} = \theta^{(2)}\overline{\mathbf{A}}_{22} + \overline{B}^{(2)}$$

(*iv*) $$k^{(1)} \le \tilde{k}^{(1)}$$

(*v*) *If* $k^{(1)} = \tilde{k}^{(1)}$ *then* $\mathbf{A}'_{11}\overline{\mathbf{A}}_{11} = \mathbf{I}_{k^{(1)}}$, $\mathbf{A}'_{22}\overline{\mathbf{A}}_{22} = \mathbf{I}_{k^{(2)}}$, *and* $\mathbf{A}_{21} = 0$.

Proof. The assertions (ii), (iii), (iv), and (v) are immediate consequences of (i).

To prove (i), observe that

$$\tilde{\theta}^{(2)} = \theta^{(1)}\overline{\mathbf{A}}_{12} + \theta^{(2)}\overline{\mathbf{A}}_{22} + \overline{B}^{(2)}. \tag{32}$$

According to assumptions, there exists some value of $\theta^{(2)}$, $\theta_0^{(2)}$ say, such that $\dim \operatorname{aff} \Theta_{\theta_0^{(2)}} = k^{(1)}$. Consequently one can find $\theta_i^{(1)}$ such that $\theta_i = (\theta_i^{(1)}, \theta_0^{(2)}) \in \Theta$, $i = 0, 1, 2, \ldots, k^{(1)}$, and such that $\theta_i^{(1)} - \theta_0^{(1)}$, $i = 1, 2, \ldots, k^{(1)}$, are linearly independent. Using (32) and assumption (*b*) one obtains

$$0 = (\theta_i^{(1)} - \theta_0^{(1)})\bar{\mathbf{A}}_{12}, \qquad i = 1, 2, \ldots, k^{(1)}$$

and hence $\bar{\mathbf{A}}_{12} = 0$. ▶

8.2 DERIVED FAMILIES

From a given statistical field $(\mathfrak{X}, \mathfrak{A}, \mathfrak{P})$ (with $\mathfrak{P}$ exponential) other statistical fields may be constructed, e.g. by margining or conditioning. Several important types of such constructions lead again to exponential families and may preserve regularity properties of the original family. This is discussed here, together with certain related results, in a sequence of subsections.

Let

$$\frac{dP_\theta}{d\mu}(x) = a(\theta)b(x)\,e^{\theta\cdot t(x)}, \qquad \theta \in \Theta \tag{1}$$

be a minimal exponential representation of $\mathfrak{P}$ with $0 \in \Theta$ and $a(0) = 1$.

(i) *Affine hypotheses.* Consider the field $(\mathfrak{X}, \mathfrak{A}, \mathfrak{P}_0)$ where $\mathfrak{P}_0$, which will occasionally be called the hypothesis, is a proper subset of $\mathfrak{P}$. Any such $\mathfrak{P}_0$ is obviously exponential. Let Θ_0 denote the subset of Θ corresponding to $\mathfrak{P}_0$. Clearly, $\operatorname{ord}\mathfrak{P}_0 = \dim\operatorname{aff}\Theta_0$. The hypothesis $\mathfrak{P}_0$ is said to be *affine* if Θ_0 is of the form $\Theta \cap L$ where L is an affine subspace of R^k.

Theorem 8.5. *Let $\mathfrak{P}_0$ be a subfamily of $\mathfrak{P}$. If $\mathfrak{P}_0$ is full then $\mathfrak{P}_0$ is affine. The converse assertion holds provided $\mathfrak{P}$ is full.*

Proof. Since the properties of being full or regular or affine are intrinsic, it causes no loss of generality to assume that $0 \in \Theta_0$ and that $\operatorname{aff}\Theta_0$ is of the form $R^{k^{(1)}} \times \{0\}$. Then, writing $\theta = (\theta^{(1)}, \theta^{(2)})$ and $t = (t^{(1)}, t^{(2)})$ with $\theta^{(1)}$ and $t^{(1)}$ of dimension $k^{(1)}$, one has that

$$\frac{dP_{(\theta^{(1)},0)}}{dP_0} = a(\theta^{(1)}, 0)\,e^{\theta^{(1)}\cdot t^{(1)}} \qquad ((\theta^{(1)}, 0) \in \Theta_0)$$

is a minimal standard representation of $\mathfrak{P}_0$. Let

$$\tilde{\Theta}_0 = \{(\zeta^{(1)}, 0) : \int e^{\zeta^{(1)}\cdot t^{(1)}}\,dP_0 < \infty\}$$

and note that

$$\Theta_0 \subset \Theta \cap \operatorname{aff}\Theta_0 \subset \tilde{\Theta}_0. \tag{2}$$

Clearly, $\mathfrak{P}_0$ full means $\Theta_0 = \tilde{\Theta}_0$, and $\mathfrak{P}$ full implies $\tilde{\Theta}_0 = \Theta \cap \operatorname{aff}\Theta_0$. The theorem now follows by (2). ▶

Corollary 8.2. *Suppose $\mathfrak{P}_0$ is regular. Then p_0 is regular if and only if it is affine.*

Let $\mathfrak{P}$ be regular and let $\mathfrak{P}_0$ be the affine hypothesis obtained by fixing $\theta^{(2)}$ at some value $\theta_0^{(2)}$. Moreover, let $\mathfrak{T}_0$ be the subset of $\mathfrak{T}$ corresponding to $\mathfrak{P}_0$. It will be shown in Section 9.1 that regularity of $\mathfrak{P}$ implies $\mathfrak{T} = \operatorname{int} C$. Using this, Theorem 8.4, and Corollary 8.2 one finds that the projections of $\mathfrak{T}$ ($= \operatorname{int} C$) and of $\mathfrak{T}_0$ on $R^{k^{(1)}} \times \{0\}$ are identical and equal to $\operatorname{int} C_0$, where C_0 denotes the convex support of the distribution of $t^{(1)}$.

It follows, furthermore, from Theorem 8.1 and equation (20) of Section 8.1 that if $\mathfrak{P}$ is steep and $\mathfrak{P}_0$ is affine then $\mathfrak{P}_0$ is steep.

Generally, of course, for any concrete model $\mathfrak{P}$ some affine hypotheses are more interesting than others.

Example 8.11. Affine hypotheses of $\mathfrak{N}_r$. For the class $\mathfrak{N}_r$ of r-dimensional normal distributions those hypotheses $\mathfrak{P}_0$ which are affine with $\xi = 0$ for some element of $\mathfrak{P}_0$ and which have the property that ξ and $\mathbf{\Delta}$ are variation independent under $\mathfrak{P}_0$ are of particular interest.

For any hypothesis $\mathfrak{P}_0$ of $\mathfrak{N}_r$ let Ξ and $\underline{\Delta}$ denote the sets of values of ξ and $\mathbf{\Delta}$ under $\mathfrak{P}_0$, and let $\Xi_{\mathbf{\Delta}}$ be the set of those ξ for which $(\xi, \mathbf{\Delta})$ corresponds to an element of $\mathfrak{P}_0$. Note that if $\mathfrak{P}_0$ is affine then the sets $\Xi_{\mathbf{\Delta}}\mathbf{\Delta}$, $\mathbf{\Delta} \in \underline{\Delta}$, are affine and parallel and $\underline{\Delta}$ is an affine subset of the set Γ_r of positive definite $r \times r$ matrices. Using this remark and equation (19) of Section 8.1 it is trivial to show:

Theorem 8.6. *Suppose $\mathfrak{P}_0$ is of the form $\{P_{(\xi,\mathbf{\Delta})}: (\xi, \mathbf{\Delta}) \in \Xi \times \underline{\Delta}\}$ with $0 \in \Xi$.*

Then $\mathfrak{P}_0$ is affine if and only if $\underline{\Delta}$ is an affine subset of Γ_r and Ξ is a linear subspace such that $\Xi\mathbf{\Delta}$ is independent of $\mathbf{\Delta}$ for $\mathbf{\Delta} \in \underline{\Delta}$.

Suppose $\mathfrak{P}_0$ is a *linear* hypothesis, i.e. $\Theta_0 = \Theta \cap L$ where L is a linear subspace of R^k. Denoting the projection onto L by $\mathbb{P}$ one has

$$\frac{dP_\theta}{d\mu}(x) = a(\theta)b(x)\,e^{\mathbb{P}\theta \cdot \mathbb{P}t(x)}, \qquad \theta \in \Theta_0. \tag{3}$$

If $\operatorname{aff}\Theta_0 = L$ then $\mathbb{P}t$ is minimal sufficient and, interpreting $\mathbb{P}\theta$ and $\mathbb{P}t$ as vectors of dimension $k_0 = \dim L$, (3) is a minimal representation of $\mathfrak{P}_0$.

Example 8.12. Let x be an r-dimensional normal variate whose family of distributions is of the affine type discussed in the previous example, i.e. $\mathfrak{P}_0 = \{P_{(\xi,\mathbf{\Delta})} : (\xi, \mathbf{\Delta}) = \Xi \times \underline{\Delta}\}$ with $0 \in \Xi$ and $\mathfrak{P}_0$ is affine.

A minimal canonical statistic is then given by

$$(\mathbb{P}^{(1)}x, \mathbb{P}^{(2)}(\tfrac{1}{2}x_{11}^2, \ldots, \tfrac{1}{2}x_{rr}^2, x_1x_2, \ldots, x_1x_r, \ldots, x_{r-1}x_r))$$

where $\mathbb{P}^{(1)}$ is the projection in R^r onto Ξ and $\mathbb{P}^{(2)}$ is the projection in $R^{\binom{r+1}{2}}$ onto the linear subspace parallel to $\operatorname{aff}\underline{\Delta}$. ▶

(ii) *Product families.* For any $n \in N$, the product family $\mathfrak{P}^{(n)}$ is exponential with ord $\mathfrak{P}^{(n)} = \text{ord}\,\mathfrak{P}$, minimal canonical parametrization $\{P_\theta^{(n)}: \theta \in \Theta\}$ and minimal standard representation

$$\frac{dP_\theta^{(n)}}{dP_0^{(n)}}(x^{(n)}) = a(\theta)^n \, e^{\theta \cdot (t(x_1) + \cdots + t(x_n))}$$

where $x^{(n)} = (x_1, \ldots, x_n)$. Thus, for a sample $x_1, \ldots, x_n$, a minimal canonical (and minimal sufficient) statistic is given by $t(x_1) + \cdots + t(x_n)$.

The family $\mathfrak{P}^{(n)}$ is full if and only if $\mathfrak{P}$ is full.

(iii) *Marginality.* The family $\mathfrak{P}t$ of marginal distributions of t is exponential with $\text{ord}(\mathfrak{P}t) = \text{ord}\,\mathfrak{P}$ and minimal standard representation

$$\frac{dP_\theta t}{dP_0 t} = a(\theta)\, e^{\theta \cdot t}.$$

The restriction $\mathfrak{P}_{t^{(1)}}$ of $\mathfrak{P}$ to the σ-field generated by the component $t^{(1)}$ of t has minimal representation

$$\frac{dP_{\theta t^{(1)}}}{dP_{0 t^{(1)}}} = a(\theta) b(t^{(1)}; \theta^{(2)})\, e^{\theta^{(1)} \cdot t^{(1)}} \tag{4}$$

where

$$b(t^{(1)}; \theta^{(2)}) = E_0(e^{\theta^{(2)} \cdot t^{(2)}} | t^{(1)})$$

(cf. equation (4) of Section 1.1). This family, and hence the family of marginal distributions of $t^{(1)}$, is not in general exponential (although, of course, each of its subfamilies determined by fixing the value of $\theta^{(2)}$ is exponential).

Example 8.13. Let u be a random variable following a Poisson distribution with parameter λ and suppose $y_1, y_2, \ldots$ are stochastically independent, mutually and of u, each following the logarithmic distribution with point probabilities

$$(-\ln(1 - \pi))^{-1} y^{-1} \pi^y \qquad y = 1, 2, \ldots .$$

Set

$$v = y_1 + \cdots + y_u.$$

The distribution of $t = (u, v)$ has point probabilities

$$(1 - \pi)^\chi b(u, v) \chi^u \pi^v;$$

here

$$\chi = \lambda / \{-\ln(1 - \pi)\}$$

while $b(u, v)$ is the coefficient of χ^u in the power series expansion of $\binom{\chi + v - 1}{v}$. Thus the family $\mathfrak{P}$ of distributions of t as (λ, π) varies in $(0, \infty) \times (0, 1)$ is exponential (of order 2 and regular). However, the marginal distribution of v is the negative binomial

$$(1-\pi)^{\chi}\binom{\chi+v-1}{v}\pi^{v}$$

(Jones and Mollison 1948) and hence $\mathfrak{P}v$ is not exponential. ▶

It may be noted that if $t^{(1)}$ is a cut, then $\mathfrak{P}t^{(1)}$ is exponential. On the other hand, if $\mathfrak{P}t^{(1)}$ is exponential of order $k^{(1)}$ and if $\mathfrak{P}$ is regular then $t^{(1)}$ is a cut. To see the latter, fix $\theta^{(2)}$ at some value and denote the corresponding subfamily of $\mathfrak{P}$ by $\mathfrak{P}_{\theta^{(2)}}$. On account of Theorem 8.5, $\mathfrak{P}_{\theta^{(2)}}$ is full and hence $\mathfrak{P}_{\theta^{(2)}}t^{(1)}$ must be full. Using again Theorem 8.5, one finds that $\mathfrak{P}_{\theta^{(2)}}t^{(1)}$ is an affine subfamily of $\mathfrak{P}t^{(1)}$, and since both are of order $k^{(1)}$ one has $\mathfrak{P}_{\theta^{(2)}}t^{(1)} = \mathfrak{P}t^{(1)}$, whatever the value of $\theta^{(2)}$. Thus $t^{(1)}$ is a cut.

The densities of $\mathfrak{P}t$ with respect to a σ-finite measure μ (typically Lebesgue measure or counting measure) are given by

$$p(t;\theta) = a(\theta)p(t)\,e^{\theta\cdot t} \tag{5}$$

where

$$p(\cdot) = \frac{dP_0 t}{d\mu}.$$

When in concrete cases the densities are sought one will naturally look for a P_0 which makes the determination of $p(\cdot)$ relatively simple.

Example 8.14. *Wishart-density*. In order to derive the density of the Wishart-distribution $W_r(f,\Sigma)$, i.e. the density for $t = x_1'x_1 + \cdots + x_f'x_f$ where $x_1,\ldots,x_f$ is a sample from $N_r(0,\Sigma)$, one may conveniently proceed as follows. Letting $\mathfrak{P}$ be the family of distributions of $(x_1,\ldots,x_f)$ and letting P_0 correspond to $\Sigma = \mathbf{I}$ one obtains from (18) of Section 8.1 and (5) that the density of $W_r(f,\Sigma)$ with respect to Lebesgue measure on $R^{\binom{r+1}{2}}$ is

$$|\Delta|^{f/2}\,e^{-\frac{1}{2}\mathrm{tr}(\Delta t)}\,e^{\frac{1}{2}\mathrm{tr}\,t}p(t),$$

so that the problem is reduced to that of finding $p(t)$, the density of $W_r(f,\mathbf{I})$. In deriving the latter one may draw on the independence properties which hold when $\Sigma = \mathbf{I}$ (cf., for instance, p. 597–598 in Rao (1973)). ▶

Example 8.15. Resultant density. For a sample $v_1,\ldots,v_n$ from the von Mises–Fisher distribution (equation (8) of Section 8.1) the density of the so-called *resultant* $v_.$ with respect to Lebesgue measure on R^k is

$$a(\chi)^n\,e^{\chi\mu\cdot v_.}p(v_.)$$

where p is the density under the uniform distribution on the unit sphere S_d. Clearly, $p(v_.)$ must be proportional to $|v_.|^{-d+1}$ times the density of $|v_.|$ under the uniform distribution. An expression for this latter density in terms of the Bessel function $J_{\frac{1}{2}d-1}$ is available, cf. Mardia (1975). ▶

More generally than (5) one has that for any statistic u on $\mathfrak{X}$ the densities of $\mathfrak{P}u$ with respect to a dominating measure μ are determined by

$$p(u;\theta) = a(\theta)E_0(e^{\theta\cdot t}|u)p(u) \tag{6}$$

where

$$p(\cdot) = \frac{dP_0u}{d\mu}.$$

This is an immediate consequence of equation (4) of Section 1.1 and equation (1) of the present section.

Example 8.16. Resultant length density. In the situation of Example 8.15 the density of the resultant length $r = |v_.|$ with respect to Lebesgue measure is

$$a(\chi)^n a(\chi r)^{-1} p(r) \qquad (0 < r < n),$$

$p(r)$ being the density for the isotropic case $\chi = 0$, which was mentioned in Example 8.15. To see this all one has to note is that for $\chi = 0$ the distribution of $r^{-1}v_.$ must be uniform on S_d, whence

$$E_0(e^{\chi\mu\cdot v_.}|r) = E_0(e^{\chi r\mu\cdot(r^{-1}v_.)}|r) = a(\chi r)^{-1}.$$ ▶

Example 8.17. Non-central χ^2-density. Let $x_1, \ldots, x_n$ be independent and normally distributed with mean values $\xi_1, \ldots, \xi_n$ and a common variance σ^2. To determine the density (with respect to Lebesgue measure) for $u = x_1^2 + \cdots + x_n^2$, note first that the density of $x_* = (x_1, \ldots, x_n)$ may be written

$$\sigma^{-n} e^{-\frac{1}{2}|\xi_*|^2/\sigma^2} \varphi(x_1) \cdots \varphi(x_n)\, e^{(1/\sigma^2)\xi_*\cdot x_* - \frac{1}{2}(\sigma^{-2}-1)u}$$

where φ denotes the density of $N(0,1)$. Hence, it is possible to apply (6) with $t = (x_*, u)$ and P_0 corresponding to $(\xi_*, \sigma^2) = (0, \ldots, 0, 1)$. Under P_0 the statistic u follows the χ^2-distribution with n degrees of freedom and for general (ξ_*, σ^2) the density of u is therefore

$$\sigma^{-n} e^{-\frac{1}{2}|\xi_*|^2/\sigma^2} \Gamma(n/2)^{-1} 2^{-n/2} u^{n/2-1} e^{-u/(2\sigma^2)} E_0(e^{(1/\sigma^2)\xi_*\cdot x_*}|u).$$

Under P_0 and given u, the statistic $u^{-1}x_*$ is uniformly distributed on the unit sphere S_n and so

$$E_0(e^{(1/\sigma^2)\xi_*\cdot x_*}|u) = a\left(\frac{|\xi_*|}{\sigma^2}u\right)^{-1}$$

the latter quantity being the normalizing constant for the von Mises–Fisher distribution on S_n with $\chi = (|\xi_*|/\sigma^2)u$ (cf. Example 8.1). ▶

A further application of formula (6) may be found in Example 9.32.

(iv) *Conditionality.* On account of (5) in Section 1.1, one has

$$\frac{dP_\theta(\cdot|t^{(1)})}{dP_0(\cdot|t^{(1)})} = a(\theta^{(2)}|t^{(1)})\, e^{\theta^{(2)}\cdot t^{(2)}} \tag{7}$$

where

$$a(\theta^{(2)}|t^{(1)})^{-1} = E_0(e^{\theta^{(2)}\cdot t^{(2)}}|t^{(1)}).$$

Since the conditional distribution of $t^{(2)}$ given $t^{(1)}$ depends on θ through $\theta^{(2)}$ only one may write $P_{\theta^{(2)}}(\cdot|t^{(1)})$ instead of $P_\theta(\cdot|t^{(1)})$. Let $\Theta^{(2)}$ be the set of values of $\theta^{(2)}$ for θ varying in Θ. It follows from (7) that for a fixed value $t^{(1)}$ the family

$$\mathfrak{P}(\cdot|t^{(1)}) = \{P_{\theta^{(2)}}(\cdot|t^{(1)}): \theta^{(2)} \in \Theta^{(2)}\}$$

is exponential with $\theta^{(2)}$ and $t^{(2)}$ as canonical quantities.

Example 8.18. Let x_i follow the $N(\xi_i, \sigma^2)$ distribution, $i = 1, \ldots, k$, and suppose $x_1, \ldots, x_k$ are independent. The minimal canonical statistic is $(x_*, |x_*|^2)$, and $(\sigma^{-2}\xi_*, -\frac{1}{2}\sigma^{-2})$ is the corresponding canonical parameter. Taking $t^{(1)} = |x_*|^2$ and observing that for $\xi_1 = \cdots = \xi_k = 0$ the conditional distribution of $t^{(2)} = x_*$ given $|x_*|^2 = 1$ is the uniform distribution on the unit sphere S_k, one sees at once from (7) that the class of conditional distributions of x_* given $|x_*| = 1$ and $\sigma^2 = 1$ is equal to the family of von Mises–Fisher distributions on S_k (cf. Example 8.1). ▶

Fullness of $\mathfrak{P}$ does not necessarily imply fullness of $\mathfrak{P}(\cdot|t^{(1)})$, as is easily shown by example.

If the mapping $\theta^{(2)} \to P_{\theta^{(2)}}(\cdot|T^{(1)})$ is one-to-one then $\theta^{(2)}$ is said to parametrize the conditional distributions given $T^{(1)}$.

Lemma 8.4. *Assume that the conditional distribution of $T^{(2)}$ given $T^{(1)}$ is nonsingular.*

Then $\theta^{(2)}$ parametrizes the conditional distributions given $T^{(1)}$.

Proof. Suppose

$$P_{\theta^{(2)}}(\cdot|T^{(1)}) = P_{\tilde{\theta}^{(2)}}(\cdot|T^{(1)}).$$

Then (cf. the definition of singular conditional distributions given in Section 1.2) there exists a value $t^{(1)}$ of $T^{(1)}$ such that under $P_0(\cdot|t^{(1)})$ the coordinates of $T^{(2)}$ are affinely independent and

$$a(\theta^{(2)}|t^{(1)})\, e^{\theta^{(2)}\cdot T^{(2)}} = a(\tilde{\theta}^{(2)}|t^{(1)})\, e^{\tilde{\theta}^{(2)}\cdot T^{(2)}}$$

whence, by the remark preceding Example 8.1, $\theta^{(2)}$ must equal $\tilde{\theta}^{(2)}$. ▶

(v) *Truncation and censoring.* Let $A \in \mathfrak{A}$ and suppose that $0 < P_0(A) < 1$. Then the family $\mathfrak{P}_\Theta^A = \{P_\theta^A: \theta \in \Theta\}$, where P_θ^A is the truncation of P_θ to A, is exponential and has standard representation

$$\frac{dP_\theta^A}{dP_0^A} = a(\theta; A)\, e^{\theta\cdot t}$$

where

$$a(\theta; A)^{-1} = \int e^{\theta \cdot t} \, dP_0^A.$$

Obviously, the order of $\mathfrak{P}^A$ may be less than the order of $\mathfrak{P}$ since even though $t_1, \ldots, t_k$ are affinely independent their restrictions to A may well be affinely dependent. Note, furthermore, that $\mathfrak{P}$ full does not necessarily imply $\mathfrak{P}^A$ full (even if ord $\mathfrak{P}^A$ = ord $\mathfrak{P}$). However, if $\Theta = R^k$ then every truncation of $\mathfrak{P}$ is full.

Example 8.19. Doubly truncated normal family. The class $\mathfrak{N}$ of one-dimensional normal distributions has minimal canonical parameter $(\xi/\sigma^2, \sigma^{-2})$ and the domain of this parameter is $R \times (0, \infty)$. Truncation to a finite interval yields a non-full family. Indeed, since the range of the minimal canonical statistic (x, x^2) is bounded when x is restricted to (a, b), where $-\infty < a < b < \infty$, any minimal canonical parameter domain for the full family generated by the truncated distributions must equal R^2. The situation is quite similar if a whole sample $x_1, \ldots, x_n$ of observations is considered. ▶

Let P_θ^{cA} denote the censoring of P_θ to A. The family $\mathfrak{P}^{cA} = \{P_\theta^{cA} : \theta \in \Theta\}$ is exponential of order $\leq$ ord $\mathfrak{P} + 1$ and

$$\frac{dP_\theta^{cA}}{dP_0^{cA}} = a(\theta) \, e^{\theta \cdot t 1_A + \sigma(\theta) 1_{A^c}}$$

where

$$\sigma(\theta) = \ln \frac{P_\theta(A^c)}{a(\theta) P_0(A^c)}.$$

Example 8.20. Censored exponential lifetimes. Suppose an individual or item has lifetime t and that this lifetime is assumed to be exponentially distributed with hazard rate parameter λ. If, as is rather often the case in practice, observation consists in recording t provided this value is less than or equal to some prefixed time t_0 and otherwise noting merely that $t > t_0$ then one has an instance of censoring, of the exponential distribution to the interval $A = [0, t_0]$, and the family of censored distributions is exponential of order 2. ▶

(vi) *Conjugate families.* Suppose $\mathfrak{P}$ is full and let

$$a(\theta) b(x) \, e^{\theta \cdot t(x)} \tag{8}$$

be a minimal canonical representation of $\mathfrak{P}$. Consider the *conjugate* family, i.e. the family of distributions on Θ having densities with respect to Lebesgue measure

$$d(\gamma, \chi) a(\theta)^\gamma \, e^{\chi \cdot \theta} \tag{9}$$

where $\gamma \in R$ and $\chi \in R^k$ are parameters and $d(\gamma, \chi)$ is a normalizing constant; the domain of variation of (γ, χ) is the set

$$E = \{(\gamma, \chi): \int_\Theta a(\theta)^\gamma e^{\chi \cdot \theta} \, d\theta < \infty\}.$$

Note that the conjugate family is not determined by $\mathfrak{P}$ alone but depends on the chosen minimal canonical representation. However, by transferring the conjugate distributions to distributions on $\mathfrak{P}$ via the one-to-one mapping $\theta \to P_\theta$ one obtains a family of distributions $\mathfrak{P}^*$ which is the same whichever representation is considered. Clearly, $\mathfrak{P}^*$ is a full exponential family.

From Theorem 6.1(i) and (iv) and from the relation int dom $\kappa^* =$ int C, which follows from Theorem 9.1 later, one finds for the section of E at γ the expressions

$$E_\gamma = \begin{cases} \gamma \operatorname{int} C & \text{for } \gamma > 0 \\ \\ \operatorname{int} \operatorname{bar} \Theta & \text{for } \gamma = 0. \end{cases} \tag{10}$$

It is possible, in general, to have E_γ non-empty for some or all negative values of γ, but no simple, general expression for E_γ is available for $\gamma < 0$. It follows, however, from (10) that if $\Theta = R^k$ then $E = \{(\gamma, \chi): \gamma > 0, \gamma^{-1}\chi \in \operatorname{int} C\}$. Note also that (10) implies ord $\mathfrak{P}^* = k + 1$.

Example 8.21. If $\mathfrak{P}$ is the family of one-dimensional normal distributions with known variance σ^2 then $\mathfrak{P}^*$ in effect equals $\mathfrak{N}$. ▶

Example 8.22. Let $\mathfrak{P}$ be the Poisson family having model function

$$e^{-\lambda} \frac{\lambda^x}{x!}.$$

The family of distributions of λ induced by $\mathfrak{P}^*$ is the class of gamma distributions

$$\frac{1}{\Gamma(\chi)\beta^\chi} \lambda^{\chi - 1} e^{-\lambda/\beta}, \qquad (\chi, \beta) \in (0, \infty)^2.$$ ▶

Example 8.23. Take $\mathfrak{P}$ to be the family of multinomial distributions with model function

$$\frac{n!}{x_1! \cdot \cdots x_k!(n - x_.)!} \pi_1^{x_1} \cdots \pi_k^{x_k}(1 - \pi_.)^{n - x_.}.$$

Under $\mathfrak{P}^*$ the parameter $\pi = (\pi_1, \ldots, \pi_k)$ follows the family of Dirichlet distributions

$$\frac{\Gamma(\lambda_1 + \cdots + \lambda_{k+1})}{\Gamma(\lambda_1) \cdots \Gamma(\lambda_{k+1})} \pi_1^{\lambda_1 - 1} \cdots \pi_k^{\lambda_k - 1}(1 - \pi_.)^{\lambda_{k+1} - 1}. \qquad \lambda_* \in (0, \infty)^{k+1}.$$ ▶

Further examples may be found in Raiffa and Schlaifer (1961), and Ando and Kaufman (1965).

It is simple to see that $\mathfrak{P}^{(n)*} = \mathfrak{P}^*$.

If θ is considered a random variable with density (9) and if $x_1,\ldots,x_n$ is a random sample from $\mathfrak{X}$ then the probability function of the marginal distribution of $x^{(n)} = (x_1,\ldots,x_n)$ is

$$p(x^{(n)}) = d(\gamma,\chi)\left(\prod_{i=1}^{n} b(x_i)\right)/d(\gamma+n,\chi+t_.) \tag{11}$$

where $t_. = t(x_1) + \cdots + t(x_n)$. Furthermore, the probability function of θ given $x^{(n)}$ is

$$p(\theta|x^{(n)}) = d(\gamma+n,\chi+t_.)a(\theta)^{\gamma+n}\,e^{(\chi+t_.)\cdot\theta}. \tag{12}$$

In other words, (12) is the posterior distribution of θ, with (9) as prior. One notes that this posterior distribution, as well as the prior, belongs to the conjugate family. However, the family of distributions (12) as (γ,χ) varies over E may be a proper subset of the conjugate family.

8.3 COMPLEMENTS

(i) Let $\mathfrak{P}$ be exponential of order k and with minimal standard representation as equation (1) of Section 8.2, and let u be a statistic. Suppose it is desired to find the conditional mean value of u given t, such as may be the case, for instance, in a control of the model $\mathfrak{P}$. Owing to the sufficiency of t, this conditional mean value does not depend on θ and may thus be denoted by $E(u|t)$.

A derivation of $E(u|t)$ via a determination of the conditional distribution of u may seem too complicated. However, if the mean value $E_\theta u$ is a known function of θ, $E(u|t)$ can sometimes be obtained fairly simply by Laplace transform inversion since

$$\int E(u|t)a(\theta)\,e^{\theta\cdot t}\,dP_0 = E_\theta u. \tag{1}$$

(Recourse may of course be made to published lists of Laplace transforms and their inverses, such as Gradshteyn and Ryzhik (1965) and Roberts and Kaufman (1966).)

Example 8.24. Assume that $x = (x_1,\ldots,x_n)$ where $x_1,\ldots,x_n$ are independent, identically distributed Poisson variates. Here $t = x_.$ and (1) takes the form

$$\sum_{x_.=0}^{\infty} E(u|x_.)\,e^{-n\lambda}\frac{(n\lambda)^{x_.}}{x_.!} = E_\lambda u.$$

The conditional mean value $E(u|x_.)$ may be found by expanding $\exp(n\lambda)\,E_\lambda u$ in a power series in λ and identifying coefficients. For instance, for $u = \Sigma(x_i - \bar{x})^2$ and hence $E_\lambda u = (n-1)\lambda$ one sees that

$$E(\Sigma(x_i - \bar{x})^2 | x_.) = \frac{n-1}{n} x_. .$$

Thus

$$E(s^2/\bar{x} | x_.) = 1, \qquad x_. > 0,$$

where $s^2/\bar{x}$ is the dispersion index. The conditional variance of the dispersion index is, by this technique, obtained as

$$V(s^2/\bar{x} | x_.) = 2(1 - x_.^{-1})/(n-1), \qquad x_. > 0.$$

(Further formulas for conditional moments of k-statistics (sample cumulants) for the Poisson distribution are given in Gart and Pettigrew (1970).) ▶

Many other examples may be drawn from the literature on unbiased estimation, $E(u|t)$ being the minimum variance unbiased estimator of $E_\theta u$. See in particular Washio, Morimoto, and Ikeda (1956).

Note that if u, t, and P_0 is any set of two statistics and a probability measure, and if $E_0(u|t)$ is sought then the above method is potentially applicable, through introduction of the exponential family generated by P_0 and t.

The techniques described here were indicated by Tweedie (1946, 1947), though his papers have been largely unnoticed.

(ii) *Factorial series families.* A parametrized family $\mathfrak{Q} = \{Q_\eta : \eta \in H\}$ where $H \subset N_0^k$ is called a *factorial series family* if for $\eta \in H$ the support of Q_η is contained in N_0^k and if there exist a positive function g on H and a non-negative function b on N_0^k such that the probability of $x \in N_0^k$ under Q_η is of the form

$$b(x)\eta^{(x)}/g(\eta) \tag{2}$$

where $\eta^{(x)} = \eta_1^{(x_1)}\eta_2^{(x_2)}\cdots\eta_k^{(x_k)}$ (and the notation $n^{(m)}$ indicates the descending factorial, i.e. $n^{(m)} = n(n-1)\cdots(n-m+1)$). With i denoting a point in N_0^k, let $S = \{i : b(i) > 0\}$ and $\bar{i} = \{j \in N_0^k : 0 \le j_1 \le i_1, \ldots, 0 \le j_k \le i_k\}$. Note that the support S_η of Q_η in general varies with η and that $S_\eta = \bar{\eta} \cap S$. For simplicity, suppose that there exists a k-dimensional cube of unit side length whose vertices all belong to S, and that the family $\mathfrak{Q}$ is *full* in the sense that $H = \{\eta \in N_0^k : \bar{\eta} \cap S \ne \varnothing\}$. Defining $g(\eta)$ for all $\eta \in N_0^k$ by $g(\eta) = \Sigma b(x)\eta^{(x)}$, one has

$$b(x) = \Delta^x g(0)/x!$$

and

$$E_\eta x^{(i)} = \eta^{(i)}\nabla^i g(\eta)/g(\eta)$$

where ∇ and Δ denote, respectively, the descending and ascending difference operator (of dimension k).

The factorial series families were introduced by Berg (1974, 1977), who noted their similarity to power series families, studied some of their properties, and showed how they arise as the result of certain sampling procedures employed to obtain data for inference on the sizes of various classes of elements or individuals, the parameters $\eta_1,\ldots,\eta_k$ denoting these sizes.

Example 8.25. Multivariate hypergeometric distributions. The probability function of the multivariate hypergeometric distribution with indices m and $n \le m$ (where $m, n \in N$) and size parameters $\eta_1,\ldots,\eta_k$ is given by

$$\frac{\binom{\eta_1}{x_1}\cdots\binom{\eta_k}{x_k}\binom{m}{n-x_.}}{\binom{\eta_. + m}{n}}$$

and this is of the form (2) with

$$g(\eta) = \binom{\eta_. + m}{n}.$$ ▶

Example 8.26. Binomial distributions. For a fixed value π of the probability parameter, the family of binomial distributions with trial parameter $n \in N_0$ is a full factorial series family for which $g(n) = (1-\pi)^{-n}$. ▶

Example 8.27. Matching. Suppose n balls marked $1, 2, \ldots, n$ are distributed at random among n urns, also numbered $1, 2, \ldots, n$, in such a way that each urn contains exactly one ball. The probability of observing x non matches is

$$\sum_{i=0}^{x} \{(-1)^i/i!\} n^{(x)}/n!$$

which is of the form (2). ▶

Example 8.28. Multiple-recapture census. On l successive occasions a random sample is removed from an animal population and each time the individuals removed are furnished with a tag, unless they have already been tagged at one of the previous occasions, and are then returned to the population. Here the tags are supposed to be identical. Let n_i and u_i denote, respectively, the sample size and the number of unmarked individuals in the sample at the ith occasion, and suppose $n_1,\ldots,n_l$ can be considered given. The $u_. = u_1 + \ldots + u_l$ is a minimal sufficient statistic with respect to size η of the population and the probability of $u_.$ is

$$\left[\sum_{j=0}^{u_.} (-1)^j \Big/ \left\{ j! \prod_{i=1}^{l} (u_. - n_i - j)! \right\}\right] \eta^{(u_.)} \Big/ \prod_{i=1}^{l} \eta^{(n_i)},$$

cf. Berg (1974). Thus $u_.$ follows a distribution from a factorial series family with $g(\eta) = \Pi \eta^{(n_i)}$. ▶

If the function g in (2) depends on η through $\eta_{.} = \eta_1 + \cdots + \eta_k$ only then the factorial series family is said to be *sum-symmetric*. The multivariate hypergeometric family of Example 8.25 is an instance of this.

(iii) *Infinite divisibility*. Let $\mathfrak{P}$ be a full and linear exponential family of distributions on R^k and suppose ord $\mathfrak{P} = k$.

If one member P of $\mathfrak{P}$ is infinitely divisible then so is every other member.

To see this, let P_n be the probability measure on R^k whose nth convolution equals P and let $\mathfrak{P}_n$ be the full exponential family generated by P_n and the identity mapping x on R^k. The family obtained by convoluting each member of $\mathfrak{P}_n$ with itself n times equals $\mathfrak{P}$, cf. Section 8.2(ii), and this establishes the result.

8.4 NOTES

Much of the material in this chapter belongs to the folklore of statistics. Previous accounts of fundamental properties of exponential families may be found in Lehmann (1959), Witting (1966), Chentsov (1966, 1972), and Barndorff-Nielsen (1970). The concepts of regular exponential families and mixed parametrization were introduced in Barndorff-Nielsen (1970), and Theorem 8.3 is from Barndorff-Nielsen (1969). For discussions of the role of conjugate families (cf. Section 8.2(vi)) in Bayes statistics, see Raiffa and Schlaifer (1961) and Lindley (1971). (In statistical mechanics and certain probability theoretical contexts, the term conjugate family is used in a sense different from that given in Section 8.2(vi), namely to denote the exponential family generated by a probability measure P_0 and a random variable, cf. Section 8.1. See, for instance, Keilson (1965), Feller (1966), and references mentioned there; Feller employs the word associated instead of conjugate.)

As mentioned in Section 1.2, the notion of exponential families first occurred in Fisher (1934). Fisher argued—in his characteristic, mathematically somewhat imprecise, manner–that families of (one-dimensional) distributions which admit a sufficient reduction have to be exponential, and he also indicated that these families were the only ones supplying uniformly most powerful tests in the sense of Neyman and Pearson. Many subsequent papers, by other authors, have been concerned with the mathematical questions left open by Fisher in making the former of the two claims. The first of these papers were by Darmois (1935), Koopman (1936), and Pitman (1936) (and exponential families have sometimes been referred to as Darmois–Koopman or Fisher–Darmois–Koopman–Pitman families.) For discrete type families no mathematically and statistically satisfactory formulation has been found, whereas a fairly adequate discussion can be given in the continuous type case, see Dynkin (1951), Brown (1964), and Hipp (1974) (generalization to higher dimensions is considered in Barndorff-Nielsen and Pedersen (1968)). The second of Fisher's claims has been treated fairly recently by Pfanzagl (1968); see also the comments by Neyman and Pearson (1936) and Bartlett (1937).

Exponential families, or at least certain types of such families, are also arrived at from various starting points other than those of sufficient reduction and uniformly most powerful test. Boltzmann's law in statistical mechanics (see e.g. Khinchin 1949) is a result to this effect. Martin-Löf (1970, 1974a,b,c, 1975) has proposed an adaptation and development of the reasoning connected with Boltzmann's law for use in statistics. A related way of constructing families of distributions is that of maximum entropy or minimum discrimination information estimation; this in turn has been shown to be formally translatable into the construction, originating with Gauss, of distribution families from the requirement that the arithmetic mean of a specified statistic should be the maximum likelihood estimator, and the families yielded by these two methods are exponential (see Campbell 1970 and the references therein). Finally, it may be recalled that the Cramér–Rao lower bound for the variance of an unbiased estimator is attained only for exponential families, see Barankin (1951), Chentsov (1972), and Wijsman (1973). (All the above-mentioned derivations of exponential families are, of course, subject to regularity conditions.)

Various generalizations of exponential families are discussed in Crain (1974), Johansen (1977), Lauritzen (1975), Nelder and Wedderburn (1972), and Soler (1977).

CHAPTER 9

Duality and Exponential Families

For exponential families, sample-hypothesis duality and lods function theory (Chapter 3) combine intimately with the mathematical theory of convex duality (Chapters 5 and 6) and as a result it is possible, employing those theories, to establish a considerable number of statistically useful, general properties of exponential families, in a unified way.

Throughout the present chapter the sample space $\mathfrak{X}$ is, for convenience, taken to be the whole Euclidean space R^k, the exponential family $\mathfrak{P}$ is assumed to be of order k and linear, and only minimal representations of $\mathfrak{P}$ are considered—unless explicitly stated otherwise. With the canonical statistic t being the identity mapping on $\mathfrak{X} = R^k$, the representations are thus of the form

$$(\dagger) \qquad p(t; \theta) = a(\theta)b(t)\,e^{\theta \cdot t} \qquad (t \in R^k).$$

(The motivation for using t, and not x, to denote sample points and the identity mapping on the sample space is that in applications of the results of this chapter the family $\mathfrak{P}$ will often have been arrived at by margining to the minimal sufficient and canonical statistic of an underlying exponential family.) The dominating measure with respect to which the densities (†) are derived will, as usual, be denoted by μ. Further, it will be supposed that $b(t) = 0$ for $t \notin S$ (the support of $\mathfrak{P}$), and that $0 \in \Theta$ and $a(0) = 1$. Thus $b(t) = p(t; 0)$ and this function will also occasionally be denoted by $p(t)$. Set $\varphi(t) = -\ln b(t) = -\ln p(t)$.

The basic parts of the convex duality theory for exponential families are presented in Section 9.1.

Section 9.2 contains some results on stochastic independence and likelihood independence in exponential families.

The lods functions to be studied, in Sections 9.3–9.6, are the log-likelihood function

$$l(\theta) = l(\theta; t) = \theta \cdot t - \kappa(\theta) - \delta(\theta|\Theta) \qquad (\theta \in R^k)$$

(where $\kappa = -\ln a$), the log-probability function

$$\theta \cdot t - \varphi(t) \qquad (t \in R^k)$$

(where $\varphi = -\ln b$), and the log-plausibility function

$$\pi(\theta) = \pi(\theta; t) = \theta \cdot t - \varphi^*(\theta) - \delta(\theta|\Theta) \qquad (\theta \in R^k)$$

(where φ^* is the conjugate of φ).

Note that both $\theta \cdot t - \kappa(\theta)$ and $\theta \cdot t - \varphi^*(\theta)$ are closed concave functions on R^k, the first because κ is the logarithm of a Laplace transform (cf. Theorem 7.1) and the second due to the fact that the conjugation operation always yields closed convex functions (see the beginning of Section 5.3). It will be shown in Section 9.5 that $\theta \cdot t - \varphi(t)$ coincides on dom φ with a concave function if and only if $\mathfrak{P}$ is universal.

(For the study of the likelihood functions of an exponential family the assumptions made here obviously cause no loss of generality. It may also be noted that even for a nonlinear exponential family with representation $a(\theta)b(x)\exp\{\theta \cdot t(x)\}$ the function $\varphi^*(\theta) = \sup_{x \in \mathfrak{X}}\{\theta \cdot t(x) + \ln b(x)\}$ is closed convex.)

Prediction functions under exponential models are briefly discussed in Section 9.7.

9.1 CONVEX DUALITY AND EXPONENTIAL FAMILIES

In this section $\mathfrak{P}$ is assumed to be full. Recall the notations $C = \operatorname{cl}\operatorname{conv} S$ and $\mathfrak{T} = \tau(\operatorname{int}\Theta)$. The cumulant transform κ is a closed convex function and the conjugate of κ is the 'sup-log-likelihood function', to be denoted by $\hat{l}$. Thus

$$\kappa^*(t) = \sup_\theta \{\theta \cdot t - \kappa(\theta)\} = \sup_\theta l(\theta; t) = \hat{l}(t), \qquad t \in R^k.$$

Similarly, φ^* is closed convex and its conjugate is, under mild regularity conditions, equal to the 'sup-log-plausibility function' $\breve{\pi}$, i.e.

$$\varphi^{**}(t) = \sup_\theta \{\theta \cdot t - \varphi^*(\theta)\} = \sup_\theta \pi(\theta; t) = \breve{\pi}(t), \qquad t \in R^k.$$

The discussion in the present section centres around the two pairs of functions κ and $\hat{l}$, φ^* and $\breve{\pi}$, and the pair of sets Θ and C. Each of these pairs is investigated for statistically interesting properties, under the viewpoint of convex duality. The results derived will be drawn upon in subsequent sections.

It may be noted that, letting q_L and q_Π denote the likelihood and plausibility ratio test statistics for the hypothesis $\theta = 0$, one has $\hat{l} = -\ln q_L$ and $\breve{\pi} = -\ln q_\Pi + d$ where d is a constant ($d = -\ln \sup_t b(t)$). Thus $\hat{l}$ and $\breve{\pi}$ both have an immediate statistical interpretation.

The 'likelihood pair' κ and $\hat{l}$ will be studied first.

Theorem 9.1. *One has*

(i) $\kappa^* = \hat{l}$ *and* $\hat{l}^* = \kappa$.

(ii) κ *is a closed and strictly convex function with* dom $\kappa = \Theta$.

(ii)* $\hat{l}$ *is a closed and essentially smooth convex function with* $\operatorname{int} C \subset \operatorname{dom} \hat{l} \subset C$.

Proof. Assertion (ii) is just a reiteration of Theorem 7.1.

The equality $\kappa^* = \hat{l}$ was pointed out above, and since κ is closed $\hat{l}^* = \kappa^{**} = \kappa$.

The essential smoothness of $\hat{l}$ is a consequence of Theorem 5.30.

To prove dom $\hat{l} \subset C$, consider a point $t \notin C$. Let H be a hyperplane separating C and t strongly, and let e be the unit vector in R^k which is normal to H and such that C lies in the negative halfspace determined by H and e. Setting $d = e \cdot t$ one has $d > \delta^*(e|C)$ and hence, by (5) of Section 7.1,

$$l(re; t) = rd - \ln c(re) \to \infty \quad \text{as} \quad r \to \infty.$$

Consequently, $\hat{l}(t) = \infty$ and $t \notin \text{dom}\, \hat{l}$.

If t is a point in int C then $t \in \text{dom}\, \hat{l}$, i.e. $\theta \cdot t - \kappa(\theta)$, considered as a function of θ, is bounded above. This follows immediately from the next two lemmas. ▶

Lemma 9.1. *For any $\theta \in R^k$, $t \in R^k$*

$$\theta \cdot t - \kappa(\theta) \leq -\ln \rho(t)$$

where

$$\rho(t) = \inf_e P_0\{e \cdot T \geq e \cdot t\}$$

the infimum being taken over all unit vectors in R^k. Hence $t \in \text{dom}\, \hat{l}$ *if* $\rho(t) > 0$.

The proof of this lemma is simple and will not be given. ▶

Lemma 9.2. *Suppose $t \in \text{int}\, C$. Then* $\inf P_0\{e \cdot T \geq e \cdot t\} > 0$, *where the infimum is taken over all unit vectors e in R^k.*

Proof. By Theorem 5.3 there exists a natural number $m \leq 2k$ and m points $s_1, \ldots, s_m$ in S such that $t \in \text{int conv}\{s_1, \ldots, s_m\}$. Let $H(e)$ denote the hyperplane through t with normal e, let $\delta_i(e)$ be the distance from s_i to $H(e)$ and set $\delta(e) = \max\{\delta_1(e), \ldots, \delta_m(e)\}$. The mapping $\delta: e \to \delta(e)$ defined on the set of unit vectors is a continuous mapping on a compact set; thus it attains its infimum δ_0, say, and clearly δ_0 must be positive. Consequently, for every e the closed positive halfspace determined by $H(e)$ and e contains at least one of the spheres $B(s_i, \delta_0)$ with centre s_i and radius δ_0, $i = 1, 2, \ldots, m$, and each of these spheres have positive P_0 measure since $s_i \in S$. Hence

$$\inf_e P_0\{e \cdot T \geq e \cdot t\} \geq \min\{P_0(B(s_i, \delta_0)) : i = 1, \ldots, m\} > 0. \quad ▶$$

Note that if θ and t satisfy

$$\theta = D\hat{l}(t) \quad \text{or, equivalently,} \quad t \in \partial\kappa(\theta) \tag{1}$$

then (cf. (2) of Section 5.4)

$$\kappa(\theta) + \hat{l}(t) = \theta \cdot t.$$

Condition (1) is fulfilled, in particular, if

$$\theta \in \text{int}\, \Theta, \quad t = \tau$$

(where $\tau = \tau(\theta) = E_\theta t$) and then

(2) $$\frac{\partial \kappa}{\partial \theta} = \tau, \qquad \frac{\partial \hat{l}}{\partial \tau} = \theta$$

(3) $$\frac{\partial^2 \kappa}{\partial \theta \partial \theta'} = V_\theta t, \qquad \frac{\partial^2 \hat{l}}{\partial \tau \partial \tau'} = (V_\theta t)^{-1}.$$

It is obvious, then, that $\hat{l}$ must be a very 'smooth' functiόn. In certain cases (but not always) $\hat{l}$ is, even, a cumulant transform.

Example 9.1. For $\mathfrak{P}$ the family of Poisson distributions and $\theta = \ln \lambda$ one has

$$\Theta = R, \qquad C = [0, \infty)$$

and

$$\kappa(\theta) = e^\theta - 1, \qquad \theta \in \Theta$$

$$\hat{l}(t) = t \ln t - t + 1, \qquad t \in C.$$

It was mentioned in Example 7.1 that the function t^t, $t \in [0, \infty)$, is the Laplace transform of an extreme stable distribution with index $\alpha = 1$, support R and density of the form (3) of Section 7.1. Hence $\hat{l}$ is the logarithm of a Laplace transform. ▶

The duality relation between κ and $\hat{l}$ is especially strong when κ is essentially smooth which is equivalent to κ being steep. As noted earlier, if Θ is open, i.e. if $\mathfrak{P}$ is regular, then κ and $\mathfrak{P}$ are steep.

Theorem 9.2. *κ is steep if and only if* $\mathfrak{T} = \text{int}\, C$.

Proof. If κ is steep then $(\text{int}\,\Theta, \kappa)$ is of Legendre type and hence, by Theorem 5.33 and Theorem 9.1 (ii)*, $(\text{int}\; C, \hat{l})$ is also of Legendre type and $\text{int}\; C = D\kappa(\text{int}\,\Theta) = \mathfrak{T}$.

Next, suppose $\mathfrak{T} = \text{int}\, C$. Then, in view of formula (3), $\hat{l}$ is strictly convex on $\text{int}\, C$. The fact that $\hat{l}$ is essentially smooth implies that $\text{dom}\, \partial\hat{l} = \text{int}\, C$ and hence $\hat{l}$ is essentially strictly convex. The steepness of κ now follows by Theorem 5.30. ▶

The first part of this proof is, in effect, also a verification of:

Theorem 9.3. *Suppose κ is steep. Then* $(\text{int}\,\Theta, \kappa)$ *and* $(\mathfrak{T}, \hat{l})$ *are both of Legendre type, and they are each other's Legendre transform.*

Example 9.2. Beta family. The beta distribution with shape parameters $\lambda_1 (>0)$ and $\lambda_2 (>0)$ has density

$$\frac{1}{B(\lambda_1, \lambda_2)} x^{\lambda_1 - 1}(1 - x)^{\lambda_2 - 1}, \qquad 0 < x < 1.$$

The family of these distributions, for (λ_1, λ_2) varying in $(0, \infty)^2$, is regular exponential, with $t(x) = (\ln x, \ln(1 - x))$ as minimal canonical statistic and

$$C = \{t: t_1 < 0, t_2 \leq \ln(1 - e^{t_1})\}.$$

Moreover, the mean value of t is

$$\tau = (\psi(\lambda_1) - \psi(\lambda.), \psi(\lambda_2) - \psi(\lambda.))$$

where ψ denotes the digamma function. It follows from Theorems 9.2 and 9.3 that the range of τ equals int C. (A direct proof of this from (4) would not be trivial.) ►

By Theorem 9.1(ii)*, $\operatorname{dom} \hat{l} \subset C$ and $\operatorname{int}(\operatorname{dom} \hat{l}) = C$. The results mentioned next throw light on the question of when a boundary point of C belongs to $\operatorname{dom} \hat{l}$.

Theorem 9.4. *Suppose S is a finite or countable set. Then* $\operatorname{conv} S \subset \operatorname{dom} \hat{l}$.
Hence, in particular, $\operatorname{dom} \hat{l} = C$ *if S is finite.*

Proof. Any point $t \in \operatorname{conv} S$ can be written as a convex combination of $k + 1$ points $s_1, \ldots, s_{k+1}$ from S (Theorem 5.3) and every closed halfspace H, whose corresponding hyperplane contains t, must contain at least one of the point $s_1, \ldots, s_{k+1}$. Thus

$$\rho(t) = \inf_e P_0\{e \cdot \mathrm{T} \geq e \cdot t\} \geq \min\{P_0\{s_i\}: i = 1, \ldots, k + 1\} > 0$$

and hence, by Lemma 9.1, $t \in \operatorname{dom} \hat{l}$.

When S is finite, $C = \operatorname{conv} S$ which implies $\operatorname{dom} \hat{l} = C$ ►

Theorem 9.5. *Let t be a boundary point of C. If there exists a hyperplane H supporting C at t and satisfying $P_0(H) = 0$ then $t \notin \operatorname{dom} \hat{l}$.*

Thus, in particular, $\operatorname{dom} \hat{l} = \operatorname{int} C$ *if P_0 is absolutely continuous with respect to Lebesgue measure.*

Proof. This result follows easily from formula (5) of Section 7.1. ►

Next, the 'plausibility pair' φ^* and $\check{\pi}$ will be discussed. For this discussion $\mathfrak{P}$ is supposed to be of either discrete or continuous type, and in the latter case the probability functions $p(\cdot; \theta)$, $\theta \in \Theta$, are assumed to be densities with respect to Lebesgue measure

Note that, by formula (2) of Section 5.3,

$$\varphi^* = (\operatorname{conv} \varphi)^* \quad \text{and} \quad \varphi^{**} = \operatorname{cl} \operatorname{conv} \varphi,$$

and that if S is finite then $\operatorname{conv} \varphi$ is closed because it is finitely generated, cf. Rockafellar (1970), Corollary 19.1.2.

One finds in partial analogy to Theorem 9.1:

Theorem 9.6. *Suppose that* $\sup_t p(t; \theta) < \infty$ *for every $\theta \in \Theta$, and that either S is finite or a subset of Z^k or $\mathfrak{P}$ is of continuous type. Then*

(i) $\varphi^{**} = \check{\pi}$ *and* $\check{\pi}^* = \varphi^*$.

(ii) φ^* *is a closed convex function with* $\Theta \subset \operatorname{dom} \varphi^* \subset \operatorname{cl} \Theta$.
(ii)* $\breve{\pi}$ *is a closed convex function with* $S \subset \operatorname{dom} \breve{\pi} \subset C$.

Proof. That the domain of the closed convex function φ^* includes Θ is the same as the first of the stated suppositions. To prove $\operatorname{dom} \varphi^* \subset \operatorname{cl} \Theta$ (which is trivial for S finite), suppose $\theta \notin \Theta$, then

$$\int e^{\theta \cdot t - \varphi(t)} d\mu = \infty$$

and, since $\varphi^{**} = \operatorname{cl} \operatorname{conv} \varphi \le \varphi$, this implies

$$\int e^{\theta \cdot t - \varphi^{**}(t)} d\mu = \infty.$$

In the continuous type case one may conclude from this, on account of Theorem 6.1(i) and (iv), that $\theta \notin \operatorname{int} \operatorname{dom} \varphi^*$. The same conclusion is reachable for $S \subset Z^k$ by invoking Theorems 6.2 and 6.1. This establishes assertion (ii), and (i) and (ii)* follow simply. ▶

Theorem 9.7. *Suppose* $\operatorname{conv} \varphi$ *is closed and let* $\mathfrak{P}$ *be of either c-discrete or continuous type.*

If $\mathfrak{P}$ *is strongly unimodal then*

$$\Theta = \operatorname{int} \operatorname{dom} \varphi^*$$

and hence $\mathfrak{P}$ *is regular.*

Proof. Apply, again, Theorems 6.1 and 6.2. ▶

By the same reasoning as that yielding $\operatorname{int} \operatorname{dom} \varphi^* = \Theta$ one finds, more generally, that if $\mathfrak{P}$ is c-discrete and if $\tilde{\varphi}$ is any closed convex function which coincides with φ on S (whence $\mathfrak{P}$ is strongly unimodal) and for which $S = Z^k \cap \operatorname{dom} \tilde{\varphi}$ then

$$\Theta = \operatorname{int} \operatorname{dom} \tilde{\varphi}^*. \tag{5}$$

Theorem 9.8. *Suppose* $\mathfrak{P}$ *is of continuous type and that* φ *is a closed convex function (whence* $\mathfrak{P}$ *is strongly unimodal and regular).*

If the convexity of φ *is strict then* φ^* *is differentiable on* Θ *and steep.*

If φ *is differentiable on* $\operatorname{int} C$ *and steep then* φ^* *is essentially strictly convex.*

Proof. Use Theorem 5.30. ▶

Further results on κ, $\hat{l}$, φ^* and $\breve{\pi}$ are given in the following sections.

The theorem below gives some information on the relation between Θ and C.

Theorem 9.9. *Let* P_0 *be a probability measure on* $\mathfrak{X}$ *and let* t *be a k-dimensional statistic. Furthermore, let* $\mathfrak{P}$ *denote the exponential family generated by* P_0 *and* t, *let C be the convex support of P_0t, and set*

$$\Theta = \{\theta : \int e^{\theta \cdot t} \, dP_0 < \infty\}.$$

Then

$$\text{bar}\,\Theta \cap \text{bar}\,C = \{0\}. \tag{6}$$

Furthermore

$$\text{bar}\,C \subset 0^+\Theta \tag{7}$$

and, dually,

$$\text{bar}\,\Theta \subset 0^+C. \tag{8}$$

Proof. Suppose e is a unit vector contained in bar Θ. Then there exists a positive number r such that

$$\int e^{re \cdot t} \, dP_0 = \infty.$$

But this is possible only if the set $\{e \cdot t : t \in C\}$ is unbounded above, i.e. $e \notin \text{bar}\,C$.

To prove (7) suppose e is a unit vector in bar C and let θ_0 be an arbitrary point in Θ. Then $\delta^*(e|C) < \infty$ and

$$\int e^{(\theta_0 + re) \cdot t} \, dP_0 \leq e^{r\delta^*(e|C)} \int e^{\theta_0 \cdot t} \, dP_0 < \infty \qquad \text{for all } r \geq 0.$$

Thus $e \in 0^+\Theta$.

From (7) one obtains

$$(\text{bar}\,C)^0 \supset (0^+\Theta)^0.$$

By (5) of Section 5.1, $(\text{bar}\ C)^0 = 0^+C$ and by (5), (4), and (3) of Section 5.1,

$$\begin{aligned}(0^+\Theta)^0 &\supset (0^+\,\text{cl}\,\Theta)^0 \\ &= (\text{bar}\,\text{cl}\,\Theta)^{00} \\ &= \text{cl}\,\text{bar}\,\text{cl}\,\Theta \\ &\supset \text{bar}\,\Theta\end{aligned}$$

whence (8) follows.

(If it has been assumed that ord $\mathfrak{P} = k$ then the theorem could be proved by appealing to Theorems 5.19 and 9.1.) ▶

In many instances the inclusions in (7) and (8) may be replaced by equalities, or nearly so. That this is not always the case is shown by the third of the examples mentioned next.

Example 9.3. If S is bounded, in particular finite, then $\operatorname{bar} C = R^k = 0^+\Theta$ and $\operatorname{bar}\Theta = \{0\} = 0^+C$. ▶

Example 9.4. Gamma family. The density of the gamma distribution with shape parameter λ and scale parameter β is

$$\frac{1}{\Gamma(\lambda)\beta^{\lambda}} x^{\lambda-1} e^{-x/\beta} \qquad (x > 0)$$

where $(\lambda, \beta) \in (0, \infty)^2$. The family constituted by these distributions is regular with $t = (x, \ln x)$ and $\theta = (1 - \beta^{-1}, \lambda - 1)$ as a pair of minimal canonical variates. The sets Θ and C are as indicated in Figures 9.1 and 9.2.

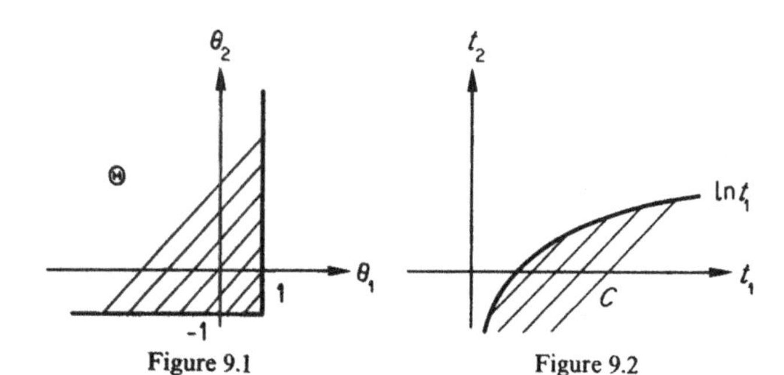

Figure 9.1 Figure 9.2

Here

$$\operatorname{bar}\Theta = \{\theta: \theta_1 \geq 0, \theta_2 \leq 0\}$$
$$0^+\Theta = \{\theta: \theta_1 \leq 0, \theta_2 \geq 0\}$$
$$\operatorname{bar} C = \{\tau: \tau_1 < 0, \tau_2 \geq 0\} \cup \{0\}$$
$$0^+C = \{\tau: \tau_1 \geq 0, \tau_2 \leq 0\}$$

so that

$$\operatorname{bar} C \subsetneq 0^+\Theta, \qquad \operatorname{cl} \operatorname{bar} C = 0^+\Theta \qquad \operatorname{bar}\Theta = 0^+C$$ ▶

Example 9.5. Suppose $\mathfrak{P}$ is the family of those normal distributions on R^k for which the coordinate random variables are independent and have variance 1. This family is full exponential with the identity mapping as minimal canonical statistic, and the mean value θ of the distribution in $\mathfrak{P}$ afford a minimal canonical parametrization of $\mathfrak{P}$ with $\Theta = R^k$. In this case $C = R^k$ and $\operatorname{bar}\Theta = \{0\}$, $0^+C = R^k$, $\operatorname{bar} C = \{0\}, 0^+\Theta = R^k$. ▶

***Corollary* 9.1.** *Under the setup of Theorem 9.9, if* $\dim C = k$ and $\operatorname{int}(\operatorname{bar} C) \neq \varnothing$ *then* $\operatorname{ord} \mathfrak{P} = k$.

Proof. The condition $\dim C = k$ implies that $t_1, \ldots, t_k$ are affinely independent, and $\operatorname{int}(\operatorname{bar} C) \neq \varnothing$ and (8) together imply $\operatorname{int}\Theta \neq \varnothing$. The result now follows from Corollary 8.1. ▶

9.2 INDEPENDENCE AND EXPONENTIAL FAMILIES

In this section $\mathfrak{P}$ is not assumed to be linear, as is otherwise standard in this chapter.

Theorem 9.10. *Let P_0 be a probability measure, let t be a statistic and let $\mathfrak{P}$ be the exponential family generated by P_0 and t. Furthermore, let $(t^{(1)}, \dots, t^{(m)})$ be a partition of t.*

If $t^{(1)}, \dots, t^{(m)}$ are independent under P_0 then they are independent under every element of $\mathfrak{P}$.

Proof. $\mathfrak{P}$ has the representation

$$\frac{dP_\theta}{dP_0} = a(\theta)\, e^{\theta \cdot t}$$

and, due to the independence assumption, one has

$$a(\theta) = a_1(\theta^{(1)}) \cdots a_m(\theta^{(m)})$$

where

$$a_j(\theta^{(j)})^{-1} = \int e^{\theta^{(j)} \cdot t^{(j)}}\, dP_0.$$

Hence, if $B^{(j)}$ belongs to the σ-algebra generated by $t^{(j)}$ $(j = 1, \dots, m)$,

$$\begin{aligned} P_\theta(B^{(1)} \cap \cdots \cap B^{(m)}) &= E_0(1_{B^{(1)} \cap \cdots \cap B^{(m)}} a(\theta)\, e^{\theta \cdot t}) \\ &= \prod_{j=1}^{m} E_0(1_{B^{(j)}} a_j(\theta^{(j)})\, e^{\theta^{(j)} \cdot t^{(j)}}) \\ &= \prod_{j=1}^{m} P_\theta(B^{(j)}). \end{aligned}$$

▶

Corollary 9.2. *Let $\mathfrak{P}$ be the exponential family generated by P_0 and t, and let $(t^{(0)}, t^{(1)}, \dots, t^{(m)})$ be a partition of t.*

If $t^{(1)}, \dots, t^{(m)}$ are conditionally independent given $t^{(0)}$ under P_0 then they have the same property under every element of $\mathfrak{P}$.

Proof. For any value $t^{(0)}$, the conditional distribution of $(t^{(1)}, \dots, t^{(m)})$, given $t^{(0)}$ and under P_0, belongs to the exponential family generated by $P_0(\cdot | t^{(0)})$ and $(t^{(1)}, \dots, t^{(m)})$.

▶

Theorem 9.11. *Suppose $\mathfrak{P}$ has open kernel, let t be a minimal canonical statistic for $\mathfrak{P}$, and let $(t^{(1)}, \dots, t^{(m)})$ be a partition of t.*

If $t^{(1)}, \dots, t^{(m)}$ are uncorrelated under $\mathfrak{P}$ then $t^{(1)}, \dots, t^{(m)}$ are independent under $\mathfrak{P}$.

Proof. It suffices to prove the result for $m = 2$ and, in view of Theorem 9.10, for Θ a product set $\Theta = \Theta^{(1)} \times \Theta^{(2)}$ with $\Theta^{(1)}$ and $\Theta^{(2)}$ being regions and Θ containing 0. From (12) and (13) of Section 8.1 one obtains

$$\frac{\partial \tau^{(1)}}{\partial \theta^{(2)}}(\theta) = 0, \qquad \theta \in \Theta,$$

whence

$$\tau^{(1)}(\theta^{(1)}, 0) = \tau^{(1)}(\theta), \qquad \theta \in \Theta$$

which in turn implies

$$\kappa(\theta) = \kappa(\theta^{(1)}, 0) + \kappa(0, \theta^{(2)}).$$

So, the Laplace transform $c = \exp \kappa$ factorizes and $t^{(1)}$ and $t^{(2)}$ are therefore independent under P_0, and hence under $\mathfrak{P}$.

Lemma 9.3. *Let $\{P_\omega : \omega \in \Omega\}$ be a parametrization of $\mathfrak{P}$, let π be an element of $\mathfrak{P}$ and let $(\omega^{(1)}, \ldots, \omega^{(m)})$ be a partition of ω.*

If $\omega^{(1)}, \ldots, \omega^{(m)}$ are L-independent then any minimal representation of the densities of $\mathfrak{P}$ with respect to π is of the form

$$\frac{dP_\omega}{d\pi} = a_1(\omega^{(1)}) \cdots a_m(\omega^{(m)}) \, e^{(\alpha_1(\omega^{(1)}) + \cdots + \alpha_m(\omega^{(m)})) \cdot t}.$$

Proof. This result is simple to obtain from the definition of L-independence and the affine independence of the coordinates of a minimal canonical statistic. ▶

The lemma is practical in a verification of the following assertion which is also simple to show.

Let $\{P_\theta : \theta \in \Theta\}$ be a minimal parametrization of $\mathfrak{P}$, let $(\theta^{(1)}, \ldots, \theta^{(m)})$ and $(t^{(1)}, \ldots, t^{(m)})$ be similar partitions of θ and t, and suppose $\mathfrak{P}$ is full. Then $t^{(1)}, \ldots, t^{(m)}$ are (stochastically) independent if and only if $\theta^{(1)}, \ldots \theta^{(m)}$ are L-independent.

Thus L-independence of the components of a partition of a canonical parameter is a rather trivial phenomenon. It will appear from Section 10.2 that L-independence in exponential families is almost exclusively tied to L-independence of the mean value component and the canonical component of mixed parametrizations. The next theorem contains a criterion for this latter property. Consider a minimal representation of $\mathfrak{P}$ and suppose $\mathfrak{P}$ is interior.

Theorem 9.12. *If $\tau^{(1)}$ and $\theta^{(2)}$ are L-independent then $\theta^{(1)}$, considered as a function of $(\tau^{(1)}, \theta^{(2)})$, is of the form*

$$\theta^{(1)} = \varphi(\tau^{(1)}) + \psi(\theta^{(2)}) \tag{1}$$

for some functions φ and ψ.

The converse assertion is valid provided $\mathfrak{P}$ is open and connected, and provided $\tau^{(1)}$ and $\theta^{(2)}$ are variation independent.

Proof. The first assertion follows from Lemma 9.1, the second from Theorem 10.4. ▶

Corollary 9.3. *Suppose $\mathfrak{P}$ is regular. Then the following three conditions are equivalent.*

(i) $\tau^{(1)}$ *and* $\theta^{(2)}$ *are L-independent*

(ii) $\theta^{(1)}$ *is of the form* $\theta^{(1)} = \varphi(\tau^{(1)}) + \psi(\theta^{(2)})$

(iii) $\tau^{(1)}$ *and* $\theta^{(2)}$, *considered as random variables on* Θ, *are stochastically independent under the conjugate family.*

Proof. The equivalence of (i) and (ii) is obvious from Theorem 9.12 and Theorem 8.4.

To show (i) $\Rightarrow$ (iii) $\Rightarrow$ (ii) note first that, since $\mathfrak{P}$ is regular, the domain of variation of $(\tau^{(1)}, \theta^{(2)})$ is of the form $\mathfrak{T}^{(1)} \times \Theta^{(2)}$. Moreover, the density of $(\tau^{(1)}, \theta^{(2)})$, with respect to Lebesgue measure on $\mathfrak{T}^{(1)} \times \Theta^{(2)}$ and under the element of the conjugate family given by (9) of Section 8.2, is

$$d(\gamma, \chi)a(\tau^{(1)}, \theta^{(2)})^{\gamma}\, e^{\chi(\theta^{(1)}(\tau^{(1)}, \theta^{(2)}), \theta^{(2)})} \left|\frac{\partial \theta^{(1)}}{\partial \tau^{(1)'}}\right|. \tag{2}$$

If $\tau^{(1)}$ and $\theta^{(2)}$ are L-independent then, by Lemma 9.3, this density may be written

$$d(\gamma, \chi)a_1(\tau^{(1)})^{\gamma} a_2(\theta^{(2)})^{\gamma}\, e^{\chi^{(1)} \cdot \varphi(\tau^{(1)}) + \chi \cdot (\psi(\theta^{(2)}), \theta^{(2)})} \left|\frac{\partial \varphi}{\partial \tau^{(1)'}}\right|,$$

i.e. the density factorizes into a function of $\tau^{(1)}$ times a function of $\theta^{(2)}$. Thus $\tau^{(1)}$ and $\theta^{(2)}$ are stochastically independent.

On the other hand, stochastic independence of $\tau^{(1)}$ and $\theta^{(2)}$ implies that (2) factorizes and this has, as is simple to see, the consequence that $\theta^{(1)}$ is of the form $\theta^{(1)} = \varphi(\tau^{(1)}) + \psi(\theta^{(2)})$. ▶

Example 9.6. Let x_1 and x_2 be independent and Poisson distributed with mean values λ_1 and λ_2. Then $t = (x_., x_2)$ and $\theta = (\ln \lambda_1, \ln\{\lambda_2/\lambda_1\})$ is a corresponding pair of canonical variates, and $\tau^{(1)} = \lambda_.$. Now

$$\theta^{(1)} = \ln \tau^{(1)} - \ln(1 + e^{\theta_2})$$

and Corollary 9.3 yields the two well-known results that $\lambda_.$ and λ_2/λ_1 are L-independent and that if λ_1 and λ_2 are stochastically independent and follow gamma-distributions with a common scale parameter, as is the case under $\mathfrak{P}^*$, then $\lambda_.$ and λ_2/λ_1 are also stochastically independent. ▶

Example 9.7. Independence in the Wishart family. Suppose t is a $W_r(f, \mathbf{\Sigma})$-distributed variate so that the density of t is

$$w_r(f)|\mathbf{\Delta}|^{f/2}|t|^{(f-r-1)/2} \exp\{-\tfrac{1}{2}\mathrm{tr}(\mathbf{\Delta}t)\} \tag{3}$$

where $\Delta = \Sigma^{-1}$ (cf. Example 6.6). Partition t

$$t = \begin{pmatrix} t_{11} & t_{12} \\ t_{21} & t_{22} \end{pmatrix}$$

such that t_{11} and t_{22} are square matrices and set $t^{(1)} = t_{11}$. Correspondingly one has $\tau^{(1)} = \Sigma_{11}$, $\theta^{(1)} = \Delta_{11}$ and

$$\theta^{(2)} = \begin{pmatrix} 0 & \Delta_{12} \\ \Delta_{21} & \Delta_{22} \end{pmatrix}.$$

The standard formula $\Sigma_{11}^{-1} = \Delta_{11} - \Delta_{12}\Delta_{22}^{-1}\Delta_{21}$ shows that condition (ii) of Corollary 9.3 is satisfied and hence Σ_{11} and $(\Delta_{12}, \Delta_{22})$ are L-independent and, moreover, stochastically independent under $\mathfrak{P}^*$.

It is evident from the expression (3) that the class of marginal distributions of Δ under $\mathfrak{P}^*$ includes all the Wishart distributions on $R^{\binom{r+1}{2}}$. Thus the important and well-known fact that $t_{11} - t_{12}t_{22}^{-1}t_{21}$ and (t_{12}, t_{22}) are stochastically independent is an immediate consequence of the above result. ▶

9.3 LIKELIHOOD FUNCTIONS FOR FULL EXPONENTIAL FAMILIES

The present section is devoted to the study of likelihood functions for full exponential families, in particular the question of existence and uniqueness of maximum likelihood estimates. Nearly all such families met in statistics are regular, or at least steep. The most important part of the section is Corollary 9.6 which summarizes the main results for regular and steep families.

The log-likelihood function

$$l(\theta) = \theta \cdot t - \kappa(\theta) \tag{1}$$

is, for any $t \in R^k$, a closed concave function on R^k, and

$$-\frac{\partial^2 l}{\partial\theta\partial\theta'} = V_\theta T = i(\theta) \qquad (\theta \in \operatorname{int} \Theta)$$

where $i(\theta)$ denotes Fisher's information function.

For a fixed t, consider the levels sets of $l(\cdot)$, i.e. the sets

$$C_d = \{\theta: l(\theta) \geq d\}, \qquad d \in R.$$

By Theorem 5.14, the non-empty sets among the C_d, $d \in R$, are closed and convex, and they all have the same recession cone. Hence they are either all unbounded or all bounded, and the boundedness case occurs if and only if $t \in \operatorname{int} C$. To obtain the latter assertion, note that

$$[-l(\cdot\,; t)]^*(\cdot) = \hat{l}(t + \cdot)$$

and invoke Theorems 9.10 and 5.20.

It is convenient to take the maximum likelihood estimator $\hat{\theta}$ to be the function defined for every $t \in R^k$ as the set of points θ which maximize l, rather than the restriction of this function to the support S of $\mathfrak{P}$. Note that with this general definition, $\hat{\theta}$ is the solution to the maximum likelihood estimation problem not only for $\mathfrak{P}$ but for every $\mathfrak{P}^{(n)}$ (cf. Section 8.2(ii)).

Theorem 9.13. *The log-likehood function l has a maximum if and only if $t \in \text{int}\, C$, and then the maximum is unique.*

The maximum likelihood estimator $\hat{\theta}$ equals $D\hat{l}$, the gradient mapping of $\hat{l}$, and the inverse mapping $\hat{\theta}^{-1}$ equals $\partial\kappa$.

Proof. Let $\bar{\theta} \in R^k$, $t \in R^k$. From the remark following immediately after equation (2) of Section 5.4 it is seen that $l(\cdot) = l(\cdot\,; t)$ attains its supremum at $\bar{\theta}$ if and only if $\bar{\theta} \in \partial\hat{l}(t)$. But $\hat{l}$ is essentially smooth and thus $\partial\hat{l}$ equals the gradient mapping $D\hat{l}$ of $\hat{l}$, which is single-valued and has domain $\text{int}(\text{dom}\, \hat{l}) = \text{int}\, C$, cf. Theorems 9.1 and 5.28. ▶

Example 9.8. Let $x_1, \ldots, x_n$ be a sample of gamma-distributed variates. Then

$$t = \frac{1}{n}\sum_{i=1}^{n}(x_i, \ln x_i)$$

is a minimal canonical statistic and the convex support of the distribution of t is as shown in Figure 9.2. Clearly, $t \in \text{bd}\, C$ for $n = 1$ while for $n > 1$ one has $t \in \text{int}\, C$ with probability 1, i.e. for samples of size 2 or more the maximum likelihood estimate of the parameter (λ, β) of the gamma-distribution exists with probability 1. ▶

Corollary 9.4. *It is a necessary and sufficient condition for $\hat{\theta}$ to be defined with probability one that* bd C, *the boundary of* C, *has probability zero.*

Corollary 9.5. *Let $\mathfrak{P}_0$ be an affine hypothesis of $\mathfrak{P}$ and consider a $t \in R^k$.*

If the maximum likelihood estimate exists under $\mathfrak{P}$ (i.e. if $t \in \text{int}\, C$) then it also exists under $\mathfrak{P}_0$.

Proof. Since $\mathfrak{P}$ is full so is $\mathfrak{P}_0$, and if t_0 is a minimal canonical statistic for $\mathfrak{P}_0$ then t_0 is an affine function of t. Now apply Lemma 5.1. ▶

Example 9.9. Suppose $\mathfrak{P}_0$ is an affine hypothesis of $\mathfrak{N}_r$ of the form $\{P_{(\xi, \Delta)}: (\xi, \Delta) \in \Xi \times \Delta\}$ with $0 \in \Xi$. (Such hypotheses were considered in Examples 8.11 and 8.12.) For simplicity, assume moreover that Δ contains the identity matrix. Let $\hat{\theta}_0$ denote the maximum likelihood estimator under the model $\mathfrak{P}_0$. It seems rather temping to conjecture that dom $\hat{\theta}_0$ has probability either 1 or 0.

That this is not the case is shown by the example $r = 3$, $\Xi = \{0\}$, and Δ equal to the set of positive definite 3×3 matrices $\mathbf{\Delta}$ whose diagonal elements δ_{ii} satisfy $\delta_{11} + \delta_{22} - \delta_{33} = 1$. This example is due to Erlandsen (1975).

However, if Δ is a cone and $\Sigma = \underline{\Delta}$ is the set of variance matrices corresponding to $\mathfrak{P}_0$, then the conjecture is true. The latter requirements are equivalent to Σ being a cone and affine. See Erlandsen (1975), and also Jensen (1975). The paper by Jensen is a deep and comprehensive study of the structure of this kind of model (with $\Xi = \{0\}$) and of the distributions of the maximum likelihood estimators and likelihood ratio testors for such models. Of particular importance among these models are those for which $\mathfrak{P}_0$ is, in a suitable coordinate system for the observation vector, describable as a family of distributions of a set of independent normal vectors such that the set may be partitioned into subsets within each of which the vectors are identically distributed, while, otherwise, Δ varies freely. Most of the models $\mathfrak{P}_0$ which have practical interest are of this latter type. ▶

It is immediate from (1) that the likelihood equation is

$$E_\theta t = t \qquad (\theta \in \operatorname{int} \Theta).$$

If this has a solution, i.e. if $t \in \mathfrak{T}$, then the solution is unique and, by (2) of Section 9.1, equal to $\hat{\theta}(t)$. In the case $\mathfrak{T} = \operatorname{int} C$ (which is equivalent to κ being steep) the maximum likelihood estimate can therefore, whenever it exists, be found as the (unique) solution to the likelihood equation.

Theorem 9.14. *The maximum likelihood estimator $\hat{\theta}$ is a one-to-one function if and only if κ is steep.*

Proof. $\hat{\theta}$ is one-to-one if and only if $\partial\kappa$ is single-valued and, on account of Theorem 5.28, this means steepness of κ. ▶

Theorem 9.14 and Corollary 9.4 imply that if κ is steep and $\operatorname{bd} C$ has probability zero then $\hat{\theta}$ is sufficient.

On the other hand, suppose θ is a boundary point of Θ and that κ is not steep at θ. Then, by Theorem 5.26, $\partial\kappa(\theta)$ ($\subset \operatorname{int} C$) contains a non-empty cone, and $\hat{\theta}$ is constant on $\partial\kappa(\theta)$. For $k = 1$, $\partial\kappa(\theta)$ is a halfline and must contain infinitely many support points, so $\hat{\theta}$ is not sufficient.

Example 9.10. Let $\mathfrak{X} = \chi > 1$ be a constant and let P_0 be the continuous-type probability measure on R with density

$$\chi x^{-\chi-1}, \qquad x \in (1, \infty)$$

(i.e. P_0 is the so-called Pareto distribution with support $(1, \infty)$ and shape parameter χ). As noted by Chentsov (1966), the exponential family generated by P_0 and t, where $t(x) = x$, has

$$\Theta = (-\infty, 0]$$

$$C = [1, \infty)$$

$$\mathfrak{T} = \left(1, 1 + \frac{1}{\chi - 1}\right),$$

so that the maximum likelihood estimation problem is not, as it were, solved by writing down the likelihood equation.

In this case the cumulant transform κ is strictly increasing and convex on $(-\infty,0]$ with $\kappa(\theta)\downarrow -\infty$ and $\kappa'(\theta)\downarrow 1$ for $\theta\downarrow -\infty$, $\kappa(0)=0$ and $\kappa'(0)=1+(\chi-1)^{-1}$, while $\kappa(\theta)=\infty$ for $\theta>0$. Clearly, κ is not steep (and $\mathfrak{P}$ is not regular). The conjugate $\kappa^*=\hat{l}$ of κ is ∞ on $(-\infty,1]$, strictly decreasing and convex on $(1,1+(\chi-1)^{-1}]$ with $\hat{l}(\tau)\uparrow\infty$ for $\tau\downarrow 1$, $\hat{l}(1+(\chi-1)^{-1})=0$, $\hat{l}'(1+(\chi-1)^{-1})=0$ and $\hat{l}(\tau)=0$ for $\theta>(1+(\chi-1)^{-1})$. Figures 9.3 and 9.4 show the graphs of κ and $\hat{l}$.

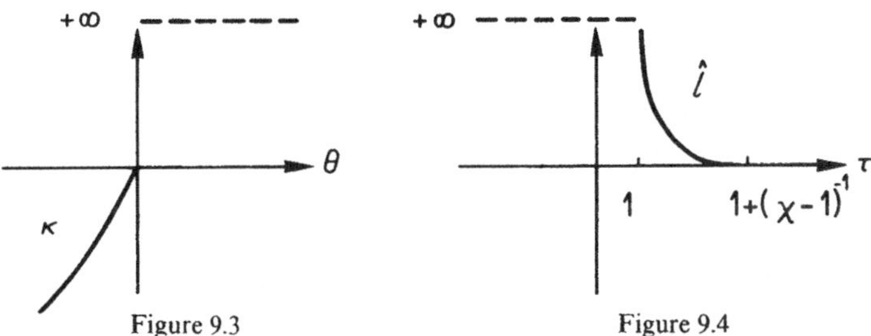

Figure 9.3 Figure 9.4

The maximum likelihood estimator $\hat{\theta}$ is defined with probability one but is not sufficient; indeed $\hat{\theta}(t)=0$ for $t\in[1+(\chi-1)^{-1},\infty)$. ▶

It is clear from the above discussion that steepness of κ is a very essential property of a full exponential family. Notice however that, in the case κ is steep, boundary points of Θ which belong to Θ do not occur among the values of $\hat{\theta}$ and are in this sense superfluous. But there are no such boundary points if Θ is open, i.e. if $\mathfrak{P}$ is regular. The main conclusions about steep exponential families, which are contained in the previous discussion, may be summarized as follows.

Corollary 9.6. *Suppose $\mathfrak{P}$ is steep, which is true in particular if $\mathfrak{P}$ is regular. The maximum likelihood estimate exists if and only if $t\in \operatorname{int} C$, and then it is unique. Furthermore, $\mathfrak{T}=\operatorname{int} C$ and the maximum likelihood estimator $\hat{\theta}$ is the one-to-one mapping on $\mathfrak{T}$ and onto* int Θ *whose inverse is τ (where $\tau(\theta)=E_\theta t$).*

Example 9.11. Let $x_1,\ldots,x_n$ be a sample from the $N_r(\xi,\Sigma)$-distribution. By Corollary 9.6, the maximum likelihood estimate of (ξ,Σ) exists if and only if $t=\left(x.,\sum_{i=1}^{n}x_i'x_i\right)\in\mathfrak{T}$ which happens precisely when $\Sigma(x_i-\bar{x}').(x_i-\bar{x})$ is positive definite. (It is straightforward to see that the latter condition is satisfied with probability 0 or 1 according as $n\le r$ or $>r$.) ▶

From Corollary 9.6 and the final remark in Section 8.2(i) one finds:

Corollary 9.7. *Suppose $\mathfrak{P}$ is regular, let $\mathfrak{P}_0$ be a linear subfamily of $\mathfrak{P}$, i.e. $\mathfrak{P}_0=\{P_\theta:\theta\in\Theta_0\}$ where $\Theta_0=\Theta\cap L$ and L is a linear subspace of R^k, and let $\mathbb{P}$ denote the projection onto L. Then, under the model $\mathfrak{P}_0$, the maximum likelihood* *estimate $\hat{\theta}_0(t)$ exists (and is unique) if and only if the likelihood equation $\mathbb{P}t = \mathbb{P}\tau(\theta)$ has a solution in Θ_0.* ▶

Example 9.12. Let $\mathfrak{P}$ be the model for an m-dimensional contingency table t of independent Poisson variates whose mean values vary freely, and let $\mathfrak{P}_0$ be a hierarchic submodel of $\mathfrak{P}$. Here $\mathbb{P}t$ stands in one-to-one, affine correspondence with the, so-called, minimal set of fitted marginals, and thus the maximum likelihood estimate exists if and only if these estimates can indeed be fitted. (For details, see Andersen 1974.) ▶

Example 9.13. Suppose x is r-dimensional and normally distributed and let $\mathfrak{P}_0$ be an affine hypothesis of $\mathfrak{N}_r$ of the form $\mathfrak{P}_0 = \{P_{(0,\Delta)}: \Delta \in \boldsymbol{\Delta}\}$ and such that $\boldsymbol{\Delta}$ is a cone and $\boldsymbol{\Sigma} = \boldsymbol{\Delta}$, cf. Example 9.9. The likelihood equation under $\mathfrak{P}_0$ may be written $\mathbb{P}x'x = \mathbb{P}\Sigma$, but the relation $\boldsymbol{\Sigma} = \boldsymbol{\Delta}$ implies $\mathbb{P}\Sigma = \Sigma$ and hence

$$\hat{\Sigma} = \mathbb{P}x'x.$$ ▶

It is of some interest to explore the situation when bd C has positive probability (cf. Corollary 9.4). Thus, one may ask whether it is possible to enlarge the family $\mathfrak{P}$ in a natural way such that the maximum likelihood estimator becomes defined with probability one. The discussion of this topic will be confined to the case where the support S of $\mathfrak{P}$ consists of finitely many points. This case is perhaps the most important from a practical point of view; at the same time it is fairly simple and admits of a complete solution.

The mean value parametrization of $\mathfrak{P}$ establishes a one-to-one correspondence between int C and $\mathfrak{P}$ and it follows from Theorem 8.3 that this correspondence is a homeomorphism ($\mathfrak{P}$ being endowed the weak topology). Let ψ denote the mapping $\tau \to P_\tau$. It will now be shown, in a constructive way, that there exists a family of probability measures $\bar{\mathfrak{P}}$ on R^k and a mapping $\bar{\psi}$ on C and onto $\bar{\mathfrak{P}}$ such that $\mathfrak{P} \subset \bar{\mathfrak{P}}$, $\bar{\psi}$ is an extension of ψ and $\bar{\psi}$ is a homeomorphism. Clearly, $\bar{\psi}$ and $\bar{\mathfrak{P}}$ are uniquely determined by these properties.

Let F be a proper face of C. Then $F = \operatorname{conv} S_F$ where S_F is the set of those points of S which belong to F (Theorem 5.8). Thus, in particular, $P_0(F) > 0$ and the conditional distribution $P_0(\cdot|F)$ is well-defined. The exponential family $\mathfrak{P}^F$ generated by $P_0(\cdot|F)$ and the identity mapping t on R^k has support S_F and convex support F. Moreover, $\mathfrak{P}^F$ is regular and hence the set of mean values of t under the probability measures in $\mathfrak{P}^F$ is equal to ri F and $\mathfrak{P}^F$ is parametrized (homeomorphically) by the mean values. Note that

$$\mathfrak{P}^F = \{P(\cdot|F): P \in \mathfrak{P}\}$$

(cf. Section 8.2(v)).

Now, define $\bar{\psi}$ as the mapping on C into the set of probability measures on R^k which coincides with ψ on int C and which to any point t in bd C lets correspond the element of $\mathfrak{P}^F$ having mean value t, F being the uniquely determined face of C with $t \in \operatorname{ri} F$. The entities $\bar{\psi}$ and $\bar{\mathfrak{P}} = \bar{\psi}(C)$ will be called the *completion* of ψ and $\mathfrak{P}$, respectively.

***Theorem* 9.15** *Suppose that the support S of $\mathfrak{P}$ consists of only finitely many points. Let ψ denote the mean value parametrization of $\mathfrak{P}$ and let $\bar{\psi}$ and $\bar{\mathfrak{P}}$ be the completions of ψ and $\mathfrak{P}$.*

Then $\bar{\psi}$ is a homeomorphism on C onto $\bar{\mathfrak{P}}$ ($\bar{\mathfrak{P}}$ being endowed with the weak topology).

Proof. $\bar{\psi}$ is clearly one-to-one and the restriction ψ of $\bar{\psi}$ to $\operatorname{int} C$ is a homeomorphism. Therefore it suffices to show that if $\tau_0 \subset \operatorname{bd} C$ and if $\tau_1, \ldots, \tau_n, \ldots$ is a sequence of points in $\operatorname{int} C$ such that $\tau_n \to \tau_0$ then $\bar{\psi}(\tau_n) \to \bar{\psi}(\tau_0)$. This, namely, implies that $\bar{\psi}$ is continuous which in turn implies continuity of $\bar{\psi}^{-1}$ since C is compact.

Let θ_n be the value of the canonical parameter for $\mathfrak{P}$ which corresponds to τ_n $(n = 1, 2, \ldots)$ and let F be the uniquely determined proper face of C which contains τ_0 in its relative interior. Using $\tau_n \to \tau_0$ and the finiteness of S it is simple to see that $P_{\theta_n}(F) \to 1$ and hence, for any bounded continuous function f on R^k,

$$\int f \, dP_{\theta_n} = \int f \, dP_{\theta_n}(\cdot|F) + O(1) \tag{2}$$

(as $n \to \infty$). In particular, therefore, the mean value of $P_{\theta_n}(\cdot|F)$ converges to τ_0 and, in view of Theorem 8.3, this implies that $P_{\theta_n}(\cdot|F)$, which is an element of $\mathfrak{P}^F$, converges (weakly) to the member of $\mathfrak{P}^F$ having mean value τ_0, i.e. to $\bar{\psi}(\tau_0)$. Formula (2) shows that $P_{\theta_n} = \bar{\psi}(\tau_n)$ must too converge to $\bar{\psi}(\tau_0)$. ▶

$\{P_\tau : \tau \in C\}$, where $P_\tau = \bar{\psi}^{-1}(\tau)$, will be called the *mean value parametrization* of $\bar{\mathfrak{P}}$. The likelihood function corresponding to this parametrization and to an observation t, where t may be any point in C, is the function with domain C and whose value at $\tau \in C$ is

$$\frac{dP_\tau}{dP_0}(t) = \frac{e^{\theta \cdot t}}{\int e^{\theta \cdot t} \, dP_0(\cdot|F)} \frac{dP_0(\cdot|F)}{dP_0}(t). \tag{3}$$

Here F is the uniquely determined face of C such that $\tau \in \operatorname{ri} F$ while θ is a point in R^k such that the probability measure determined by the right-hand side has mean value τ. The value (real) of the quantity on the right-hand side is the same for all such points θ.

***Theorem* 9.16.** *Suppose S is finite and let $\bar{\mathfrak{P}} = \{P_\tau : \tau \in C\}$ be the mean value parametrization of the completion $\bar{\mathfrak{P}}$ of $\mathfrak{P}$.*

For each $t \in S$ the likelihood function (3) is a continuous function of $\tau \in C$ and attains its maximum at exactly one point, namely $\tau = t$.

Proof. The continuity of the likelihood function follows from the continuity of $\bar{\psi}$ and the fact that S is finite.

If $\tau \in C$ is such that the face F determined by τ does not contain t then the value of the likelihood function is 0 and the maximum, which exists since C is compact, is not attained at τ. Also, the maximum is not attained at τ if $t \in \operatorname{rbd} F$ since this would imply that the likelihood function of the family $\mathfrak{P}^F$ assumed its supremum which is impossible by Theorem 9.13. Thus the maximum point is to be sought among those τ for which $t \in \operatorname{ri} F$ or, equivalently, among the relatively interior points of the face F_0 determined by t; i.e. one is faced with the maximum likelihood estimation problem for the family $\mathfrak{P}^{F_0}$, the solution of which is $\tau = t$. ▶

The question of maximum likelihood estimation for an observation t on the boundary of C may also be looked at from a somewhat different angle. Although the maximum likelihood estimate of θ does not exist it is still possible that certain subparameters have a maximum likelihood estimate according to the definition given in section 4.7(i), and, in fact, if d is the dimension of the proper face F of C for which $t \in \operatorname{ri} F$ then a $(k - d)$-dimensinal affine transformation of θ is estimable. This follows from the above discussion. To illustrate, let $(t^{(1)}, t^{(2)})$ and $(\theta^{(1)}, \theta^{(2)})$ be similar partitions of t and θ into components of dimensions d and $k - d$, and suppose that the face F may be expressed as $F = \{t \in C: t^{(1)} = 0\}$, which can always be obtained by suitable choice of the exponential representation of $\mathfrak{P}$. Then the maximum likelihood estimate of $\theta^{(2)}$ exists and is the unique solution of

$$E_{\theta^{(2)}}(t^{(2)}|t^{(1)}) = t^{(2)}$$

(cf. formula 3).

Example 9.14. Logistic dose-(binomial) response model. In the statistical theory for analysis of bio-assays with quantal response a prominent role is played by the logistic response model. Data corresponding to this model consist of a set of real numbers $x_1 < x_2 < \cdots < x_d$, which normally are the logarithms of the various doses applied in the experiment, and to each x_i is associated a pair of integers (n_i, a_i) where $n_i(>0)$ is the number of trials performed with the ith dose while a_i is the number of positive outcomes among the n_i (in the present context death is often the positive outcome). The model describes the a_i, $i = 1, \ldots, d$, as independent observations, a_i following the binomial distribution with numbering parameter n_i and probability parameter

$$p_i = \frac{1}{1 + e^{-(\alpha + \beta x_i)}}$$

where the parameters α and β vary freely in R. (The graph of the function

$$x \to \frac{1}{1 + e^{-(\alpha + \beta x)}}, \qquad x \in R$$

is the logistic curve). Hence, the probability of observing $a_1, \ldots, a_d$ is

$$\prod_{i=1}^{d} (1 + e^{\alpha + \beta x_i})^{-n_i} \prod_{i=1}^{d} \binom{n_i}{a_i} e^{\alpha s + \beta w}$$

where

$$s = \sum_{i=1}^{d} a_i, \qquad w = \sum_{i=1}^{d} x_i a_i. \tag{4}$$

The model is thus regular exponential of order 2 with $\Theta = R^2$ and

$$C = \text{cl conv}\{(s, w) \colon s = \sum_{i=1}^{d} a_i, w = \sum_{i=1}^{d} x_i a_i, \ 0 \le a_i \le n_i, i = 1, \ldots, d\}.$$

Let $\mathfrak{P}$ denote the family of distributions of (s, w).

From the results developed above it follows that the problem of estimating (α, β) on the basis of observations $a_1, \ldots, a_d$ is solvable if and only if $(s, w) \in \mathfrak{T} = \text{int}\, C$ and that the estimate may be obtained as the unique solution to the likelihood equations

$$\sum n_i p_i = s$$
$$\sum x_i n_i p_i = w$$

(which have to be solved by numerical iteration). It follows furthermore that for the completion $\bar{\mathfrak{P}}$ endowed with the mean value parametrization the estimation problem is solvable for every $(s, w) \in C$.

Consider more closely the simplest case, namely the one for which $d = 3$, $x_1 = -1$, $x_2 = 0$, $x_3 = 1$ and $n_1 = n_2 = n_3 = n$. Here

$$s = a_1 + a_2 + a_3, \qquad w = a_3 - a_1$$

and the convex support C has appearance as shown in Figure 9.5.

Let (σ, ω) denote the mean value parameter for $\bar{\mathfrak{P}}$, so that $\bar{\mathfrak{P}}$ may be written $\{P_{(\sigma,\omega)} \colon (\sigma, \omega) \in C\}$. If, for instance, (σ, ω) is a boundary point of C of the form

$$(\sigma, \omega) = (\lambda, \lambda)$$

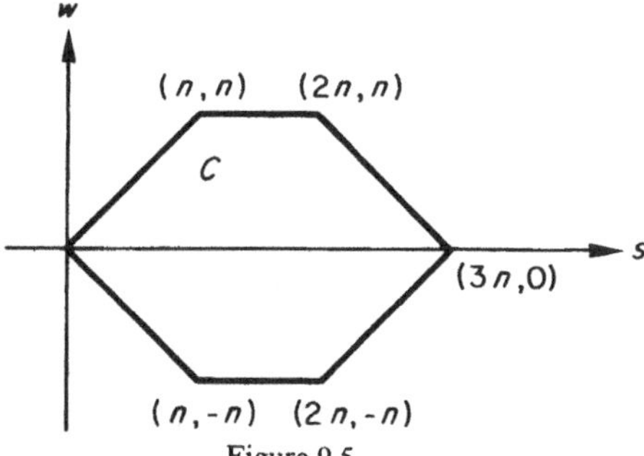

Figure 9.5

where $\lambda \in (0, n)$ then, by definition,

(5) $$P_{(\lambda,\lambda)}\{S = s, W = w\} = \binom{n}{s}\left(\frac{\lambda}{n}\right)^s\left(1 - \frac{\lambda}{n}\right)^{n-s} \quad \text{for } w = s, s = 0, 1 \ldots, n,$$

while the probability is 0 when $w \neq s$. To show the validity of this formula, note first that the face F determined by the point (λ, λ) is given by $F = C \cap \{(s, w): s = w\}$ and that $s = w$ implies $a_1 = a_2 = 0$, $a_3 = s$. The mean value parameter corresponding to $\alpha = 0$, $\beta = 0$ is $(\sigma_0, \omega_0) = (3n/2, 0)$ and

$$P_{(\sigma_0,\omega_0)}\{S = s = W\} = 2^{-3n}\binom{n}{s}$$

$$P_{(\sigma_0,\omega_0)}(F) = P_{(\sigma_0,\omega_0)}\{S = W\} = 2^{-2n}$$

so

$$P_{(\sigma_0,\omega_0)}\{S = s = W|F\} = \binom{n}{s}2^{-n}$$

whence (5) follows. The elements of $\mathfrak{P}$ corresponding to the extreme points of C are the one-point probability measures at these points. (Silverstone (1957) discussed another kind of compactification of this model).

Note furthermore that if $0 < s = w < n$, for example, then $\alpha + \beta$ is estimable by the method of maximum likelihood, even though (α, β) is not. In the case $n < s < 2n$, the parameter α is estimable irrespectively of the value of w. Thus, for $n = 2$, $a_1 = 0$, $a_2 = 1$, $a_3 = 2$ one has $\hat{\alpha} = 0$, while $\hat{\beta}$ does not exist. ▶

9.4 LIKELIHOOD FUNCTIONS FOR CONVEX EXPONENTIAL FAMILIES†

In the present section $\mathfrak{P}_0$ denotes a convex subfamily of the full exponential family $\mathfrak{P}$ and Θ_0 stands for the (convex) subset of Θ corresponding to $\mathfrak{P}_0$. Unless explicitely stated otherwise, $\mathfrak{P}_0$ is assumed to be of order k which is equivalent to $\operatorname{int}\Theta_0 \neq \varnothing$. Full exponential families are convex and results obtained here generalize some of those given in Sections 9.1 and 9.3.

It will be convenient, and will cause no loss of generality, to think of Θ_0 as given by a relation of the form $\Theta_0 = \Theta \cap D$ where D is a convex subset of R^k (necessarily of dimension k). Set $\mathfrak{T}_0 = \tau(\Theta_0 \cap \operatorname{int}\Theta)(= \tau(D \cap \operatorname{int}\Theta))$ and, for $t \in R^k$,

$$l_0(\theta) = l_0(\theta; t) = \theta \cdot t - \kappa(\theta) - \delta(\theta|D), \qquad \theta \in R^k,$$

and

$$\hat{l}_0(t) = \sup_\theta l_0(\theta).$$

† This section treats a rather special topic. The results discussed are not used elsewhere in the book.

The function l_0 is the log-likelihood function of $\mathfrak{P}_0$. When necessary, the dependence of $\hat{l}_0$ on D will be indicated by writing $\hat{l}_0(\cdot|D)$. Since

$$\text{int}\,\Theta_0 = (\text{int}\,\Theta) \cap (\text{int}\,D) \tag{1}$$

one has

$$\begin{aligned}\hat{l}_0(t|D) &= \sup\{\theta \cdot t - \kappa(\theta) : \theta \in \Theta_0\} \\ &= \sup\{\theta \cdot t - \kappa(\theta) \in (\text{int}\,\Theta) \cap (\text{int}\,D)\}\end{aligned}$$

and consequently

$$\hat{l}_0(\cdot|D) = \hat{l}_0(\cdot|\text{cl}\,D). \tag{2}$$

The following theorem is analogous to Theorem 9.1 and is similarly important in discussing maximum likelihood estimation.

Theorem 9.17. *One has*

(i) $(\kappa + \delta(\cdot|D))^* = \hat{l}_0$ *and* $\hat{l}_0 = \kappa + \delta(\cdot|\text{cl}\,D)$.
(ii) $\kappa + \delta(\cdot|D)$ *is a strictly convex function with* $\text{dom}\,(\kappa + \delta(\cdot|D)) = \Theta_0$, *and this function is closed if* D *is closed.*
(ii)* $\hat{l}_0$ *is a closed and essentially smooth convex function with* $\text{int}\,C \subset \text{dom}\,\hat{l}_0$.

Proof. Assertions (i) and (ii) are immediate. That $\hat{l}_0$ is essentially smooth follows from (ii), formula (2), and Theorem 5.30, and the inclusion $\text{int}\,C \subset \text{dom}\,\hat{l}_0$ is a consequence of the inequality $\hat{l}_0 \leq \hat{l}$ and Theorem 9.1. ▶

By formula (3) of Section 5.3

$$\hat{l}_0 = \hat{l} \,\square\, \delta^*(\cdot|D)$$

and hence

$$\text{dom}\,\hat{l}_0 = \text{dom}\,\hat{l} + \text{bar}\,D. \tag{3}$$

Let $t \in R^k$ and consider the level sets of $l_0(\cdot;t)$

$$C_{0d} = \{\theta : l_0(\theta;t) \geq d\}, \qquad d \in R.$$

If D is closed a statement completely analogous to that made at the beginning of Section 9.3 holds for the collection C_{0d}, $d \in R$. Note also that, whether D is closed or not,

$$C_{0d} = D \cap C_d$$

(where $C_d = \{\theta : l(\cdot;t) \geq d\}$).

As in Section 9.3, the maximum likelihood estimator is taken to be defined for every $t \in R^k$. The maximum likelihood estimators under $\mathfrak{P}_0$ and $\mathfrak{P}$ will be denoted respectively by $\hat{\theta}_0$ and $\hat{\theta}$.

For the next theorem, which generalizes Theorem 9.13, note that by Theorem 5.24 and Example 5.6 one has

$$\partial(\kappa + \delta(\cdot|D)) = \partial\kappa + K \tag{4}$$

where K is the normal cone mapping for D.

Theorem 9.18. *The log-likelihood function l_0 has a maximum if and only if $t \in (\partial\kappa + K)(\Theta_0)$, and then the maximum is unique.*

The mapping inverse to the maximum likelihood estimator $\hat{\theta}_0$ is $\partial\kappa + K$.

If D is closed then $\hat{\theta}_0$ equals $D\hat{l}_0$ and $\operatorname{dom}\hat{\theta}_0 = \operatorname{int} C + \operatorname{bar} D$.

If D is open then $\hat{\theta}_0$ equals the restriction of $D\hat{l}_0$ to $\partial\kappa(\Theta_0)(\subset \operatorname{int} C)$.

Proof. A point $\bar{\theta} \in R^k$ is a maximum point for the log-likelihood function l_0 corresponding to $t \in R^k$ if and only if $t \in (\partial\kappa + K)(\bar{\theta})$, cf. formulas (2) and (4) of Section 5.4, and in this case $\bar{\theta} \in \Theta_0$ since $l_0(\theta) = -\infty$ for $\theta \notin \Theta_0$. To see that the maximum is unique, note first that for every $\theta \in R^k$

$$\begin{aligned} \partial(\kappa + \delta(\cdot|D))(\theta) &= \partial\kappa(\theta) + \partial\delta(\cdot|D)(\theta) \\ &\subset \partial\kappa(\theta) + \partial\delta(\cdot|\operatorname{cl} D)(\theta) \\ &= \partial(\kappa + \delta(\cdot|\operatorname{cl} D))(\theta). \end{aligned} \tag{5}$$

The assertion of uniqueness is equivalent to

$$\partial(\kappa + \delta(\cdot|D))(\theta) \cap \partial(\kappa + \partial(\cdot|D))(\bar{\theta}) = \emptyset \tag{6}$$

for every pair $\theta, \bar{\theta}$ with $\theta \neq \bar{\theta}$. In view of (5) it suffices to verify (6) for D closed, and on this assumption $\kappa + \delta(\cdot|D)$ is closed. Application of Theorem 5.29 now yields the result.

Obviously, then, $\hat{\theta}_0^{-1} = \partial\kappa + K$.

If D is closed then $D\hat{l}_0$ is the inverse of $\partial(\kappa + \delta(\cdot|D)) = \hat{\theta}_0^{-1}$. Furthermore, by (3)

$$\begin{aligned} \operatorname{dom}\hat{\theta}_0 &= \operatorname{int}\operatorname{dom}\hat{l}_0 \\ &= \operatorname{int} C + \operatorname{bar} D. \end{aligned}$$

For D open one has $\kappa = \delta(\cdot|D)$ and hence $\hat{\theta}_0^{-1}$ is the restriction of $\partial\kappa$ to D or, equivalently, to Θ_0. But $\partial\kappa$ is the inverse of $D\hat{l}$ and therefore the last statement of the theorem is true. ▶

The likelihood equation is

$$E_\theta T = t \qquad (\theta \in \Theta_0).$$

If this has a solution $\bar{\theta}$, i.e. if $t \in \mathfrak{T}_0$, then $\bar{\theta} = \hat{\theta}_0(t)$, because $(\partial\kappa + K)(\bar{\theta}) \supset \partial\kappa(\bar{\theta}) = \tau(\bar{\theta})$.

In the case $\mathfrak{P}$ is regular Theorem 9.18 specializes to

Corollary 9.8. *Suppose $\mathfrak{P}$ is regular. The maximum likelihood estimate exists if and only if $t \in (\tau + K)(\Theta_0)$, and then it is unique. Furthermore, $\hat{\theta}_0^{-1} = \tau + K$ and range $\hat{\theta}_0 = \Theta_0$.*

If D is closed then $\operatorname{dom}\hat{\theta}_0 = \operatorname{int} C + \operatorname{bar} D$.

If D is open then $\hat{\theta}_0$ equals the restriction of $\hat{\theta}$ to $\mathfrak{T}_0$.

Consider the situation where $\mathfrak{P}$ is regular and D has the simplest possible form, namely that of a half-space

$$D = \{\theta : \theta \cdot e \leq d\}$$

where e denotes a unit vector. Furthermore, let $t \in \operatorname{dom} \hat{\theta}_0 \backslash \mathfrak{T}_0$. Then the maximum likelihood estimate $\bar{\theta} = \hat{\theta}_0(t)$ belongs to $\Theta \cap \{\theta : \theta \cdot e = d\}$ and

$$t - \tau(\bar{\theta}) \in \{\lambda e : \lambda > 0\}.$$

$\bar{\theta}$ can therefore be found as follows. First one projects t in the direction $-e$ onto the boundary of $\mathfrak{T}_0$. This operation yields a uniquely determined point $\bar{\tau} \in (\operatorname{bd} \mathfrak{T}_0) \cap \mathfrak{T}$, and $\bar{\theta}$ is obtained as solution to the equation $\tau(\theta) = \bar{\tau}$.

Example 9.15. Logistic dose-(binomial) response model with $\beta > 0$ or $\beta \geq 0$. The terminology employed is that introduced in Example 9.14. The family $\mathfrak{P}$ is regular with $\Theta = R^2$ so that for any D, $\Theta_0 = D = \operatorname{range} \hat{\theta}_0$. Assume for simplicity that $d = 3$, $x_1 = -1$, $x_2 = 0$, $x_3 = 1$ and $n_1 = n_2 = n_3 = n$. In this case the line $\{(\alpha, \beta) : \beta = 0\}$ in Θ is mapped by the mean value mapping onto $\{(s, w) : 0 < s < 3n, w = 0\}$.

If $D = \{(\alpha, \beta) : \beta > 0\}$ then, by Corollary 9.8, $\hat{\theta}_0$ is the restriction to $\mathfrak{T}_0$ of $\hat{\theta}$ and $\mathfrak{T}_0$ is the part of $\operatorname{int} C$ contained in the open first quadrant, cf. Figure 9.5.

If $D = \{(\alpha, \beta) : \beta \geq 0\}$ then

$$\operatorname{dom} \hat{\theta}_0 = \mathfrak{T}_0 \cup \{(s, w) : 0 < s < 3n, w < 0\}$$

where $\mathfrak{T}_0 = \{(s, w) : (s, w) \in \operatorname{int} C, w \geq 0\}$, and if (s, w) is a point of $(\operatorname{dom} \hat{\theta}) \cap C$ with $w \leq 0$ then (s, w) gives rise to the same maximum likelihood estimate as does the point $(s, 0)$. ▶

Example 9.16. Doubly truncated normal distribution. Let a and b be real numbers with $a < b$. The normal distribution $N(\xi, \sigma^2)$ truncated to the interval (a, b) has density with respect to Lebesgue measure given by

$$\left[\sqrt{(2\pi)}\sigma\left\{\Phi\left(\frac{b-\xi}{\sigma}\right) - \Phi\left(\frac{a-\xi}{\sigma}\right)\right\}\right]^{-1} e^{-\frac{1}{2}(x-\xi)^2/\sigma^2}, \qquad a < x < b. \tag{7}$$

Let $x_1, x_2, \ldots, x_n (n > 1)$ be independent, identically distributed random variates following the distribution (7) and consider the problem of estimating $(\xi, \sigma^2) \in R \times (0, \infty)$ on the basis of an observation $(x_1, \ldots, x_n)$.

This problem is, of course, equivalent to that of estimating

$$\theta = \left(\frac{\xi}{\sigma^2}, 1 - \frac{1}{\sigma^2}\right) \in \Theta_0 = R \times (-\infty, 1)$$

on the basis of the minimal sufficient statistic

$$t = \left(\sum_{i=1}^{n} x_i, \frac{1}{2}\sum_{i=1}^{n} x_i^2\right).$$

Let P_θ denote the probability measure of the marginal distribution of t and let $\mathfrak{P}_0 = \{P_\theta : \theta \in \Theta_0\}$. One has

$$\frac{dP_\theta}{dP_0}(t) = a(\theta)\, e^{\theta \cdot t}$$

where

$$a(\theta)^{1/n} = \{\Phi(b) - \Phi(a)\}\left[\sigma\left\{\Phi\left(\frac{b-\xi}{\sigma}\right) - \Phi\left(\frac{a-\xi}{\sigma}\right)\right\}\right] e^{-\frac{1}{2}\xi^2/\sigma^2}$$

$$= \{\Phi(b) - \Phi(a)\}\sqrt{(1-\theta_2)}\left\{\Phi\left(\frac{b(1-\theta_2)-\theta_1}{\sqrt{(1-\theta_2)}}\right) - \Phi\left(\frac{a(1-\theta_2)-\theta_1}{\sqrt{(1-\theta_2)}}\right)\right\}^{-1} \times e^{-\frac{1}{2}\theta_1^2/(1-\theta_2)}.$$

The parametrization $\{P_\theta : \theta \in \Theta_0\}$ is minimal and may be extended, in unique manner, to a minimal parametrization $\{P_\theta : \theta \in \Theta\}$ of $\mathfrak{P}$, the canonical family generated by $\mathfrak{P}_0$. The convex support of $\mathfrak{P}$ is the set

$$C = \{t : na \le t_1 \le nb, \tfrac{1}{2}t_1^2 \le nt_2 \le \tfrac{1}{2}(na)^2 + \tfrac{1}{2}(t_1 - na)(na + nb)\}$$

and, since C is bounded, $\Theta = R^2$. $\mathfrak{P}$ is therefore regular.

Invoking Corollary 9.8 one may conclude that the maximum likelihood estimate of (ξ, σ^2) exists if and only if t is contained in the open subset $\mathfrak{T}_0 = \tau(\Theta_0)$ of int C. In order to get an impression of how $\mathfrak{T}_0$ looks, the mapping τ is studied next.

For arbitrary $\theta \in \Theta$

$$c(\theta)^{1/n} = \frac{1}{\sqrt{(2\pi)}\{\Phi(b) - \Phi(a)\}} \int_a^b e^{\theta_1 x + (\theta_2 - 1)\frac{1}{2}x^2}\, dx$$

and hence τ is given by

$$\tau(\theta) = \frac{n}{\int_a^b e^{\theta_1 x + (\theta_2 - 1)\frac{1}{2}x^2}\, dx}\left(\int_a^b x\, e^{\theta_1 x + (\theta_2 - 1)\frac{1}{2}x^2}\, dx,\ \frac{1}{2}\int_a^b x^2\, e^{\theta_1 x + (\theta_2 - 1)\frac{1}{2}x^2}\, dx\right). \tag{8}$$

That part of the boundary of $\mathfrak{T}_0$ which lies in int C is the image under τ of bd $\Theta_0 = \{\theta : \theta_2 = 1\}$. For $\theta_2 = 1$ all three integrations in (8) can be performed and one obtains

$$\tau_1(\theta_1, 1) = n\left(\frac{b\, e^{b\theta_1} - a\, e^{a\theta_1}}{e^{b\theta_1} - e^{a\theta_1}} - \frac{1}{\theta_1}\right) \tag{9}$$

$$\tau_2(\theta_1, 1) = n\frac{1}{2}\frac{b^2 e^{b\theta_1} - a^2 e^{a\theta_1}}{e^{b\theta_1} - e^{a\theta_1}} - \frac{1}{\theta_1}\tau_1(\theta_1, 1), \tag{10}$$

where for $\theta_1 = 0$ the two right hand sides should be interpreted as the limiting values for $\theta_1 \to 0$, i.e.

$$\tau_1(0, 1) = n\frac{a + b}{2}$$

$$\tau_2(0, 1) = n\tfrac{1}{6}(a^2 + ab + b^2).$$

The following limiting relations hold

$$\tau(\theta_1, 1) \to n(a, \tfrac{1}{2}a^2) \qquad \text{as } \theta_1 \to -\infty$$

$$\tau(\theta_1, 1) \to n(b, \tfrac{1}{2}b^2) \qquad \text{as } \theta_1 \to +\infty$$

and, as is not difficult to see, $\mathfrak{T}_0$ is the region bounded by the two curves

$$\{\tau(\theta_1, 1): -\infty < \theta_1 < +\infty\} \tag{11}$$

$$\{t: nt_2 = \tfrac{1}{2}t_1^2, na \le t_1 \le nb\}. \tag{12}$$

The curves are tangent to each other at the endpoints $n(a, \frac{1}{2}a^2)$ and $n(b, \frac{1}{2}b^2)$. To show this, form the difference quotient

$$\frac{n\frac{1}{2}b^2 - \tau_2(\theta_1, 1)}{nb - \tau_1(\theta_1, 1)}$$

which, in view of (9) and (10), may be written

$$\frac{-\frac{1}{2}n(b^2 - a^2)(\theta_1/\{e^{(b-a)\theta_1} - 1\} + \tau_1(\theta_1, 1)}{-n(b - a)(\theta_1/e^{(b-a)\theta_1} - 1) + n}.$$

It is apparent from the latter expression that as $\theta_1 \to \infty$ the difference quotient tends to b which is equal to the slope of the tangent to (12) at $n(b, \frac{1}{2}b^2)$. Figure 9.6 indicates the appearance of the curves (11) and (12) as well as the sets C and $\mathfrak{T}_0$. ▶

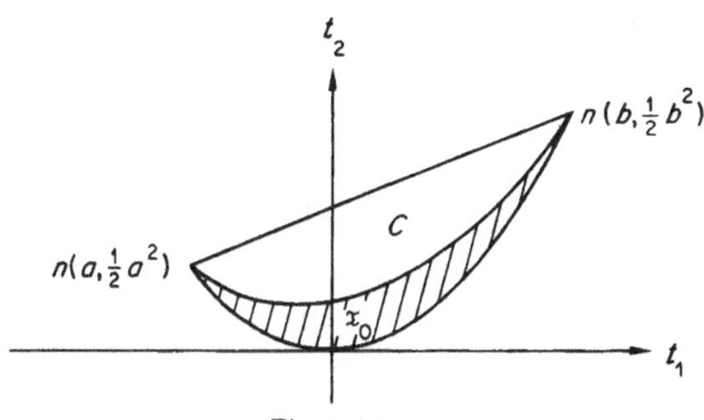

Figure 9.6

If bd C has positive probability then θ_0 will not, in general, exist with probability one. However, by enlarging the family $\mathfrak{P}_0$ suitably, as was done for full families in Section 9.3, it is possible to remedy the matter. Specifically, suppose S is finite and let $\bar{\mathfrak{P}}_0$ denote the closure of $\mathfrak{P}_0$ in the weak topology. Then $\bar{\mathfrak{P}}_0 \subset \bar{\mathfrak{P}}$, the completion of $\mathfrak{P}$ and under $\bar{\mathfrak{P}}_0$ the maximum likelihood estimate exists and is unique for every $t \in S$. Space will not be taken here to provide a proof of this fact.

Theorem 9.19 below, and also Theorem 9.32 and Corollary 9.11 in Section 9.8(ix), contain necessary and sufficient conditions for a pair t, $\breve{\theta} \in R^k$ to satisfy $t \in \operatorname{dom} \hat{\theta}_0$ and $\breve{\theta} = \hat{\theta}_0(t)$. These conditions are useful for determining whether a proposed $\breve{\theta}$ is in fact the maximum likelihood estimate corresponding to t.

Theorem 9.19. *Suppose κ is steep and let $t \in R^k$, $\breve{\theta} \in R^k$. Then $t \in \operatorname{dom} \hat{\theta}_0$ and $\breve{\theta} = \hat{\theta}_0(t)$ if and only if $\breve{\theta} \in D \cap \operatorname{int} \Theta$ and*

$$(\theta - \breve{\theta}) \cdot (\tau(\breve{\theta}) - t) \geq 0, \qquad \theta \in \Theta_0. \tag{13}$$

Proof. If $t \in \operatorname{dom} \hat{\theta}_0$ and $\breve{\theta} = \hat{\theta}_0(t)$ then, by Theorem 9.17, $\breve{\theta} \in D \cap \operatorname{int} \Theta$ *and*

$$t \in \tau(\breve{\theta}) + K(\breve{\theta}). \tag{14}$$

From the definition of a normal cone it follows that (14) is equivalent to

$$(\theta - \breve{\theta}) \cdot (t - \tau(\breve{\theta})) \leq 0, \qquad \theta \in D,$$

which clearly implies (13).

On the other hand, suppose that $\breve{\theta} \in D \cap \operatorname{int} \Theta$ and that (13) holds. Let θ be an arbitrary element of D. On account of the convexity of D one has $\theta_\lambda = \breve{\theta} + \lambda(\theta - \breve{\theta}) \in D$ for every $\lambda \in (0, 1)$. If λ is sufficiently small then $\theta_\lambda \in \operatorname{int} \Theta$ and hence $\theta_\lambda \in \Theta_0$ and

$$(\theta_\lambda - \breve{\theta})(\tau(\breve{\theta}) - t) \geq 0.$$

Thus

$$(\theta - \breve{\theta})(\tau(\breve{\theta}) - t) \geq 0$$

and one may conclude that $t \in \tau(\breve{\theta}) + K(\breve{\theta})$ or, equivalently, $t \in \operatorname{dom} \hat{\theta}_0$ and $\breve{\theta} = \hat{\theta}_0(t)$. ▶

9.5 PROBABILITY FUNCTIONS FOR EXPONENTIAL FAMILIES

It will be shown in this section that for exponential families the concepts of strong unimodality and universality are virtually equivalent. Moreover, certain mild regularity conditions will be specified which are sufficient to ensure that a strongly unimodal or universal family is strictly universal, and that unimodality of $\mathfrak{P}$ implies strong unimodality.

The relation between unimodality and strong unimodality will be discussed first.

Note that, since the densities of $\mathfrak{P}$ are of the form

$$a(\theta)b(t)\,e^{\theta\cdot t},$$

the family $\mathfrak{P}$ is strongly unimodal if just one element of $\mathfrak{P}$ is strongly unimodal.

The analogous proposition for unimodality is not true, as is shown by the following example.

Example 9.17. Let P_0 be the probability measure on R having support $S = [0, 1]$ and density

$$b(t) = \tfrac{1}{2}t^{-\frac{1}{2}}, \qquad t \in [0, 1],$$

and let $\mathfrak{P}$ be the corresponding full exponential family. Then

$$\varphi(t) - \theta t = \tfrac{1}{2}\ln t - \theta t + \ln 2.$$

For $\theta = 1$ this function is not quasiconvex on S and hence $\mathfrak{P}$ is not unimodal. ▶

However, one has

Theorem 9.20. *Suppose $\mathfrak{P}$ is full and $\Theta = R^k$.*

Then $\mathfrak{P}$ is unimodal if and only if it is strongly unimodal.

Proof. The if assertion is trivial and for $\mathfrak{P}$ of continuous type the converse follows immediately from Theorem 5.12.

In the discrete case, suppose $\mathfrak{P}$ is not strongly unimodal. Then for some $\theta_0 \in R^k$, P_{θ_0} is not strongly unimodal. Without loss of generality it may be assumed that $\theta_0 = 0$. Let $\tilde{\varphi} = \operatorname{conv} \varphi$. Clearly, then, there must exist a point $t_0 \in S$ for which

$$\varphi(t_0) > \tilde{\varphi}(t_0)$$

and this, by Theorem 5.16, implies the existence of $k + 1$ points $s_1, \ldots, s_{k+1}$ in S and of $k + 1$ non-negative scalars $\chi_1, \ldots, \chi_{k+1}$ with $\chi_1 + \cdots + \chi_{k+1} = 1$ such that

$$t_0 = \chi_1 s_1 + \cdots + \chi_{k+1} s_{k+1}$$

and

$$\varphi(t_0) > \chi_1 \varphi(s_1) + \cdots + \chi_{k+1}\varphi(s_{k+1}). \tag{1}$$

Set

$$S_0 = \{t_0, s_1, \ldots, s_{k+1}\},$$

let φ_0 be the function on R^k which coincides with φ on S_0 and is $+\infty$ elsewhere, and set $\tilde{\varphi}_0 = \operatorname{conv} \varphi_0$. On account of (1) one has

$$\varphi_0(t_0) > \tilde{\varphi}_0(t_0).$$

Let L be a non-vertical supporting hyperplane to epi $\tilde{\varphi}_0$ at $(t_0, \tilde{\varphi}_0(t_0))$. Such a hyperplane exists on account of Theorem 5.23 and it has the form

$$L = \{(t, \eta) : t \in R^k, \eta \in R, -\omega \cdot t + \eta = \alpha\}$$

for some $\omega \in R^k$, $\alpha \in R$. Set $F = L \cap \text{epi}\,\tilde{\varphi}_0$. F is a face of epi $\tilde{\varphi}_0$ and epi $\tilde{\varphi}_0 = \text{conv}\,\mathcal{S}_0$ where $\mathcal{S}_0$ consists of the points $(s, \varphi(s))$, $s \in S_0$ and the direction $(0, 1)$, cf. Theorem 5.17. Therefore (cf. Rockafellar 1970, Theorem 18.3) there exist points $t_1, \ldots, t_m$ in S_0 and non-negative scalars $\lambda_1, \ldots, \lambda_m$ with $\lambda_1 + \cdots + \lambda_m = 1$ such that

$$(t_i, \varphi_0(t_i)) \in F, \qquad i = 1, \ldots, m,$$

and

$$(t_0, \tilde{\varphi}_0(t_0)) = \lambda_1(t_1, \varphi_0(t_1)) + \cdots + \lambda_m(t_m, \varphi_0(t_m)).$$

It follows that

$$\varphi(t_0) - \omega \cdot t_0 > \alpha$$

$$\varphi(t_i) - \omega \cdot t_i = \alpha \qquad\qquad i = 1, \ldots, m$$

$$t_0 = \lambda_1 t_1 + \cdots + \lambda_m t_m.$$

Thus P_ω is not unimodal. ▶

The conclusion of the theorem does not hold in general without the assumption that $\Theta = R^k$. Thus, for instance, the family of gamma-distributions with fixed shape parameter λ, and $\lambda < 1$ is regular and unimodal, but not strongly unimodal (cf. Example 6.2).

A number of examples of strongly unimodal exponential families have, in fact, already been given in Sections 6.2 and 6.3.

Theorem 9.21. *The following conditions are equivalent.*

(i) $\mathfrak{P}$ *is universal*

(ii) $\varphi(t) \leq \sup_{\theta \in \Theta} \{\theta \cdot t - \varphi^*(\theta)\}$ *for* $t \in \text{dom}\,\varphi$

(iii) $\varphi(t) = \sup_{\theta \in \Theta} \{\theta \cdot t - \varphi^*(\theta)\}$ *for* $t \in \text{dom}\,\varphi$

(iv) $\varphi(t) = (\text{conv}\,\varphi)(t)$ *for* $t \in \text{dom}\,\varphi$

(v) *On* dom φ, φ *coincides with a convex function on* R^k.

Proof. By definition, $\mathfrak{P}$ is universal if for every $\varepsilon > 0$ and every $t \in \text{dom}\,\varphi$ there exists a $\theta \in \Theta$ such that

$$(1 + \varepsilon)a(\theta)b(t)\,e^{\theta \cdot t} \geq a(\theta)b(\tilde{t})\,e^{\theta \cdot \tilde{t}}, \qquad \tilde{t} \in R^k.$$

This inequality may be written

$$\varphi(t) \leq \theta \cdot t - (\theta \cdot \tilde{t} - \varphi(\tilde{t})) + \delta, \qquad \tilde{t} \in R^k$$

where $\delta = \ln(1 + \varepsilon) > 0$.

The equivalence between (i) and (ii) is now evident, and the other equivalences follow from the relation

$$\sup_{\theta \in \Theta} \{\theta \cdot t - \varphi^*(\theta)\} \leq \varphi^{**}(t) \leq (\text{conv}\,\varphi)(t) \leq \varphi(t), \qquad t \in \text{dom}\,\varphi. \quad ▶$$

Corollary 9.9. *Let $\mathfrak{P}$ be of c-discrete type. Then $\mathfrak{P}$ is universal if and only if it is strongly unimodal.*

Corollary 9.10. *Let $\mathfrak{P}$ be of continuous type and suppose the set is convex. Then $\mathfrak{P}$ is universal if and only if it is strongly unimodal.*

Example 9.18. Let x_{**} be the $r \times c$ contingency table of independent Poisson variates whose parameters vary freely. This model is universal and it follows at once from Corollaries 9.9 and 2.1 that the conditional distribution of x_{**} given the marginals $(x_{\cdot *}, x_{* \cdot})$ is strongly unimodal.

In particular, the conditional distribution under the hypothesis of no interaction, which has point probability

$$\frac{\Pi x_{i.}!\,\Pi x_{.j}!}{x_{..}!\,\Pi x_{ij}!},$$

is strongly unimodal. (This includes, for $c = 2$, the multivariate hypergeometric distribution.) ▶

Theorem 9.22. *Suppose $\mathfrak{P}$ is full and universal, and that S is finite.*

Then range $\hat{\tau} = S$, i.e. $\mathfrak{P}$ is strictly universal.

Proof. Set $\tilde{\varphi} = \operatorname{conv} \varphi$. Since $\varphi(s) = \tilde{\varphi}(s)$ for $s \in S$, it suffices to show that to every $t \in S$ there exists a $\theta \in R^k$ for which

$$\theta \cdot t - \tilde{\varphi}(t) = \sup_{u \in R^k} (\theta \cdot u - \tilde{\varphi}(u))(= \quad {}^*(\theta))$$

or, equivalently, that $\operatorname{dom} \partial\tilde{\varphi} \supset S$. The latter relation is valid in consequence of Theorem 5.23. ▶

Theorem 9.23. *Suppose $\mathfrak{P}$ is full and c-discrete. Let $\tilde{\varphi}$ be a closed convex function on R^k such that $S = Z^k \cap \operatorname{dom} \tilde{\varphi}$, $S \subset \operatorname{dom} \partial\tilde{\varphi}$, and range $\partial\tilde{\varphi}$ is open, and assume that φ and $\tilde{\varphi}$ coincide on S (whence $\mathfrak{P}$ is strongly unimodal and regular).*

Then range $\hat{\tau} = S$, i.e. $\mathfrak{P}$ is strictly universal.

Proof. If $t \in S$ then $t \in \operatorname{dom} \cdot\tilde{\varphi}$ and hence for some $\theta \in \operatorname{range} \partial\tilde{\varphi}$

$$\theta \cdot t - \tilde{\varphi}(t) = \sup_{u \in R^k} (\theta \cdot u - \tilde{\varphi}(u))$$

which implies $t \in \operatorname{range} \hat{\tau}$, since $\theta \in \Theta$ on account of (5) of Section 9.1. ▶

Example 9.19. Let the distributions in $\mathfrak{P}$, where $\mathfrak{P}$ is full, be one-dimensional and c-discrete and suppose that $\mathfrak{P}$ is strongly unimodal, i.e.

$$b(t-1)b(t+1) \le b(t)^2, \qquad t \in Z. \tag{2}$$

If the inequality in (2) is strict for every $t \in S$ the suppositions in Theorem 9.23 can be fulfilled ($\tilde{\varphi}$ may be chosen to be strictly convex whence range $\partial\tilde{\varphi}$ is open).

On the other hand, for the family of geometric distributions equality holds in (2), except for $t = 0$, and range $\hat{\tau} = \{0\} \neq S$. ▶

In the case where $\mathfrak{P}$ is full and of continuous type and where φ is a closed convex function, the mode mapping $\check{t}$ is clearly equal to the restriction of $\partial\varphi^*$ to Θ; moreover, by Theorem 9.8, one has $\Theta = \operatorname{int}\operatorname{dom}\varphi^*$. In particular, $\check{t} = \partial\varphi^*$ if $\operatorname{dom}\partial\varphi^* = \Theta$ or, in other words, if range $\partial\varphi$ is open, and then $\check{t}^{-1} = \partial\varphi$ and range $\check{t} = \operatorname{dom}\partial\varphi \supset \operatorname{int} C$. Thus:

Theorem 9.24. *Suppose $\mathfrak{P}$ is full, of continuous type, and strongly unimodal with φ closed. (Hence $\mathfrak{P}$ is regular.)*

If range $\partial\varphi$ is open, which is the case in particular if φ is essentially strictly convex, then range $\check{t} \supset \operatorname{int} C$, i.e. $\mathfrak{P}$ is strictly universal.

The assumption of openness of range $\partial\varphi$ is essential for the conclusion that range $\check{t} \supset \operatorname{int} C$, as is apparent from:

Example 9.20. If P_0 is the Laplace distribution on R then

$$\tilde{\varphi}(t) = |t| + \ln 2$$

$$\text{range}\,\partial\varphi = [-1, 1]$$

and

$$\Theta = (-1, 1).$$

But

$$\text{range}\,\check{t} = \{0\}.$$

because

$$\begin{aligned}\text{range}\,\check{t} &= \partial\varphi^*(\Theta)\\ &= \{t: \cdot\varphi(t) \cap (-1, 1) \neq \varnothing\}\\ &= \{0\}.\end{aligned}$$

▶

9.6 PLAUSIBILITY FUNCTIONS FOR FULL EXPONENTIAL FAMILIES

In the present section it is assumed that $\sup_t p(t; \theta) < \infty$ for every $\theta \in \Theta$, and that $\mathfrak{P}$ is full. The structure of the plausibility functions of $\mathfrak{P}$, including properties of the maximum plausibility estimator, will be studied. To some extent the discussion parallels that of Section 9.3.

The log-plausibility function

$$\pi(\theta) = \theta \cdot t - \varphi^*(\theta) - \delta(\theta|\Theta)$$

is, for any $t \in R^k$, a closed concave function on R^k.

For a fixed t, consider the level sets

$$C_d = \{\theta: \pi(\theta) \geq d\}, \qquad d \in R,$$

of π. These convex sets are bounded if $t \in \operatorname{int} C$ (cf. Theorem 5.20).

In particular, when the maximum plausibility estimate $\check{\theta}(t)$ exists it forms a convex set, which is bounded if $t \in \operatorname{int} C$.

Recall that, whether $\mathfrak{P}$ is exponential or not, the range of the maximum plausibility estimator is, as a rule and certainly for $\mathfrak{P}$ of finite discrete type, equal to the whole parameter domain.

For $\mathfrak{P}$ discrete the maximum plausibility estimate has, ordinarily, a non-empty interior, i.e. its dimension is k. However, the estimates $\check{\theta}(t')$ and $\check{\theta}(t'')$ corresponding to two different points t' and t'' have at most boundary points in common. This follows from an application of a general argument indicated in Section 2.2.

Thus, if $\mathfrak{P}$ is discrete, $\operatorname{int} \check{\theta}(t)$ is non-empty for every $t \in S$ and if $\Theta = \check{\theta}(S)$ then there exists a partition $\{A_s : s \in S\}$ of Θ such that $\operatorname{int} \check{\theta}(s) \subset A_s$ for every $s \in S$. Adding the assumption that $\mathfrak{P}$ is universal, whence $\theta \in \check{\theta}(t) \Leftrightarrow t \in \check{t}(\theta)$, one obtains that the plausibility function may be written in the form

$$\Pi(\theta) = \frac{b(t)}{b(s)} e^{\theta \cdot (t - s)}, \qquad \theta \in A_s, s \in S.$$

(Compare Figure 2.1.)

Theorem 9.25. *If $\mathfrak{P}$ is of finite discrete type then the maximum plausibility estimate $\check{\theta}(t)$ exists (and is a convex set) for every $t \in S$, and $\check{\theta}(S) = \Theta = R^k$.*

Proof. Since S is finite, π is a polyhedral concave function. Moreover, π is bounded above and hence it attains its supremum, cf. Rockafellar (1970), Corollary 27.3.2. ▶

Example 9.21. Suppose $\mathfrak{P}$ is discrete with $S \subset Z$, and strongly unimodal, i.e. S is a set of consecutive integers and

$$b(t-1)b(t+1) \leq b(t)^2, \qquad t \in Z. \tag{1}$$

In the case $\mathfrak{P}$ is the family of geometric distributions, $\check{\theta}(0) = \Theta$ and $\check{\theta}(t) = \varnothing$ for $t \in S \setminus \{0\} (= \{1, 2, \ldots\})$.

However, if the inequality in (1) is strict then $b(t-1)/b(t)$ is strictly increasing on $S + \{0, 1\}$ and

$$\theta(t) = \left[\ln \frac{b(t-1)}{b(t)}, \ln \frac{b(t)}{b(t+1)} \right], \qquad t \in S,$$

(where this interval is to be interpreted respectively as $(-\infty, \ln \{b(t)/b(t+1)\}]$ or $[\ln \{b(t-1)/b(t)\}, \infty)$ if $t - 1 \notin S$ or $t + 1 \notin S$.) Furthermore, $\check{\theta}(S) = \Theta$. To show this one may, for instance, use Theorems 6.2 and 6.1. ▶

Incidentally, the results mentioned in Examples 4.18 and 4.19 are extendable to distribution families of the type considered in the above example.

Theorem 9.26. *Suppose $\mathfrak{P}$ is of continuous type and that φ is a closed and strictly convex function which is differentiable on* $\operatorname{int} C$.

Then $\check{\theta}(t) = D\varphi(t)$ for $t \in \operatorname{int} C$.

Proof. By the remark following formula (2) of Section 5.4. the function

$$\theta \cdot t - \varphi^*(\theta) \tag{2}$$

has a maximum at a $\bar{\theta} \in R^k$ if and only if $\bar{\theta} \in \partial\varphi^{**}(t)$. But here $\varphi^{**} = \varphi$ and, provided $t \in \text{int } C$, one has $\partial\varphi(t) = D\varphi(t)$. Moreover, on account of Theorem 9.7, range $\partial\varphi = \text{dom } \partial\varphi^* = \Theta$ and hence $D\varphi(t) \in \Theta$. Since $D\varphi(t)$ is a maximum point of (2), it also maximizes $\pi(\theta) = \theta \cdot t - \varphi^*(\theta) - \delta(\theta|\Theta)$. ▶

Finally, for the continuous-type case, a set of requirements can now be stipulated which are sufficient to ensure that the conditions (iv) and (v) of Theorem 4.7 are fulfilled when t, $\check{t}$, ψ and $\check{\psi}$ of those conditions are specified, in the present context, as $t^{(1)}$, $\check{t}^{(1)}$, $\theta^{(2)}$, and $\check{\theta}^{(2)}$, where these latter variables are components of similar partitions of t, $\check{t}$, θ, and $\check{\theta}$.

Theorem 9.27. *Let $\mathfrak{P}$ be of continuous type.*

If φ is closed and strictly convex, and differentiable on int C, *and if* dom φ = int C *then $\check{\theta} = D\varphi = \check{t}^{-1}$ and $\check{\theta}$ is a homeomorphism on* int C *onto Θ. Moreover, $t^{(1)}$ and $\check{\theta}^{(2)}$ are variation independent, and $\check{t}^{(1)}$ and $\theta^{(2)}$ are variation independent.*

Proof. By Theorem 9.22, $\mathfrak{P}$ is universal and hence $\check{\theta} = \check{t}^{-1}$. Theorem 9.8 shows that $\Theta = \text{int dom } \varphi^*$ and therefore a maximum of $\pi(\theta)$ is also a maximum of $\theta \cdot t - \varphi^*(\theta)$. This latter function has a maximum only if $t \in \text{int } C$ and it now follows from Theorem 9.26 that $\check{\theta} = D\varphi$. The remaining conclusions may be obtained from Theorems 5.33 and 5.34.

9.7 PREDICTION FUNCTIONS FOR FULL EXPONENTIAL FAMILIES

Suppose an observation t, with distribution

$$a(\theta)b(t)\,e^{\theta \cdot t},$$

has been taken, and that it is desired to make inference on the unknown outcome u of another, independent experiment for which u is assumed to follow an exponential family which is also of order k and has a minimal representation

$$a^\dagger(\theta)d(u)\,e^{\theta \cdot u}.$$

The support and convex support for u will be denoted by $S^\dagger$ and $C^\dagger$ while the variation domain and the value of the parameter θ are supposed to be the same for t and u. In typical situations where this is the case, t and u are both minimal canonical statistics based on independent samples $x_1, \ldots, x_m$ and $x_1^\dagger, \ldots, x_n^\dagger$ respectively, the common distribution of the xs being from an exponential family (cf. Section 8.2(ii)).

Furthermore, let the distributions of t and u be of finite discrete or c-discrete or continuous type, and consider the likelihood prediction function

$$\vec{L}(u|t) = \sup_{\theta \in \Theta} a(\theta)b(t)\,e^{\theta \cdot t}\,\frac{d(u)\,e^{\theta \cdot u}}{\sup_u\{d(u)\,e^{\theta \cdot u}\}}$$

and the plausibility prediction function

$$\vec{\Pi}\,(u|t) = \sup_{\theta \in \Theta} \frac{b(t)\,e^{\theta \cdot t}\,d(u)\,e^{\theta \cdot u}}{\sup_t\{b(t)\,e^{\theta \cdot t}\}\,\sup_u\{d(u)\,e^{\theta \cdot u}\}}.$$

With the notations $\vec{l} = \ln \vec{L}$, $\vec{\pi} = \ln \vec{\Pi}$, $\kappa = -\ln a$, $\varphi = -\ln b$ and $\psi = -\ln d$ one has

$$\vec{l}(u|t) = \sup_{\theta \in \Theta}\,\{\theta \cdot (t+u) - (\kappa + \psi^*)(\theta)\} - \varphi(t) - \psi(u) \tag{1}$$

and

$$\vec{\pi}(u|t) = \sup_{\theta \in \Theta}\,\{\theta \cdot (t+u) - (\varphi^* + \psi^*)(\theta)\} - \varphi(t) - \psi(u). \tag{2}$$

Obviously, $\hat{\theta}$ and $\check{\theta}$ (the sets of points $\theta \in \Theta$ for which the suprema in (1) and (2), respectively, are attained) depend on t and u only through $t + u$.

Finally, assume that the distribution family of u, as well as that of t, is full. Then, in view of Theorem 9.2(**ii**), the suprema over Θ, in (1) and (2), equal the suprema over all of R^k, and (1) and (2) may be written

$$\begin{aligned}\vec{l}(u|t) &= (\kappa + \psi^*)^*(t+u) - \varphi(t) - \psi(u)\\ &= (\hat{l} \,\Box\, \psi^{**})(t+u) - \varphi(t) - \psi(u)\end{aligned}$$

where $\hat{l} = \kappa^*$ (cf. Theorem 9.1), and

$$\begin{aligned}\vec{\pi}(u|t) &= (\Box^* + \psi^*)^*(t+u) - \varphi(t) - \psi(u)\\ &= (\varphi^{**} \,\Box\, \psi^{**})(t+u) - \varphi(t) - \psi(u).\end{aligned}$$

In the discussion of $\vec{l}$ and $\vec{\pi}$ below the plausibility prediction function $\vec{\pi}$, which is symmetrical in relation to the performed experiment and the predicted experiment, will be treated first.

Theorem 9.28. *Let the distribution families of t and u be of finite discrete type and universal.*

If $t + u \in \text{range}(\check{t} + \check{u})$ then

$$\vec{\pi}(u|t) = \varphi(\check{t}) + \psi(\check{u}) - \varphi(t) - \psi(u), \tag{3}$$

i.e.

$$\vec{\Pi}(u|t) = \frac{b(t)d(u)}{b(\check{t})d(\check{u})} \tag{4}$$

where, in (3) and (4), $\check{t}$ and $\check{u}$ denote mode points corresponding to a common value of θ and determined so that $\check{t} + \check{u} = t + u$.

For any value t the maximum plausibility predictate of u exists and is given by

(5) $$\check{\vec{u}}(t) = \check{u}(\check{\vec{\theta}}(t)),$$

and

$$\vec{\Pi}(\check{\vec{u}}|t) = 1.$$

Proof. The functions $\tilde{\varphi} = \text{conv}\,\varphi$ and $\tilde{\psi} = \text{conv}\,\psi$ are closed, and they coincide with φ and ψ on $\{\varphi(\cdot) > 0\}$ and $\{\psi(\cdot) > 0\}$, respectively, cf. Theorem 9.21. If $\check{t} \in \check{t}(\theta)$ and $\check{u} \in \check{u}(\theta)$ then $\check{t} \in \partial\varphi^*(\theta)$ and $\check{u} \in \partial\psi^*(\theta)$, and the first assertion of the theorem now follows from Corollary 5.2.

As discussed in Section 3.2, one always has $\check{u}(\check{\vec{\theta}}(t)) \subset \check{\vec{u}}(t)$. Thus it remains to show that $\vec{\Pi}(u|t) = 1$ implies $u \in \check{u}(\check{\vec{\theta}}(t))$.

Clearly, $\text{int}(C + C^\dagger) \subset \text{dom}\,\partial(\tilde{\varphi} \,\square\, \tilde{\psi}) \subset C + C^\dagger$, and in fact $\text{dom}\,\partial(\tilde{\varphi} \,\square\, \tilde{\psi}) = C + C^\dagger$ because $\tilde{\varphi} \,\square\, \tilde{\psi}$ is a polyhedral convex function, cf. Rockafellar (1970), Corollaries 19.1.2 and 19.3.4 and Theorem 23.10. From (2) one finds

$$\vec{\theta}(t + u) = \partial(\varphi^* + \psi^*)^*(t + u) = \partial(\tilde{\varphi}[\,]\tilde{\psi})(t + u)$$

and so, for $t \in S$, $u \in S^\dagger$ and $\vec{\theta} \in \vec{\theta}(t + u)$,

$$\begin{aligned} \vec{\Pi}(u|t) = 1 &\Rightarrow \Pi(t; \check{\vec{\theta}})\Pi(u; \check{\vec{\theta}}) = 1 \\ &\Rightarrow t \in \check{t}(\check{\vec{\theta}}) \text{ and } u \in \check{u}(\check{\vec{\theta}}) \\ &\Rightarrow u \in \check{u}(\check{\vec{\theta}}(t)). \end{aligned}$$ ▶

When the distributions of t and u are of c-discrete or continuous type a similar line of reasoning holds under mild regularity assumptions. Thus, in the c-discrete case, if besides universality it is assumed that $\tilde{\varphi} = \text{conv}\,\varphi$ and $\tilde{\psi} = \text{conv}\,\psi$ are closed then the first assertion of Theorem 9.28 stands again, and if, moreover, range $\partial\tilde{\varphi}$ and range $\partial\tilde{\psi}$ are open then (5) is true, at least for $t \in \text{int}\,C$ (but non-emptiness of $\check{u}(\check{\vec{\theta}}(t))$ is not ensured).

As usual, let $\tau = E_\theta t$. By the same technique as that used for the proof of Theorem 9.28 one obtains:

Theorem 9.29. *Let the distribution family of t be steep, and let the distribution family of u be of finite discrete type and universal.*
If $t + u \in \text{range}\,(\tau + \check{u})$ then

(6) $$\vec{l}(u|t) = \hat{l}(\hat{t}) + \psi(\hat{u}) - \varphi(t) - \psi(u)$$

i.e.

(7) $$\vec{L}(u|t) = a(\hat{\vec{\theta}})b(t)e^{\hat{\vec{\theta}}\cdot\hat{t}}\frac{d(u)}{d(\hat{u})}$$

where, in (6) and (7), $\hat{t} = \tau(\hat{\vec{\theta}})$ and $\hat{u} \in \check{u}(\hat{\vec{\theta}})$, the quantities $\hat{\vec{\theta}}$, $\hat{t}$ and $\hat{u}$ being determined such that $\hat{t} + \hat{u} = t + u$.

For any value $t \in \text{int}\,C$ the maximum likelihood predictate exists and is given by

$$\hat{\vec{u}}(t) = \check{u}(\hat{\vec{\theta}}(t)).$$

For binomial variates t and u, Theorems 9.28 and 9.29 are illustrated by Example 3.1.

Example 9.22. Suppose t is the number of successes in a fixed number m of independent Bernoulli trials, while u is the number of trials required to obtain a further n successes. Let π denote the success probability.

If $0 < t < m$ and $n > 1$, the above conditions are satisfied and

$$\vec{\Pi}(u|t) = \binom{m}{t}\binom{u-1}{u-n}\Big/\binom{m}{\check{t}}\binom{\check{u}-1}{\check{u}-n} \qquad (u = n, n+1, \ldots)$$

where $\check{t} = [m\pi\}$, $\check{u} = n + [(n-1)\pi/(1-\pi)\}$ and π is determined so that $\check{t} + \check{u} = t + u$. Furthermore, $\vec{u}(t) = \check{u}(\check{\theta}(t))$. ▶

9.8 COMPLEMENTS

(i) In relation to Theorem 9.3, the reader may wonder whether $\mathfrak{T}$ is convex for non-steep κ. Although this may happen it is not to be expected generally. To see that, note first that range $\partial\kappa = \operatorname{dom} \partial\hat{l} = \operatorname{int} C$. Hence, by Theorem 5.25, $(\operatorname{int} C)\backslash\mathfrak{T} = \partial\kappa(\operatorname{bd}\Theta)$. Invoking Theorem 5.26 one finds that $(\operatorname{int} C)\backslash\mathfrak{T}$ is a union of half-lines with start point in bd $\mathfrak{T}$, which makes it somewhat unlikely to have $\mathfrak{T}$ convex (provided, of course, that κ is non-steep).

Example 9.23. Take $\mathfrak{X} = R^2$ and t equal to the identity mapping on R^2. Let P_0 be the probability measure having support $\{(t_1, 0): t_1 \geq 0\} \cup \{(0, t_2): t_2 \geq 0\}$, giving measure $\frac{1}{2}$ to each of the two half-axes and being such that the truncation of P_0 to $\{(t_1, 0): t_1 \geq 0\}$ is the exponential distribution with density e^{-t_1}, while the truncation to $\{(0, t_2): t_2 \geq 0\}$ has density proportional to $(1 + t_2^3)^{-1}$. Defining $\mathfrak{P}$ as the exponential family generated by P_0 and t one has

$$\Theta = \{\theta: \theta_1 < 1, \theta_2 \leq 0\}, \qquad C = [0, \infty)^2,$$

and

$$\tau = \left(\frac{c_1(\theta_1)}{c_1(\theta_1) + c_2(\theta_2)}\kappa_1'(\theta_1), \frac{c_2(\theta_2)}{c_1(\theta_1) + c_2(\theta_2)}\kappa_2'(\theta_2)\right)$$

where c_i denotes the Laplace transform of the truncation of P_0 to axis i, and $\kappa_i = \ln c_i$. Now,

$$c_1(\theta_1) = (1 - \theta_1)^{-1} = \kappa_1'(\theta_1)$$

and using this it is simple to see that $\mathfrak{T}$ is the region bounded by the two positive half-axes and the curve

$$(\lambda^2/(\lambda + 1), \kappa_2'(0)/(\lambda + 1)), \qquad 0 < \lambda < \infty.$$

This region is not convex.

This example is due to Bradley Efron. ▶

(ii) The following example shows that the inclusion sign in Theorem 9.4 cannot in general be replaced by an equality.

Example 9.24. Let $k = 2$ and let $a_1, a_2, \ldots$ be a strictly increasing sequence of real numbers with $a_1 = 0$ and $a_n \to \infty$. Furthermore, let P_0 be the probability measure on R^2 defined by

$$P_0\{(-1,0)\} = \tfrac{1}{2}, P_0\{(a_n, 1)\} = 2^{-n-1}, \qquad n = 1, 2, \ldots$$

and let $\mathfrak{P}$ be the canonical exponential family generated by P_0 and the identity mapping $t = (t_1, t_2)$ on R^2. We have

$$\begin{aligned} \operatorname{conv} S = &\{(-1,0)\} \cup \{(t_1, t_2)\colon -1 < t_1 < 0, 0 < t_2 < 1 + t_1\} \\ &\cup \{(t_1, t_2)\colon 0 \le t_1, 0 < t_2 \le 1\}, \end{aligned}$$

and $\operatorname{ord} \mathfrak{P} = 2$.

It will now be shown that, provided a_n increases sufficiently fast with n, the point $t = 0 = (0,0)$ belongs to $\operatorname{dom} \hat{l}$ even though $0 \notin \operatorname{conv} S$. Note that

$$\rho(0) = \inf_e P_0\{e \cdot t \ge 0\} = 0$$

so that the relation $0 \in \operatorname{dom} \hat{l}$ cannot be obtained by invoking Lemma 9.1.

We have

$$\hat{l}\,(0) = \sup_\theta(-\kappa(\theta)) = \sup_e \sup_{\lambda \ge 0}(-\kappa(\lambda e))$$

and thus it must be shown that, with an appropriate choice of $\{a_n\}$,

$$\inf_e \inf_{\lambda > 0} \int e^{\lambda e \cdot t}\, dP_0 > 0. \tag{1}$$

Suppose (1) does not hold. Then there exists a sequence of non-negative real numbers $\lambda_1, \lambda_2, \ldots$ and a sequence of unit vectors $e_1, e_2, \ldots$ such that

$$\int e^{\lambda_i e_i \cdot t}\, dP_0 \to 0 \qquad \text{as } i \to \infty. \tag{2}$$

The integral in (2) is bounded below by $P_0\{e_i \cdot t \ge 0\}$ and this probability is greater than or equal to $\frac{1}{2}$ unless $e_{i1} > 0$ and $e_{i2} < 0$ where e_{i1} and e_{i2} are the coordinates of e_i. Hence one may assume

$$e_{i1} > 0, e_{i2} < 0, \qquad i = 1, 2, \ldots .$$

The sequence λ_i cannot be bounded from above; if, namely, $\lambda_i \le \bar{\lambda}$ for all i and some $\bar{\lambda} < \infty$ then

$$\begin{aligned} \int e^{\lambda_i e_i \cdot t}\, dP_0 &\ge \int_{\{e_i \cdot t \le 0\}} e^{\bar{\lambda} e_i \cdot t}\, dP_0 \\ &\ge e^{\bar{\lambda} e_i \cdot (-1,0)} \tfrac{1}{2} \\ &\ge e^{-\bar{\lambda}} \tfrac{1}{2} > 0 \end{aligned}$$

in contradiction to (2). It may therefore also be assumed that

$$\lambda_i \geq 1, \qquad i = 1, 2, \ldots .$$

Let n_i denote the smallest n with $e_i \cdot (a_n, 1) > 0$. Then

$$\text{(3)} \qquad \int e^{\lambda_i e_i \cdot t}\, dP_0 \geq e^{\lambda_i e_i \cdot (a_{n_i+1}, 1)} 2^{-n_i - 2}$$

$$\geq e^{e_i \cdot (a_{n_i+1}, 1)} 2^{-n_i - 2}.$$

In view of (2) and the fact that $e_{i1} > 0$ and $a_{n_i+1} > 0$ one can conclude that $n_i \to \infty$ as $i \to \infty$. Furthermore $e_i \cdot (a_{n_i+1}, 1) > \tilde{e}_i \cdot (a_{n_i+1}, 1)$ where $\tilde{e}_i$ is the unit vector in the fourth quadrant which is orthogonal to $(a_{n_i}, 1)$, i.e.

$$\tilde{e}_i = (a_{n_i}^2 + 1)^{-\frac{1}{2}}(1, -a_{n_i}).$$

Hence

$$e_i \cdot (a_{n+1}, 1) \geq \frac{a_{n_i+1} - a_{n_i}}{\sqrt{(a_{n_i}^2 + 1)}}.$$

Setting, for instance,

$$a_n = 2^{n^2}, \qquad n = 1, 2, \ldots$$

one has

$$e_i \cdot (a_{n_i+1}, 1) \geq 2^{n_i}$$

and consequently the lowest bound in (3) tends to ∞ as $i \to \infty$. This contradicts (2), and thus $0 \in \operatorname{dom} \hat{l}$.

(iii) The first assertion of Theorem 9.13 can be extended, as follows, to cover the case of exponential representations

$$a(\theta) b(t)\, e^{\theta \cdot t}$$

with $\Theta = \{\theta : c(\theta) < \infty\}$, but which are not necessarily minimal.

Theorem 9.30. *Suppose* $\Theta = \{\theta : c(\theta) < \infty\}$ *and* $\operatorname{int} \Theta \neq \emptyset$.
Then the log-likelihood function has a maximum if and only if $t \in \operatorname{ri} C$.

Proof. Set $L = \operatorname{aff} S$, let $k^{(1)}$ denote the dimension of L and let t_0 be a point in L. Furthermore, let $\mathbf{M}$ be an orthonormal matrix whose first $l^{(1)}$ columns are vectors in the subspace $L_0 = L - t_0$, set

$$\tilde{t} = (t - t_0)\boldsymbol{M}$$

$$\tilde{\theta} = \theta \boldsymbol{M}$$

and partition $\tilde{t}$ and $\tilde{\theta}$ into $(\tilde{t}^{(1)}, \tilde{t}^{(2)})$ and $(\tilde{\theta}^{(1)}, \tilde{\theta}^{(2)})$ where $\tilde{t}^{(1)}$ and $\tilde{\theta}^{(1)}$ have dimension

$k^{(1)}$. Now, for any $\theta \in R^k$.

$$\int e^{\theta \cdot t}\, dP_0 t = e^{\theta \cdot t_0} \int e^{\tilde{\theta} \cdot \tilde{t}}\, dP_0 \tilde{t}$$

$$= e^{\theta \cdot t_0} \int e^{\tilde{\theta}^{(1)} \cdot \tilde{t}^{(1)}}\, dP_0 \tilde{t}^{(1)}$$

which shows that

$$\Theta M = \tilde{\Theta}^{(1)} \times R^{k^{(2)}} \tag{4}$$

and

$$\kappa(\theta) - \theta \cdot t_0 = \kappa_1(\tilde{\theta}^{(1)}) \tag{5}$$

where $\tilde{\Theta}^{(1)}$ is the domain of the Laplace transform c_1 *of* $P_0 \tilde{t}^{(1)}$ and $\kappa_1 = \ln c_1$. Note that, with $a_1(\tilde{\theta}^{(1)}) = c_1(\tilde{\theta}^{(1)})^{-1}$, the expression

$$a_1(\tilde{\theta}^{(1)})\, e^{\tilde{\theta}^{(1)} \cdot \tilde{t}^{(1)}} \qquad \tilde{\theta}^{(1)} \in \tilde{\Theta}^{(1)}, \tag{6}$$

is a (minimal) representation of the densities of $\mathfrak{P}$ with respect to P_0.

By (5) one has

$$\begin{aligned} l(\theta) &= \theta \cdot t - \kappa(\theta) \\ &= \theta \cdot (t - t_0) - \kappa_1(\tilde{\theta}^{(1)}) \\ &= \tilde{\theta}^{(1)} \cdot \tilde{t}^{(1)} - \kappa_1(\tilde{\theta}^{(1)}) + \tilde{\theta}^{(2)} \cdot \tilde{t}^{(2)} \end{aligned} \tag{7}$$

whence, in view of (4), one finds that l has a maximum if and only if $\tilde{t}^{(2)} = 0$ and

$$l_1(\tilde{\theta}^{(1)}) = \tilde{\theta}^{(1)} \cdot \tilde{t}^{(1)} - \kappa_1(\tilde{\theta}^{(1)}) \tag{8}$$

has a maximum.

The function l_1 given by (8) is the likelihood function for $\mathfrak{P}$ corresponding to the representation (6). The assumption $\operatorname{int} \Theta \neq \emptyset$ implies $\operatorname{int} \tilde{\Theta}^{(1)} \neq \emptyset$. Moreover the affine support of the marginal distributions $\mathfrak{P}\tilde{t}^{(1)}$ of $\tilde{t}^{(1)}$ has dimension $k^{(1)}$. Hence, according to Theorem 9.13, l_1 has a maximum if and only if $\tilde{t}^{(1)}$ belongs to the interior of the convex support of $\mathfrak{P}\tilde{t}^{(1)}$.

The latter condition together with the requirement $\tilde{t}^{(2)} = 0$ is equivalent to $t \in \operatorname{ri} C$. ▶

(iv) ***Partially observed exponential situations.*** If a variate x follows an exponential model $\mathfrak{P}$ with minimal representation

$$a(\theta) b(x)\, e^{\theta \cdot t(x)}$$

but only the value of a statistic u, and not x itself, is observed then the likelihood equation for θ is

$$E_\theta t = E_\theta(t \mid u) \tag{9}$$

while the Fisher information matrix may be written as

$$(10) \qquad i(\theta) = V_\theta E_\theta(t|u) (= V_\theta t - E_\theta V_\theta(t|u)).$$

This follows at once from formula (6) of Section 8.2. (The expressions (9) and (10) were noted by Per Martin-Löf in 1966. The asymptotic likelihood theory for the present type of situation is treated by Sundberg (1974) who also provides a copious list of examples of such situations.)

Suppose now that an affine submodel $\mathfrak{P}_0 = \{P_\theta : \theta \in \Theta_0\}$ is considered. The likelihood equation may then be written

$$(11) \qquad E_\theta(t_0) = E_\theta(t_0|u) \qquad (\theta \in \Theta_0)$$

where t_0 denotes a minimal canonical statistic under $\mathfrak{P}_0$.

Example 9.25. ABO blood group system. The table below indicates the observed and theoretical distributions according to genotype at the ABO-locus for n persons sampled from a population in which the frequencies of the A, B and O genes are, respectively p, q and r.

AA	AO	BB	BO	AB	OO
x_1	x_2	x_3	x_4	x_5	x_6
p^2	$2pr$	q^2	$2qr$	$2pq$	r^2

Since A and B are both dominant relative to O, only $x_1 + x_2$, $x_3 + x_4$, x_5 and x_6 are phenotypically observable. Taking the full multinomial model for $(x_1, \ldots, x_6)$ as the original model and letting $\mathfrak{P}_0$ correspond to the hypothesis of Hardy–Weinberg distribution, as in the table, one finds from (11) with $u = (x_1 + x_2, x_3 + x_4, x_5)$ and $t_0 = (2x_1 + x_2 + x_5, 2x_3 + x_4 + x_5)$ that the likelihood equations are

$$2np = x_1 + x_2 + x_5 + (x_1 + x_2)\frac{p}{p + 2r}$$

$$2nq = x_3 + x_4 + x_5 + (x_3 + x_4)\frac{q}{q + 2r}.$$

▶

The expression (11) was, for frequency table models and u being a vector of sums over various cells, derived by Haberman (1974) who illustrated it with various genetical examples including the one given here. Clearly, many models for frequency tables for which some of the observations are only partly categorized fall within the present framework.

(v) *Spread-stabilizing and normalizing transformations.* Transformations of argument variables of lods functions which normalize, or stabilize, the spread of, the functions were touched upon in Section 3.4(i). Here the question of transformations of this kind will be further discussed in the one-dimensional case, primarily for log-likelihood and log-probability functions of regular exponential families of order 1.

Consider first the log-likelihood function of such a family,

$$l(\theta) = \theta t - \kappa(\theta),$$

and set

$$\zeta(\theta) = \int \kappa''(\theta)^{\frac{1}{2}}\, d\theta \tag{12}$$

and

$$\nu(\theta) = \int \kappa''(\theta)^{\frac{1}{3}}\, d\theta \tag{13}$$

(the integrals being taken as indefinite). The transformation ζ stabilizes the spread of the log-likelihood function in the sense that

$$\frac{d^2 l}{d\zeta^2}(\hat{\zeta}) \text{ is the same} \qquad \text{for all } t \in \operatorname{int} C,$$

and the transformation ν is normalizing in the sense that

$$\frac{d^3 l}{d\nu^3}(\hat{\nu}) = 0 \qquad \text{for all } t \in \operatorname{int} C.$$

This proposition may of course be verified by a direct check. It is however instructive to give the subsequent line of reasoning which leads to the result and shows that the transformations ζ and ν are essentially unique.

Let $\alpha(\theta)$ be a one-to-one, smooth transformation of θ. Differentiating and inserting $\theta = \hat{\theta} (= \hat{\theta}(t))$, where $t \in \operatorname{int} C$, one obtains

$$\frac{d^2 l}{d\alpha^2}(\hat{\alpha}) = \frac{d^2 l}{d\theta^2}(\hat{\theta})\, \alpha'(\hat{\theta})^{-2}$$

and

$$\frac{d^3 l}{d\alpha^3}(\hat{\alpha}) = \left\{ \frac{d^3 l}{d\theta^3}(\hat{\theta}) - 3\frac{d^2 l}{d\theta^2}(\hat{\theta}) \frac{\alpha''(\hat{\theta})}{\alpha'(\hat{\theta})} \right\} \alpha'(\hat{\theta})^{-3}.$$

Requiring these expressions to be, respectively, constant and 0 identically for $t \in \operatorname{int} C$—or, equivalently, for $\hat{\theta} \in \Theta$—one obtains differential equations for α, which are solved by (12) and (13).

It is convenient to have the transformations expressed in terms of the mean value parameter τ as well as in terms of θ. Writing V_θ and V_τ for the variance of t considered as a function of respectively θ and τ, one sees that the appropriate transformations of these variables have the form

$$\zeta(\theta) = \int V_\theta^{\frac{1}{2}} \, d\theta \tag{14}$$
$$\zeta(\tau) = \int V_\tau^{-\frac{1}{2}} \, d\tau$$

$$\nu(\theta) = \int V_\theta^{\frac{2}{3}} \, d\theta \tag{15}$$
$$\nu(\tau) = \int V_\tau^{-\frac{1}{3}} \, d\tau.$$

A familiar technique for finding a variance-stabilizing or normalizing transformation of the arithmetic mean $\bar{t} = (t_1 + \cdots + t_n)/n$ of n independent and identically distributed variates t_i, $i = 1, \ldots, n$, proceeds as follows. Let τ denote the mean value of the t_i and suppose that τ parametrizes the family of distributions of t_i, so that the central moments $\mu_2, \mu_3, \ldots$ of the t_i are expressible as functions of τ alone. Furthermore, let $\alpha(\bar{t})$ denote the transformation sought for. From the Taylor series of $\alpha(\bar{t})$ around τ one obtains the first order terms of the expansions in powers of n^{-1} of the second and third central moments of $\alpha(\bar{t})$. Requiring the first order term for the second central moment to be constant yields the differential equation $\alpha' = \mu_2^{-\frac{1}{2}}$ for a variance-stabilizing transformation α, and setting the first order term for the third central moment equal to 0 gives the differential equation $\alpha''/\alpha' = -\mu_3/(3\mu_2^2)$ for a transformation α which can be expected to be normalizing since it makes the skewness of $\alpha(\bar{t})$ approximately 0.

If the t_i follows a linear and regular exponential family, one has $\tau = \kappa'(\theta)$, $\mu_2 = \kappa''(\theta) = V_\tau$, and $\mu_3 = \kappa'''(\theta)$ and hence the solutions of the two differential equations may be written

$$\zeta(\bar{t}) = \int^{\bar{t}} V_\tau^{-\frac{1}{2}} \, d\tau \tag{16}$$

$$\nu(\bar{t}) = \int^{\bar{t}} V_\tau^{-\frac{1}{3}} \, d\tau. \tag{17}$$

That the spread-stabilizing and normalizing transformations for exponential models might be given on the form (14)–(17) was first noticed by R. W. M. Wedderburn.

Example 9.26. For the binomial distribution

$$\binom{m}{t} \pi^t (1 - \pi)^{m-t}$$

the transformations turn out as

$$\zeta(\pi) = \int \{\pi(1 - \pi)\}^{-\frac{1}{2}} \, d\pi = \arcsin \sqrt{\pi}$$

$$\nu(\pi) = \int \{\pi(1 - \pi)\}^{-\frac{1}{3}} \, d\pi$$

and, for $m = 1$,

$$\zeta(\bar{t}) = \arc\sin\sqrt{\bar{t}} \tag{18}$$

$$v(\bar{t}) = \int \{\bar{t}(1-\bar{t})\}^{-\frac{1}{6}}\, d\bar{t}. \tag{19}$$

The most well-known of these transformations is the classical arcsin transformation for $\bar{t}$. The expression given for $v(\bar{t})$ has been considered by Cox and Snell (1968), see also Borges (1970, 1971). Anscombe (1964a) and Box and Tiao (1973) derived, respectively, $v(\pi)$ and $\zeta(\pi)$, essentially in the way it has been done here. ▶

Example 9.27. In the case of the Poisson distribution

$$e^{-\lambda}\frac{\lambda^t}{t!}$$

one obtains

$$\zeta(\lambda) = \sqrt{\lambda}$$

and

$$v(\lambda) = \lambda^{\frac{1}{3}}.$$

These transformations were proposed, respectively, by Box and Tiao (1973) and Anscombe (1964a). The expressions (16) and (17) turn out as

$$\zeta(\bar{t}) = \sqrt{\bar{t}}$$

$$v(\bar{t}) = \bar{t}^{\frac{2}{3}}$$

which, for $n = 1$, are well-known variance-stabilizing and normalizing transformations for the Poisson distribution (Anscombe (1953) introduced the latter transformation). ▶

Example 9.28. Consider the gamma distribution

$$\frac{1}{\Gamma(\lambda)\beta^{\lambda}}t^{\lambda-1}e^{-t/\beta}$$

with known shape parameter $\lambda \geq 1$. Here

$$\zeta(\beta) = \ln\beta$$

$$v(\beta) = \beta^{-\frac{1}{3}}$$

$$\zeta(\bar{t}) = \ln\bar{t}$$

$$v(\bar{t}) = \bar{t}^{\frac{1}{3}}.$$

Wilson and Hilferty (1931) suggested $t^{\frac{1}{3}}$ as a normalizing transformation for the χ^2-distribution, and $v(\beta)$ was derived in Anscombe (1964a). ▶

The derivation of the formulas (12) and (13) given in the foregoing more generally shows:

Theorem 9.31. *Let $\mathfrak{X}$ be an open interval, let f be a strictly convex and three times differentiable function on $\mathfrak{X}$ and define the function $l(x) = l(x; x^*)$, for $x^* \in \mathfrak{X}^* = Df(\mathfrak{X})$, by*

$$l(x) = x^* \cdot x - f(x).$$

Setting

$$\zeta(x) = \int f''(x)^{\frac{1}{2}}\, dx \tag{20}$$

and

$$v(x) = \int f''(x)^{\frac{1}{3}}\, dx \tag{21}$$

and letting $\tilde{x}$ denote the x solution of the equation $x^* = Df(x)$, one has (for $\tilde{\zeta} = \zeta(\tilde{x})$ and $\tilde{v} = v(\tilde{x})$)

$$\frac{d^2 l}{d\zeta^2}(\tilde{\zeta}) \text{ is the same} \qquad \text{for all } x^* \in \mathfrak{X}^*$$

and

$$\frac{d^3 l}{dv^3}(\tilde{v}) = 0 \qquad \text{for all } x^* \in \mathfrak{X}^*.$$ ▶

The theorem should be thought of as a result on spread-stabilization and normalization of linear, concave lods functions.

It may be convenient to work with f as a function of $\xi^* = Df(x)$ rather than x. In terms of ξ^* the transformations are

$$\begin{aligned}\zeta(\xi^*) &= \int f''((Df)^{-1}(\xi^*))^{-\frac{1}{2}}\, d\xi^* \\ &= \int f^{*\prime\prime}(\xi^*)^{\frac{1}{2}}\, d\xi^*\end{aligned}$$

and

$$\begin{aligned}v(\xi^*) &= \int f''((Df)^{-1}(\xi^*))^{-\frac{2}{3}}\, d\xi^* \\ &= \int f^{*\prime\prime}(\xi^*)^{\frac{1}{3}}\, d\xi^*\end{aligned}$$

where f^* is the conjugate of f.

For the log-probability function of an exponential family,

$$m(t) = \theta t - \varphi(t),$$

Theorem 9.31 applies if φ coincides on S with a function $\tilde{\varphi}$ having the properties required of f in the theorem (in which case the exponential family is strongly unimodal).

Example 9.29. Considering again the binomial distribution one finds, for $\tilde{\varphi}(t) = \ln\{\Gamma(t+1)\Gamma(n-t+1)\}$,

$$\zeta(t) = \int \{\psi'(t+1) + \psi'(n-t+1)\}^{\frac{1}{2}}\, dt \tag{22}$$

$$\nu(t) = \int \{\psi'(t+1) + \psi'(n-t+1)\}^{\frac{1}{3}}\, dt \tag{23}$$

where ψ denotes the digamma function. The asymptotic expansion

$$\psi'(\lambda) \sim \frac{1}{\lambda} + \frac{1}{2\lambda^2} + \cdots$$

implies

$$\psi'(t+1) + \psi'(n-t+1) \sim n\{t(n-t)\}^{-1}$$

and with this approximation formulas (22) and (23) change to (18) and (19). ▶

Example 9.30. For the gamma distribution with known shape parameter $\lambda > 1$ one has $\varphi(t) = -(\lambda - 1)\ln t$ and formulas (20) and (21) yield the familiar transformations $\zeta(t) = \ln t$ and $\nu(t) = t^{\frac{1}{3}}$, which were obtained from a different angle in Example 9.28.

(vi) ***Infinitesimal L-independence and orthogonality.*** In this subsection Θ is assumed to be open.

Let $\{P_\omega : \omega \in \Omega\}$ be a parametrization of $\mathfrak{P}$ such that Ω is an open subset of R^k and the (one-to-one) mapping $\omega \to \theta$ has continuous partial derivatives of the second order. Using formula (3) of Section 1.1 one finds that the log-likelihood function l satisfies

$$\frac{\partial^2 l}{\partial\omega'\partial\omega} = -\frac{\partial\theta}{\partial\omega'} V_{\omega'} \frac{\partial\theta'}{\partial\omega} + (t - \tau)\cdot \frac{\partial^2\theta}{\partial\omega'\partial\omega}.$$

Hence, if $t \in \tau(\Theta)$ then (by Corollary 9.6) one has the useful equality

$$-\frac{\partial^2 l}{\partial\omega'\partial\omega}(\hat{\omega}) = i(\hat{\omega}). \tag{24}$$

Considering now a partition $(\omega^{(1)}, \ldots, \omega^{(m)})$ of ω one sees, from (24), that $\omega^{(1)}, \ldots, \omega^{(m)}$ are orthogonal under $P_{\hat{\omega}}$ if and only if $\omega^{(1)}, \ldots, \omega^{(m)}$ are infinitesimally L-independent at t, provided only that $t \in \tau(\Theta)$. Thus, on the assumption that $\mathfrak{P}$ is full (and hence regular), $\omega^{(1)}, \ldots, \omega^{(m)}$ are orthogonal under P_ω for every $\omega \in \Omega$ if and only if $\omega^{(1)}, \ldots, \omega^{(m)}$ are infinitesimally L-independent for every $t \in R^k$ for which the log-likelihood function has a maximum.

Next, the notions of orthogonality and infinitesimal L-independence will be considered in relation to canonical, mean value, and mixed parametrizations of the exponential family $\mathfrak{P}$. Let $(\theta^{(1)},\ldots,\theta^{(m)})$, $(\tau^{(1)},\ldots,\tau^{(m)})$, and $(t^{(1)},\ldots,t^{(m)})$ be similar partitions of θ, τ, and t.

The Fisher information matrix for θ is $V_\theta t$ and that for τ is $(V_\theta t)^{-1}$. From this and Theorem 9.11 it follows that

$$\begin{aligned}
&\tau^{(1)},\ldots,\tau^{(m)} \text{ are orthogonal}\\
\Leftrightarrow\ &\theta^{(1)},\ldots,\theta^{(m)} \text{ are orthogonal}\\
\Leftrightarrow\ &t^{(1)},\ldots,t^{(m)} \text{ are uncorrelated}\\
\Leftrightarrow\ &t^{(1)},\ldots,t^{(m)} \text{ are independent.}
\end{aligned}$$

Thus orthogonality and infinitesimal L-independence of $\theta^{(1)},\ldots,\theta^{(m)}$, or of $\tau^{(1)},\ldots,\tau^{(m)}$, are not properties that are of interest in themselves.

However, with $m = 2$, one has the nontrivial result that the two components of the mixed parameter $(\tau^{(1)}, \theta^{(2)})$ are always orthogonal. This follows immediately from the formula

$$\frac{\partial^2 l}{\partial \tau^{(1)\prime}\partial\theta^{(2)}} = (t^{(1)} - \tau^{(1)})\cdot \frac{\partial^2\theta^{(1)}}{\partial\tau^{(1)\prime}\partial\theta^{(2)}}. \tag{25}$$

In the two examples given below, $\hat{\tau}^{(1)}$ and $\hat{\theta}^{(2)}$ are not only asymptotically independent, as implied by the orthogonality of $\tau^{(1)}$ and $\theta^{(2)}$, but are in fact exactly independent.

Example 9.31. Let $x_1,\ldots,x_n$ be independent and following the gamma distribution with probability function

$$\frac{1}{\Gamma(x)\beta^\lambda} x^{\lambda-1} e^{-x/\beta}.$$

The pair (μ, λ), where $\mu = \beta\lambda$ is the mean value of the gamma distribution, constitutes a mixed parameter. By the above results, μ and λ are therefore orthogonal and infinitesimally L-independent.

To prove the stronger assertion that the maximum likelihood estimates $\hat{\mu}$ and $\hat{\lambda}$ are independent one may note that $\hat{\mu}$ and $\hat{\lambda}$ are determined by the equations

$$\hat{\mu} = \bar{x}$$

$$\psi(\hat{\lambda}) - \ln\hat{\lambda} = \ln(\bar{x}/\bar{\bar{x}})$$

where ψ denotes the digamma function and $\bar{\bar{x}}$ is the geometric mean of $x_1,\ldots,x_n$. Clearly, the distribution of $\bar{x}/\bar{\bar{x}}$ does not depend on β and hence, by Basu's Theorem (Corollary 4.4), $\bar{x}$ and $\bar{x}/\bar{\bar{x}}$ are independent, which implies the independence of $\hat{\mu}$ and $\hat{\lambda}$. This property of the gamma distribution was observed by Cox and Lewis (1966). ▶

Example 9.32. For the family $\mathfrak{N}^-$ of inverse Gaussian distributions, having model function as (21) of Section 8.1, the parameter (μ, λ) is mixed. Here

$$\hat{\mu} = \bar{x}$$

$$\hat{\lambda}^{-1} = \frac{1}{n}\sum\left(\frac{1}{x_1} - \frac{1}{\bar{x}}\right).$$

That $\hat{\mu}$ and $\hat{\lambda}$ are indeed not only orthogonal but independent was shown by Tweedie (1957), who moreover established the remarkable result that $n\lambda/\hat{\lambda}$ follows a χ^2 distribution with $n-1$ degrees of freedom. Thus the inverse Gaussian family allows for an immediate analogue of the analysis of variance for nested classifications of normal variates.

These results by Tweedie follow simply from formula (6) of Section 8.2. The conditional distribution of $\Sigma(1/x_i)$ given $x_.$ is exponential and does not depend on α (cf. (22) of Section 8.1). To find this distribution it therefore suffices to determine the conditional Laplace transform

$$E_0(e^{\theta\Sigma(1/x_i)}|x_.)$$

where E_0 denotes mean value with respect to the probability measure given by $\alpha = 0$, $\lambda = 1$. Now, for $\alpha = 0$ the distribution (22) of Section 8.1 is a stable distribution with characteristic exponent $\frac{1}{2}$ and hence $x_.$ is distributed according to (22) of Section 8.1 with $\alpha = 0$ and λ replaced by $n^2\lambda$. Consequently, by (6) of Section 8.2 and for $\theta = -(\lambda - 1)/2$,

$$E_0(e^{\theta\Sigma(1/x_i)}|x_.) = \lambda^{-(n-1)/2}\, e^{n^2\theta/x_.}$$

or, equivalently,

$$E_0(e^{\theta n \hat{\lambda}^{-1}}|x) = (1 - 2\theta)^{-(n-1)/2}.$$

Since the right hand side does not depend on $x_.$, the estimates $\hat{\mu}$ and $\hat{\lambda}^{-1}$ must be independent; furthermore, the right hand side is the Laplace transform of the χ^2-distribution with $n-1$ degrees of freedom, as was to be shown. ▶

Consider a family $\mathfrak{P}$ which is not necessarily exponential but yet partly exponential in the sense that its model function may be written as

$$a(\theta^{(1)}, \omega^{(2)})b(x; \omega^{(2)})\, e^{\theta^{(1)}\cdot t^{(1)}(x)},$$

i.e. for each fixed $\omega^{(2)}$, $\mathfrak{P}_{\omega^{(2)}}$ is exponential with $\theta^{(1)}$ and $t^{(1)}$ as canonical quantities. Under smoothness assumptions, (25) generalizes to

$$\frac{\partial^2 l}{\partial\tau^{(1)'}\partial\omega^{(2)}} = (t^{(1)} - \tau^{(1)})\cdot\frac{\partial^2\theta^{(1)}}{\partial\tau^{(1)'}\partial\omega^{(2)}},$$

$\tau^{(1)}$ still being the mean value of $t^{(1)}$. The parameters $\tau^{(1)}$ and $\omega^{(2)}$ are consequently orthogonal (and often infinitesimally L-independent).

Example 9.33. The family of negative binomial distributions, having model function

$$(1-\pi)^{\chi}\binom{\chi+x-1}{x}\pi^{x},$$

is partly exponential and one sees that the mean $\mu = \chi\pi/(1-\pi)$ and the shape parameter χ are orthogonal (as noted first by Anscombe (1950)). ▶

(vii) *Least squares and maximum likelihood.* For an arbitrary vector variate t with mean value $\tau = \tau(\omega)$ and variance $V = V(\omega)$, where ω is a parameter ($\omega \in \Omega \subset R^m$), the generalized weighted least squares estimate of ω is defined as the solution to the equation

$$(t-\tau)V^{-1}\frac{\partial\tau'}{\partial\omega} = 0 \tag{26}$$

(which, when τ depends linearly on ω, $\Omega = R^m$ and V is constant, equals that of the classical weighted least squares procedure of minimizing the quadratic form $(t-\tau)V^{-1}(t-\tau)'$).

Suppose, in addition, that the family of probability measures governing t is exponential with t as minimal canonical statistic. Then the left hand side of (26) equals the derivative of the log-likelihood function, so (26) is, in fact, also the likelihood equation.

In this connection it may be noted that the iterative (generalized) Gauss–Newton method for solving (26) is identical to Fisher's scoring method.

The above has been observed previously in somewhat less generality, see Bradley (1973) and Wedderburn (1974).

(viii) Suppose $\mathfrak{P}$ is regular with minimal standard representation

$$\frac{dP_\theta}{dP_0} = a(\theta)\,e^{\theta\cdot t}.$$

If θ tends to the finite or infinite boundary of Θ, the probability mass of P_θ is swept towards the corresponding, i.e. the dual, part of the boundary of C.

One of the possible precise versions of this statement is as follows.

For every unit vector e in R^k and every $d < \delta^*(e|C)$ one has

$$P_{\lambda e}\{e\cdot t \le d\} \to 0 \qquad \text{for } \lambda \uparrow \bar{\lambda}(e)$$

where λ denotes a scalar variable and

$$\bar{\lambda}(e) = \sup\{\lambda\colon \lambda e \in \Theta\}.$$

In fact, the stronger result

$$\int_{\{e\cdot t\le d\}} e^{w(d-e\cdot t)}\,dP_{\lambda e} \to 0 \qquad \text{for } \lambda \uparrow \bar{\lambda}(e)$$

holds for every $w < \bar{\lambda}(e)$. To see this note that for $\lambda > w$

$$\int_{\{e\cdot t \leq d\}} e^{w(d-e\cdot t)}\, dP_{\lambda e} \leq e^{\lambda d} a(\lambda e) P_0\{e\cdot t \leq d\}.$$

That the right hand side tends to zero for $\lambda \uparrow \bar{\lambda}(e)$ follows from formula (5) of Section 7.1 if $\bar{\lambda}(e) = \infty$ and from the first part of Theorem 7.1 otherwise.

(ix) Theorem 9.19 is valid for arbitrary convex D. If D is a closed convex cone, the criterion of the theorem may be sharpened as follows.

Theorem 9.32. *Suppose that κ is steep and that D is a closed convex cone. Let D^* denote the negative of the polar of D and let $t \in R^k$, $\bar{\theta} \in R^k$.*

Then

$$t \in \operatorname{dom} \hat{\theta}_0, \qquad \bar{\theta} = \hat{\theta}_0(t) \tag{27}$$

if and only if

$$\bar{\theta} \in D \cap \operatorname{int} \Theta \tag{28i}$$

$$\tau(\bar{\theta}) - t \in D^* \tag{28ii}$$

$$\bar{\theta}(\tau(\bar{\theta}) - t) = 0. \tag{28iii}$$

Proof. Note that since κ is steep

$$\operatorname{range} \hat{\theta}_0 = D \cap \operatorname{int} \Theta. \tag{29}$$

Let f denote the negative of the log-likelihood function l for $\mathfrak{P}$ corresponding to the observation t, i.e.

$$f(\theta) = -l(\theta; t) = \kappa(\theta) - \theta \cdot t, \qquad \theta \in R^k.$$

f is a closed convex function, $f^* = \hat{l}(t + \cdot)$ and

$$\operatorname{int}(\operatorname{dom} f) \cap (\operatorname{int} D) = (\operatorname{int} \Theta) \cap (\operatorname{int} D) \neq \varnothing.$$

Hence

$$\inf\{f(\theta) \colon \bar{\theta} \in D\} = -\inf\{f^*(t^*) \colon t^* \in D^*\}$$

where the infimum at the right hand side is attained, see Theorem 5.35.

Now, suppose $t \in \operatorname{dom} \hat{\theta}_0$ and $\bar{\theta} = \hat{\theta}_0(t)$, and let t^* be a point in R^k such that f^* attains its infimum over D^* at t^*. Then

$$f(\bar{\theta}) = \inf_D f = -\inf_{D^*} f^* = -f^*(t^*)$$

and using Theorem 5.35 one obtains

$$t^* \in \partial f(\bar{\theta}), \quad \bar{\theta} \in D, \quad t^* \in D^*, \quad \bar{\theta} \cdot t^* = 0. \tag{30}$$

By (29),

(31) $$\bar{\theta} \in D \cap \operatorname{int} \Theta.$$

Hence $\bar{\theta} \in \operatorname{int} \Theta$ and $\partial f(\bar{\theta}) = \tau(\bar{\theta}) - t$, i.e.

(32) $$t^* = \tau(\bar{\theta}) - t.$$

Formulas (31) and (32) and the third and fourth assertion in (30) yield (28i–iii).
Conversely, if (28i–iii) are satisfied then so is (30) with $t^* = \tau(\bar{\theta}) - t$. Invoking again Theorem 5.35 one finds that $f(\bar{\theta}) = \inf\{f(\theta): \theta \in D\}$ which implies (27)▶

(Conditions (28i–iii) are, in essence, the so-called Kuhn–Tucker conditions for the mathematical programme

$$\text{minimize } f(\theta) = \kappa(\theta) - \theta \cdot t \text{ subject to } \theta \in D,$$

cf. Rockafellar (1970)).

Corollary 9.11. *Suppose* $\mathfrak{P}$. *is regular and that D is of the form*

$$D = \{\theta: \theta_1 \geq 0, \ldots, \theta_j \geq 0\}$$

where $1 \leq j \leq k$.
Then

(33) $$t \in \operatorname{dom} \hat{\theta}_0, \qquad \bar{\theta} = \hat{\theta}_0(t)$$

if and only if $\bar{\theta} \in \Theta$ *and, with* $\bar{\tau} = \tau(\bar{\theta})$,

(34i) *for* $i = 1, \ldots, j$,

either $\bar{\theta}_i = 0$ and $\bar{\tau}_i \geq t_i$

or $\bar{\theta}_i \geq 0$ *and* $\bar{\tau}_i = t_i$;

(34ii) *for* $i = j + 1, \ldots, k$,

$\bar{\tau}_i = t_i$.

Proof. The assumptions of Theorem 9.32 are fulfilled and $D^* = \{\theta: \theta_1 \geq 0,,,,, \theta_j \geq 0, \theta_{j+1} = 0, \ldots, \theta_k = 0\}$. Thus (33) is equivalent to

(35) $$\bar{\theta} \in \Theta, \qquad \bar{\theta}_1 \geq 0, \ldots, \bar{\theta}_j \geq 0,$$

(36) $$\bar{\tau}_1 \geq t_1, \ldots, \bar{\tau}_j \geq t_j, \qquad \bar{\tau}_{j+1} = t_{j+1}, \ldots, \bar{\tau}_k = t_k,$$

(37) $$\bar{\theta}_1(\bar{\tau}_1 - t_1) + \cdots + \bar{\theta}_k(\bar{\tau}_k - t_k) = 0.$$

Clearly, (35)–(37) are equivalent to $\bar{\theta} \in \Theta$ and (34i–ii). ▶

(x) The following comment is incidental to the remarks preceding Theorem 9.24. It is simple to see that if φ is closed convex and $k = 1$ then range $\check{t}$ is convex (i.e. an interval). This, however, need not be so for $k > 1$.

Example 9.34. Let $k = 2$ and let φ^* be the closed convex function on R^2 determined by

$$\varphi^*(\omega) = \frac{\omega_1^2}{4(\omega_2 + 1)} + c_0, \qquad \omega = (\omega_1, \omega_2), \omega_2 > -1$$

where c_0 is a constant to be chosen later. Set $\varphi = (\varphi^*)^*$. Then φ^* is the conjugate of φ.

φ^* is not affine along any line and hence aff dom $\varphi = R^k$, cf. Rockafellar (1970), Theorem 13.4. Moreover, $0 \in \operatorname{int} \operatorname{dom} \varphi^*$. Thus, by Theorem 6.1, the integral

$$\int e^{-\varphi} \, d\lambda$$

is finite. Let c_0 be determined so that the value of the integral is 1, which is clearly possible.

With this φ one has

$$\begin{aligned} \operatorname{range} \check{t} &= \partial\varphi^*(\operatorname{int} \operatorname{dom} \varphi^*) \\ &= D\varphi^*\{\omega: \omega_2 > -1\} \\ &= \{t: t_2 = -t_1^2\}. \end{aligned}$$

Thus range $\check{t}$ is a parabola and is not convex. ▶

(xi) To some extent the roles played by the mean value mapping τ and the mode mapping $\check{t}$ in, respectively, maximum likelihood and maximum plausibility estimation are analogous. Thus $\theta^{-1} = \tau$ provided $\mathfrak{P}$ is regular, while $\check{\theta}^{-1} = \check{t}$ provided $\mathfrak{P}$ is universal. Moreover, with the same assumptions on $\mathfrak{P}$, $(\tau^{(1)}, \theta^{(2)})$ parametrizes $\mathfrak{P}$ and $\tau^{(1)}$ and $\theta^{(2)}$ are variation independent, while the same properties of $(\check{t}^{(1)}, \theta^{(2)})$ occur in important cases, cf. Theorem 9.27 and the remark following Example 9.21.

(xii) *Factorial series families.* Consider a (full) factorial series family $\mathfrak{Q}$ of one-dimensional distributions,

$$b(x)\eta^{(x)}/g(\eta),$$

cf. Section 8.3(ii), and suppose $S(= \{b(\cdot) > 0\})$ is a set of consecutive integers.

If

$$h(x) = x + b(x)/b(x+1)$$

is a nondecreasing function on S then $\mathfrak{Q}$ is unimodal.

The family $\mathfrak{Q}$ is strongly unimodal if and only if b is strongly unimodal, i.e.

$$b(x-1)b(x+1) \le b(x)^2, \qquad x \in N,$$

(which, in turn, is equivalent to $b(x)/b(x+1)$ being nondecreasing on S) and in this case $\mathfrak{Q}$ is universal.

Furthermore, strong unimodality of $\mathfrak{Q}$ implies that the maximum plausibility estimate of η exists for every $x \in S$ and is given by

$$\check{\eta}(x) = \{\eta \in N_0 \colon h(x-1) \leq \eta \leq h(x)\}.$$

Example 9.35. For the family of hypergeometric distributions

$$\binom{\eta}{x}\binom{m}{n-x}\bigg/\binom{\eta+m}{n} \tag{38}$$

one has

$$b(x)/b(x+1) = (x+1)\left(\frac{m+1}{n-x} - 1\right)$$

and hence $\check{\eta}$ is determined by

$$(m+1)\frac{x}{n-x+1} - 1 \leq \check{\eta} \leq (m+1)\frac{x+1}{n-x} - 1.$$

The maximum likelihood estimate is obtained from

$$m\frac{x}{n-x} - 1 \leq \check{\eta} \leq m\frac{x}{n-x}.$$

(If m members of a population of size $m + \eta$ are marked, and a random sample of n is taken from the population then (38) is the probability that the sample contains x unmarked members. The estimate $m + \hat{\eta}$ of the total population size, or its approximation $mn/(n-x)$, is known in the literature on capture–recapture investigations as the Petersen estimate, after Petersen (1896).) ▶

(xiii) *Discrimination information.* The *discrimination information between P and $\tilde{P}$*, where P and $\tilde{P}$ are any two probability measures on a sample space $\mathfrak{X}$, is the quantity $I(P, \tilde{P})$ defined by

$$I(P, \tilde{P}) = -\int \ln \frac{d\tilde{P}}{dP}\, dP;$$

this quantity is also known as the information of P with respect to $\tilde{P}$, cf. Savage (1950) and Kullback (1959). By Jensen's inequality (Section 5.5(iii)), $I(P, \tilde{P}) \geq 0$ with equality if and only if $\tilde{P} = P$, and $I(P, \tilde{P})$ may be thought of as a directed measure of the dissimilarity or distance between P and $\tilde{P}$. Furthermore, for a parametrized family $\mathfrak{P} = \{P_\omega \colon \omega \in \Omega\}$ of probability measures, determined by a family $\{p(\cdot; \omega) \colon \omega \in \Omega\}$ of probability functions, one has, under mild smoothness assumptions that for fixed ω the matrix of second order partial derivatives of $I(\omega, \tilde{\omega})$, with respect to $\tilde{\omega} \in \Omega$ and evaluated at $\tilde{\omega} = \omega$, is equal to Fisher's information matrix $i(\omega)$.

For an exponential family $\mathfrak{P}$ with minimal representation

$$\frac{dP_\theta}{d\mu}(x) = a(\theta)b(x)e^{\theta \cdot t(x)}$$

the discrimination information between P_θ and $P_{\tilde{\theta}}$ may, if $\theta \in \operatorname{int} \Theta$, be written

$$I(\theta, \tilde{\theta}) = (\theta - \tilde{\theta}) \cdot \tau - \kappa(\theta) + \kappa(\tilde{\theta})$$

where $\tau = \tau(\theta) = E_\theta t$. Suppose $\mathfrak{P}$ is full, let $t \in \mathfrak{T}$ $(= \tau(\operatorname{int} \Theta))$ and let $\hat{\theta}$ be the maximum likelihood estimate of θ based on t. Then, in the notations of Section 9.3,

$$\begin{aligned} I(\hat{\theta}, \theta) &= \hat{\theta} \cdot t - \kappa(\hat{\theta}) - (\theta \cdot t - \kappa(\theta)) \\ &= \hat{l}(t) - l(\theta; t), \end{aligned}$$

which is a convex function of θ. Furthermore, if $\mathfrak{P}_0 = \{P_\theta : \theta \in \Theta_0\}$ is any subfamily of $\mathfrak{P}$, the maximum likelihood estimate $\hat{\theta}_0$ of θ under the hypothesis $\mathfrak{P}_0$ can be obtained by minimizing $I(\hat{\theta}, \theta)$ over $\theta \in \Theta_0$, and $I(\hat{\theta}, \hat{\theta}_0) = -\ln q$ where q denotes the likelihood ratio test statistic of the hypothesis $\mathfrak{P}_0$. It follows that if $\mathfrak{P}_0$ is affine and if $\mathfrak{P}_{00}$ is an arbitrary subfamily of $\mathfrak{P}_0$ then

$$I(\hat{\theta}, \hat{\theta}_{00}) = I(\hat{\theta}, \hat{\theta}_0) + I(\hat{\theta}_0, \hat{\theta}_{00}),$$

$\hat{\theta}_{00}$ being the maximum likelihood estimate under $\mathfrak{P}_{00}$.

9.9 NOTES

Except for Theorem 9.6 which is new and Theorem 9.7 which was given in Barndorff-Nielsen (1973b), the results in Sections 9.1, 9.3, and 9.4 are from Barndorff-Nielsen (1970). There is some overlap between Sections 9.1 and 9.3 on the one hand and Chapter 4 of the–independent–work by Chentsov (1972) on the other. Theorem 9.11 is due to Bildikar and Patil (1968). Sections 9.5 and 9.6 are based on material in Barndorff-Nielsen (1973b, 1976b), and Section 9.7 is based on Barndorff-Nielsen (1977b) and Mathiasen (1977).

For asymptotic properties of exponential families, in particular concerning maximum likelihood estimates and maximum plausibility estimates, the reader is referred to Efron and Truax (1968), Andersen (1969), Berk (1972), Martin-Löf (1970), Höglund (1974), and Mathiasen (1977). The paper by Höglund is of special interest in the context of the present treatise as it shows, roughly speaking, that the distance between $\hat{\theta}$ and $\check{\theta}$ will ordinarily be of the order of n^{-1}, where n is the sample size; thus the distance is infinitesimal compared to the standard deviation of $\hat{\theta}$ (see also Section 10.7).

CHAPTER 10

Inferential Separation and Exponential Families

The questions of the existence and character of the ancillary and sufficient statistics, including the cuts, under a given statistical model can to a large extent be settled by general theorems, provided the model is exponential. The present chapter demonstrates this.

The same general assumptions on the exponential families and their exponential representations as were made for the previous chapter (see p. 139) will be presupposed here. Moreover, $(\theta^{(1)}, \theta^{(2)})$, $(t^{(1)}, t^{(2)})$, and $(\tau^{(1)}, \tau^{(2)})$ stand for similar partitions of θ, t, and τ, into components of dimensions $k^{(i)}$, $i = 1, 2$, and $\mathfrak{P}$ is, except in Section 10.1, assumed to have an open kernel.

10.1 QUASI-ANCILLARITY AND EXPONENTIAL FAMILIES

The concept of quasi-ancillarity, which was introduced in Section 4.5 and of which S- and M-ancillarity are special cases, is sufficiently strong to allow useful conclusions on existence and character of quasi-ancillary statistics to be drawn, when the family of distributions is exponential. Of the results given here the most important are Corollary 10.3, which states that if $\mathfrak{P}$ has open kernel then the quasi-ancillary statistics are simply the statistics of the form $t^{(1)}$, and Theorem 10.2, which, in particular, contains a condition for maximal quasi-ancillarity.

A parameter function ψ, viewed as a function on Θ, often depends only on some, r say, of the coordinates of θ. The number r will, in general, vary with the minimal canonical parametrization and the smallest possible r will be called the rank of ψ. A convenient, more formal definition of rank is the following. If f is a mapping on a subset of $\mathfrak{Y}$ of R^k then the *rank* of f is defined as the smallest integer $m \in \{0, 1, \ldots, k\}$ for which there exists an affine mapping g on R^k into R^m and a mapping $\tilde{f}$ from R^m such that

$$f(y) = \tilde{f}(g(y)), \qquad y \in \mathfrak{Y}.$$

Now, recall the definition of quasi-ancillary statistics and note that if $\mathfrak{P}$ has open kernel then every $t^{(1)}$ is quasi-ancillary, as is simple to show.

Let u be quasi-ancillary. The *size* of u is defined as $\max(\operatorname{ord}\mathfrak{P}_0)$ where the maximum is taken over all members $\mathfrak{P}_0$ of the partition of $\mathfrak{P}$ induced by the conditional distributions given u.

Suppose u is quasi-ancillary with respect to ψ. If a component $t^{(1)}$ is a function of u then $k^{(1)} + \operatorname{rank}\psi \le k$, because ψ depends on $P \in \mathfrak{P}$ only through $P(\cdot|t^{(1)})$ which has rank less than or equal to $k - k^{(1)}$. It follows, in particular, that

$$\operatorname{size} u \times \operatorname{rank}\psi \le k. \tag{1}$$

To see this, let $\mathfrak{P}_0$ be as above and such that $\operatorname{ord}\mathfrak{P}_0 = \operatorname{size} u$, and let t_0 be a minimal canonical statistic for $\mathfrak{P}_0$. Since u is sufficient for $\mathfrak{P}_0$, the statistic t_0 is a function of u and the previous inequality applies.

Theorem 10.1. *Let u be a statistic, let ψ be a parameter function and suppose that u is quasi-ancillary with respect to ψ and that ψ parametrizes the conditional distributions given u. Let d and r denote the size of u and the rank of ψ, respectively.*

Then B-sufficiency of u is equivalent to $d = k$ and also to $r = 0$, while B-ancillarity of u is equivalent to $d = 0$ and also to $r = k$. If none of these two cases occur then there exists a couple $t = (t^{(1)}, t^{(2)})$, $\theta = (\theta^{(1)}, \theta^{(2)})$ such that $d \le k^{(1)} \le k - r$ and

(i) *$t^{(1)}$ is a function of u and u is conditionally B-ancillary given $t^{(1)}$*
(ii) *ψ stands in one-to-one correspondence with $\theta^{(2)}$.*

Proof. Clearly, B-sufficiency implies $d = k$ and is implied by $r = 0$. Moreover, B-ancillarity implies $r = k$ and is, on account of the quasi-ancillarity of u, implied by $d = 0$. In view of (1) the first part of the theorem is thus established.

Hence, suppose $0 < d \le k - r < k$ and let $t^{(1)}$ be such that $t^{(1)}$ is a function of u and such that there exists no other component $\tilde{t}^{(1)}$ for which $\tilde{t}^{(1)}$ is a function of u and $\tilde{k}^{(1)} > k^{(1)}$.

By the remark preceding Theorem 10.1, one has $k^{(1)} \le k - r$, and the inequality $d \le k^{(1)}$ will be established at the end of the proof.

The conditional distribution given $t^{(1)}$ has a density which may be written in the following two ways

$$a(\theta^{(2)}|t^{(1)})\,e^{\theta^{(2)}\cdot t^{(2)}} = p(u;\theta^{(2)}|t^{(1)})p(x;\psi|u), \tag{2}$$

and ψ depends on $\theta^{(2)}$ only. Let x, $\tilde{x}$ *and* $\theta^{(2)}$, $\tilde{\theta}^{(2)}$ be any pairs satisfying $u(x) = u(\tilde{x}) = u$ and $\psi(\theta^{(2)}) = \psi(\tilde{\theta}^{(2)}) = \psi$. For the t-values, t and $\tilde{t}$, corresponding to x and $\tilde{x}$, one has $t^{(1)} = \tilde{t}^{(1)}$ and hence one obtains from (2)

$$\frac{p(x;\psi|u)}{p(\tilde{x};\psi|u)} = e^{\theta^{(2)}\cdot(t^{(2)} - \tilde{t}^{(2)})} = e^{\tilde{\theta}^{(2)}\cdot(t^{(2)} - \tilde{t}^{(2)})}$$

whence

$$(\theta^{(2)} - \tilde{\theta}^{(2)})\cdot t^{(2)} = (\theta^{(2)} - \tilde{\theta}^{(2)})\cdot\tilde{t}^{(2)}.$$

If $\theta^{(2)} \neq \tilde{\theta}^{(2)}$ this would be in contradiction to the assumed 'maximality' of $t^{(1)}$, and therefore ψ is a one-to-one function of $\theta^{(2)}$. But this, in turn, means that the conditional distribution given $t^{(1)}$ depends on ψ only so that, by the quasi-ancillarity of u, this latter statistic is conditionally B-ancillary given $t^{(1)}$. Thus (i) and (ii) have been proved, and it is now simple to see that $d \leq k^{(1)}$. ▶

Corollary 10.1. *If $k = 1$ then any quasi-ancillary statistic is either B-ancillary or B-sufficient.*

Corollary 10.2. *If $\mathfrak{P}$ has open kernel then $t^{(1)}$ is in one-to-one correspondence with u and $d = k^{(1)} = k - r$.*

Corollary 10.3. *If $\mathfrak{P}$ has open kernel then the class of quasi-ancillary statistics equals the class of components $t^{(1)}$.* ▶

To any parameter function ψ of rank r, for which $0 < r < k$, there exist minimal canonical variates $t = (t^{(1)}, t^{(2)})$ and $\theta = (\theta^{(1)}, \theta^{(2)})$ such that $k^{(1)} = k - r$ and ψ depends on $\theta^{(2)}$ only. Under mild conditions, any statistic which is quasi-ancillary with respect to ψ is an affine function of the $t^{(1)}$ thus chosen, cf. Theorem 10.2 below. However, $t^{(1)}$ itself may not be quasi-ancillary.

Example 10.1. Let x be normally distributed with (ξ, σ^2) unknown and set $\psi = \sigma^2$ and $t = (x, x^2/2)$, $\theta = (\xi/\sigma^2, -1/\sigma^2)$. Here $t^{(1)} = x$ is not quasi-ancillary with respect to ψ because ψ is not a function of the conditioning mapping $P \to P^{t^{(1)}}$ which in this case is constant, $P^{t^{(1)}}$ assigning probability 1 to a single point. ▶

Essentially, the difficulty, exposed by this example, is that $\theta^{(2)}$ need not parametrize the conditional distributions given the statistic $t^{(1)}$. This, however, can happen only if the conditional distribution of $t^{(2)}$ given $t^{(1)}$ is singular, cf. Lemma 8.4.

Lemma 10.1. *Suppose $\mathfrak{P}$ is open and convex. Let ψ be a parameter function of rank r and let $t = (t^{(1)}, t^{(2)})$ and $\tilde{t} = (\tilde{t}^{(1)}, \tilde{t}^{(2)})$ be two minimal canonical statistics such that the dimensions of $t^{(1)}$ and $\tilde{t}^{(1)}$ are, respectively, $k^{(1)}$ and $k - r$. Suppose that ψ considered as a function of $\theta = (\theta^{(1)}, \theta^{(2)})$ depends on $\theta^{(2)}$ only, and considered as a function of $\tilde{\theta} = (\tilde{\theta}^{(1)}, \tilde{\theta}^{(2)})$ depends on $\tilde{\theta}^{(2)}$ only.*

Then $k^{(1)} \leq k - r$, $t^{(1)}$ is an affine transformation of $\tilde{t}^{(1)}$, and $\tilde{\theta}^{(2)}$ is an affine transformation of $\theta^{(2)}$. Furthermore, if $k^{(1)} = k - r$ then these transformations are regular.

Proof. According to Lemma 8.1 there exist two constant $k \times k$ matrices $\mathbf{A}$ and $\overline{\mathbf{A}}$ and two constant $1 \times k$ vectors B and $\bar{B}$ such that

$$\tilde{t} = t\mathbf{A} + B$$
$$\tilde{\theta} = \theta\overline{\mathbf{A}} + \bar{B}$$

where

$$\mathbf{A}'\overline{\mathbf{A}} = \mathbf{I}_k. \tag{3}$$

Let

$$\overline{\mathbf{A}} = \begin{Bmatrix} \overline{\mathbf{A}}_{11} & \overline{\mathbf{A}}_{12} \\ \\ \overline{\mathbf{A}}_{21} & \overline{\mathbf{A}}_{22} \end{Bmatrix}$$

be the partition of $\overline{\mathbf{A}}$ such that $\overline{\mathbf{A}}_{11}$ is a $k^{(1)} \times (k - r)$ matrix. Since

$$\tilde{\theta}^{(2)} = \theta^{(1)}\overline{\mathbf{A}}_{12} + \theta^{(2)}\overline{\mathbf{A}}_{22} + \overline{B}^{(2)} \tag{4}$$

it suffices, in view of (3), to prove that $\overline{\mathbf{A}}_{12} = 0$.

Let the vector $(0, \ldots, 0, 1, 0, \ldots, 0)$ in R^k with 1 as the ith coordinate be denoted by e_i and let $E = \text{span}\{e_1, \ldots, e_{k^{(1)}}\}$, the subspace of R^k spanned by $\{e_1, \ldots, e_{k^{(1)}}\}$. Furthermore, let

$$\tilde{E} = \{(\theta^{(1)}, \theta^{(2)}) \colon \theta^{(1)}\overline{\mathbf{A}}_{12} + \theta^{(2)}\overline{\mathbf{A}}_{22} = 0\}.$$

Clearly $\tilde{E}$ is a subspace of dimension $k - r$ and if θ and θ_0 are elements in Θ such that $\theta - \theta_0 \in \tilde{E}$ then, by (4), $\tilde{\theta}^{(2)}(\theta) = \tilde{\theta}^{(2)}(\theta_0)$.

Assume $\overline{\mathbf{A}}_{12} \neq 0$ and note that this implies that E is not a subspace of $\tilde{E}$. If

$$F = \text{span}(E, \tilde{E})$$

one can conclude that $m = \dim F - d \geq 1$. Let $\theta_0 \in \Theta$ and let

$$\Theta_0 = (\theta_0 + F) \cap \Theta.$$

By assumption Θ is open and convex and hence it is possible to connect every point in Θ_0 with θ_0 using a finite number of line segments contained in Θ_0 and of one of the two forms

$$\alpha\theta + (1 - \alpha)\theta', \qquad \text{where } \theta - \theta' \in E$$

and

$$\alpha\theta + (1 - \alpha)\theta'', \qquad \text{where } \theta - \theta'' \in \tilde{E}.$$

Since

$$\theta - \theta' \in E \Leftrightarrow \theta^{(2)} = \theta'^{(2)}$$

and

$$\theta - \theta'' \in \tilde{E} \Leftrightarrow \tilde{\theta}^{(2)}(\theta) = \tilde{\theta}^{(2)}(\theta'')$$

one has

$$\psi(\theta) = \psi(\theta')$$

respectively

$$\psi(\theta) = \psi(\theta'').$$

It follows that

$$\psi(\theta) = \psi(\theta_0) \qquad \text{for all } \theta \in \Theta_0.$$

Consequently

$$\operatorname{rank} \psi \leq k - \dim \operatorname{aff} \Theta_0 = r - m$$

in contradiction to the fact that $\operatorname{rank} \psi = r$. ▶

It may be noticed that if $r - m = 0$ then the assumption that $\mathfrak{P}$ is convex can be replaced by the weaker assumption that $\mathfrak{P}$ is connected. Hence, if $\mathfrak{P}$ is open and connected and $r = 1$ the conclusions in Lemma 10.1 hold.

From the discussion in connection with Example 10.1, and from Corollary 10.3 and Lemmas 10.1 and 8.3, one obtains:

Theorem 10.2. *Let ψ be a parameter function of rank r, where $0 < r < k$, and let $t = (t^{(1)}, t^{(2)})$, $\theta = (\theta^{(1)}, \theta^{(2)})$ be such that $k^{(1)} = k - r$ and ψ depends on $\theta^{(2)}$ only.*

If $\theta^{(2)}$ parametrizes the conditional distributions of $t^{(2)}$ given $t^{(1)}$—which is the case, in particular, if the conditional distributions are non-singular—then $t^{(1)}$ is quasi-ancillary with respect to ψ.

Suppose either (a) $\mathfrak{P}$ has open kernel and ψ is one-to-one as a function of $\theta^{(2)}$; or (b) $\mathfrak{P}$ is open and convex. Then any statistic, which is quasi-ancillary with respect to ψ, is (equivalent to) an affine function of $t^{(1)}$. Hence, if $t^{(1)}$ itself is quasi-ancillary with respect to ψ then it is maximal quasi-ancillary. In the opposite case there exists no statistic, quasi-ancillary with respect to ψ, whose size equals $k - r$.

Note that if $r = 1$ then the assumption in (*b*) that $\mathfrak{P}$ is convex may be weakened to $\mathfrak{P}$ being connected, cf. the remark immediately after Lemma 10.1.

Example 10.2. Consider the regular family $\mathfrak{P}$ of trinomial distributions with model function

$$\frac{n!}{x_1!x_2!(n - x_1 - x_2)!} p_1^{x_1} p_2^{x_2} (1 - p_1 - p_2)^{n - x_1 - x_2}$$

and let the interest parameter be

$$\psi = \frac{p_2}{p_1 + p_2}.$$

Taking $t^{(1)} = x_1 + x_2$ and $\theta^{(2)} = \ln(p_2/p_1)$ one sees that $x_1 + x_2$ is quasi-ancillary with respect to ψ and is the only statistic having this property. (The statistic $x_1 + x_2$ is, in fact, both S- and M-ancillary.) ▶

Example 10.3. Let x_1 and x_2 be independent, Poisson distributed with mean values λ_1 and λ_2, and assume that $\lambda_2 \geq \lambda_1$. The statistic x_1 is the unique statistic quasi-ancillary with respect to λ_2. (However, x_1 is not ancillary with respect to λ_2, cf. Example 4.17). ▶

Example 10.4. Again, let x_1 and x_2 be independent Poisson variates, but suppose now that $\lambda = (\lambda_1, \lambda_2)$ varies in $(0,\infty) \times (0,\infty)$ and let

$$\psi(\lambda) = \begin{cases} 0 & \text{if } \lambda_1 = \lambda_2 \\ 1 & \text{if } \lambda_1 \neq \lambda_2, \end{cases}$$

i.e. ψ indicates the hypothesis that $\lambda_1 = \lambda_2$. Then $x_.$ is the only statistic quasi-ancillary with respect to ψ. (In fact, $x_.$ is both S- and M-ancillary.) ▶

Theorem 10.2 will be further drawn upon and exemplified in Sections 10.4 and 10.5.

In order to keep the theoretical developments, to be discussed in the rest of this chapter, simple while still covering most cases of interest it will be assumed henceforth that $\mathfrak{P}$ has open kernel and only statistics of the type $t^{(1)}$ will be considered. On account of Corollaries 10.2 and 10.3 little or nothing is lost by the second restriction.

10.2 CUTS IN GENERAL EXPONENTIAL FAMILIES

Certain conditions for a component $t^{(1)}$ to be a cut will be given here. One of these suggests a way of constructing exponential families with cuts, and this is discussed at the end of the section. For families of discrete type it is possible to obtain a considerable body of further results, to be described in the next section.

Lemma 10.2. *Let $t^{(1)}$ be a cut of size d and let $\omega^{(1)}$ and $\omega^{(2)}$ be a corresponding pair of L-independent parameters. Suppose that $\mathfrak{P}$ has open kernel.*

Then $\theta^{(2)}$ and $\omega^{(2)}$ are in one-to-one correspondence if and only if $d = k^{(1)}$.

Proof. The only if assertion is straightforward to derive from the definition of size. The converse assertion is a consequence of Theorem 10.1, Corollary 10.2, and Lemma 8.3(v). ▶

Theorem 10.3. *Let $\mathfrak{P}$ be interior (i.e. if $\{P_\theta : \theta \in \tilde{\Theta}\}$ is a minimal parametrization of the exponential family $\tilde{\mathfrak{P}}$ generated by $\mathfrak{P}$ and if Θ is the subject of $\tilde{\Theta}$ corresponding to $\mathfrak{P}$ then $\Theta \subset \operatorname{int} \tilde{\Theta}$), with open kernel, and suppose $t^{(1)}$ is a cut of size $k^{(1)}$.*

Then $\tau^{(1)}$ and $\theta^{(2)}$ is a corresponding pair of L-independent parameters, and $\theta^{(1)}$ and a are of the form

$$\theta^{(1)} = \varphi(\tau^{(1)}) + \chi(\theta^{(2)}) \tag{1}$$

$$a(\tau^{(1)}, \theta^{(2)}) = a_1(\tau^{(1)}) a_2(\theta^{(2)}). \tag{2}$$

Furthermore, the quantity

$$a_2(\theta^{(2)})\, e^{\chi(\theta^{(2)}) \cdot t^{(1)}} E_0(e^{\theta^{(2)} \cdot t^{(2)}} \mid t^{(1)})$$

does not depend on $\theta^{(2)}$.

Proof. That $\tau^{(1)}$ and $\theta^{(2)}$ is a corresponding pair of L-independent parameters follows at once from Lemma 10.2.

Hence, for any $\tau_0^{(1)} \in \mathfrak{T}^{(1)}$

$$\frac{p(t^{(1)}; \tau^{(1)})}{p(t^{(1)}; \tau_0^{(1)})} = \frac{a(\tau^{(1)}, \theta^{(2)})}{a(\tau_0^{(1)}, \theta^{(2)})} e^{\{\theta^{(1)}(\tau^{(1)}, \theta^{(2)}) - \theta^{(1)}(\tau_0^{(1)}, \theta^{(2)})\} \cdot t^{(1)}}$$

and using the affine independence of the coordinates of $t^{(1)}$ one obtains (1) and (2).

The last assertion is an expression of the fact that the marginal distribution of $t^{(1)}$ does not depend on $\theta^{(2)}$. ▶

Let $c(\cdot|t^{(1)})$ be the Laplace transform of the conditional distribution of $t^{(2)}$ given $t^{(1)}$, under P_0. Thus

$$c(\theta^{(2)}|t^{(1)}) = E_0(e^{\theta^{(2)} \cdot t^{(2)}}|t^{(1)}), \qquad \theta^{(2)} \in \Theta^{(2)}.$$

Furthermore, let $S^{(1)}$ denote the set of possible values of $t^{(1)}$,

$$S^{(1)} = \{s^{(1)}: (s^{(1)}, s^{(2)}) \in S \text{ for some } s^{(2)}\}, \tag{3}$$

and let $t_0^{(1)}$ be a point in $S^{(1)}$. From the last conclusion of Theorem 10.3 it follows simply that for $\eta(\theta^{(2)}) = \chi(0) - \chi(\theta^{(2)})$ one has

$$c(\theta^{(2)}|t^{(1)}) = c(\theta^{(2)}|t_0^{(1)}) \, e^{\eta(\theta^{(2)}) \cdot (t^{(1)} - t_0^{(1)})} \tag{4}$$

or, equivalently,

$$\kappa(\theta^{(2)}|t^{(1)}) = \kappa(\theta^{(2)}|t_0^{(1)}) + (t^{(1)} - t_0^{(1)}) \cdot \eta(\theta^{(2)})$$

where $\kappa(\cdot|t^{(1)}) = \ln c(\cdot|t^{(1)})$.

Suppose $\Theta^{(2)}$ is an open subset of $R^{k^{(2)}}$. The conditional cumulant transform $\kappa(\cdot|t^{(1)})$ is differentiable in $\Theta^{(2)}$ and its gradient $\tau(\theta^{(2)}|t^{(1)})$ equals the conditional mean value of $t^{(2)}$ under $P_{(\tau^{(1)}, \theta^{(2)})}$. The vector function η must therefore be differentiable too and

$$\tau(\theta^{(2)}|t^{(1)}) = \tau(\theta^{(2)}|t_0^{(1)}) + (t^{(1)} - t_0^{(1)}) \frac{\partial \eta'}{\partial \theta^{(2)}}. \tag{5}$$

Hence $t^{(2)}$ has linear regression on $t^{(1)}$. This property is thus a necessary condition for $t^{(1)}$ to be a cut, but it is not very useful as stated. However, it is possible under further mild assumptions to derive from it a more immediately applicable condition which relates to the boundary of the convex hull of S. For $\mathfrak{P}$ of discrete type this possibility is indicated in the next section.

Theorem 10.3 implies, in particular, that if $\mathfrak{P}$ is open and $t^{(1)}$ is a cut of size $k^{(1)}$ then $\tau^{(1)}$ and $\theta^{(2)}$ are variation independent and $\theta^{(1)}$ may be written as $\varphi(\tau^{(1)}) + \chi(\theta^{(2)})$. The converse assertion holds provided $\mathfrak{P}$ is not only open but also connected, as will now be verified.

Lemma 10.3. *Suppose* $\mathfrak{P}$ *is open and connected. If*

(i) $\tau^{(1)}$ *and* $\theta^{(2)}$ *are variation independent*
(ii) $\theta^{(1)}$ *is of the form*

$$\theta^{(1)}(\tau^{(1)}, \theta^{(2)}) = \varphi(\tau^{(1)}) + \chi(\theta^{(2)})$$

then φ *is homeomorphic and, for certain functions* a_1 *and* a_2,

$$a(\tau^{(1)}, \theta^{(2)}) = a_1(\tau^{(1)})a_2(\theta^{(2)}).$$

Proof. Without loss of generality one may assume that $\chi(0) = 0$. Then

$$\theta^{(1)}(\tau^{(1)}, 0) = \varphi(\tau^{(1)})$$

and φ maps $\mathfrak{T}^{(1)}$ onto $\Theta^{(1)}(0) = \{\theta^{(1)}:(\theta^{(1)}, 0)\in\Theta\}$, in one-to-one manner. Now, let $\{\tau_n^{(1)}\}$ be a sequence of elements in $\mathfrak{T}^{(1)}$ such that $\tau_n^{(1)} \to \tau_0^{(1)} \in \mathfrak{T}^{(1)}$. One has

$$\begin{aligned}\tau_n^{(1)} \to \tau_0^{(1)} &\Rightarrow (\tau_n^{(1)}, 0) \to (\tau_0^{(1)}, 0)\\ &\Rightarrow (\theta^{(1)}(\tau_n^{(1)}, 0) \to (\theta^{(1)}(\tau_0^{(1)}, 0), 0)\\ &\Rightarrow \theta^{(1)}(\tau_n^{(1)}, 0) \to \theta^{(1)}(\tau_0^{(1)}, 0)\\ &\Rightarrow \varphi(\tau_n^{(1)}) \to \varphi(\tau_0^{(1)})\end{aligned}$$

and thus φ is continuous. Since $\mathfrak{T}^{(1)}$ must be open, it may be concluded that φ is a homeomorphism which maps $\mathfrak{T}^{(1)}$ onto $\Theta^{(1)}(0)$.

Because of (i) and (ii),

$$\theta^{(1)}(\tau^{(1)}, \theta^{(2)}) - \chi(\theta^{(2)}) = \varphi(\tau^{(1)}) \in \Theta^{(1)}(0)$$

where $\tau^{(1)} = \tau^{(1)}(\theta^{(1)}, \theta^{(2)})$. Consequently

$$\tau^{(1)}(\theta^{(1)}, \theta^{(2)}) = \varphi^{-1}(\theta^{(1)} - \chi(\theta^{(2)})). \tag{6}$$

Let

$$\kappa(\theta^{(1)}, \theta^{(2)}) = -\ln a(\theta^{(1)}, \theta^{(2)})$$

and consider for fixed $\theta^{(2)}$ the function f defined on $\Theta^{(1)}(0)$ by

$$f(\theta^{(1)}) = \kappa(\theta^{(1)}, 0) - \kappa(\theta^{(1)} + \chi(\theta^{(2)}), \theta^{(2)}), \qquad \theta^{(1)} \in \Theta^{(1)}(0).$$

Using (6) one finds the gradient of f to be 0 because

$$\begin{aligned}Df(\theta^{(1)}) &= \frac{\partial}{\partial\theta^{(1)}}\kappa(\theta^{(1)}, 0) - \frac{\partial}{\partial\theta^{(1)}}\kappa(\theta^{(1)} + \chi(\theta^{(2)}), \theta^{(2)})\\ &= \tau^{(1)}(\theta^{(1)}, 0) - \tau^{(1)}(\theta^{(1)} + \chi(\theta^{(2)}), \theta^{(2)})\\ &= \varphi^{-1}(\theta^{(1)}) - \varphi^{-1}(\theta^{(1)} + \chi(\theta^{(2)}) - \chi(\theta^{(2)}))\\ &= \varphi^{-1}(\theta^{(1)}) - \varphi^{-1}(\theta^{(1)}) = 0.\end{aligned}$$

Since Θ is assumed to be open, $\Theta^{(1)}(0)$ is open. Moreover, $\Theta^{(1)}(0)$ is connected, because φ is a homeomorphism which maps $\mathfrak{T}^{(1)}$ onto $\Theta^{(1)}(0)$ and because the connectedness of Θ and the variation independence of $\tau^{(1)}$ and $\theta^{(2)}$ imply that $\mathfrak{T}^{(1)}$ is connected. Let $\tau^{(1)}(0,0) = \tau_0^{(1)}$. Since $Df = 0$ one has

$$f(\theta^{(1)}) = f(0) \qquad \forall\ \theta^{(1)} \in \Theta^{(1)}(0),$$

i.e.

$$\begin{aligned}\kappa(\theta^{(1)},0) - \kappa(\theta^{(1)} + \chi(\theta^{(2)}),\theta^{(2)}) &= \kappa(0,0) - \kappa(\chi(\theta^{(2)}),\theta^{(2)})\\ &= -\kappa(\chi(\theta^{(2)}),\theta^{(2)})\end{aligned}$$

and, considering κ as a function of $(\tau^{(1)},\theta^{(2)})$, one obtains

$$\kappa(\tau^{(1)},0) - \kappa(\tau^{(1)},\theta^{(2)}) = -\kappa(\tau_0^{(1)},\theta^{(2)}).$$

Thus, letting

$$\kappa_1(\tau^{(1)}) = \kappa(\tau^{(1)},0) \quad \text{and} \quad \kappa_2(\theta^{(2)}) = \kappa(\tau_0^{(1)},\theta^{(2)}),$$

one has

$$\kappa(\tau^{(1)},\theta^{(2)}) = \kappa_1(\tau^{(1)}) + \kappa_2(\theta^{(2)})$$

from which the conclusion of the lemma follows immediately. ▶

Theorem 10.4. *Suppose $\mathfrak{P}$ is open and connected. Then $t^{(1)}$ is a proper cut of size $d = k^{(1)}$ if and only if*

(i) *$\tau^{(1)}$ and $\theta^{(2)}$ are variation independent.*
(ii) *$\theta^{(1)}$ is of the form*

$$\theta^{(1)} = \varphi(\tau^{(1)}) + \chi(\theta^{(2)}).$$

Proof. The necessity of this assertion follows from Theorem 10.3.

Suppose (i) and (ii) are fulfilled. By Lemma 10.3, $\mathfrak{P}$ has a minimal representation of the form

$$\frac{dP_\theta}{dP_0} = a_1(\tau^{(1)})a_2(\theta^{(2)})\,e^{(\varphi(\tau^{(1)})+\chi(\theta^{(2)}))\cdot t^{(1)}+\theta^{(2)}\cdot t^{(2)}}.$$

The distribution of $t^{(1)}$ has density

$$p(t^{(1)};(\tau^{(1)},\theta^{(2)})) = a_1(\tau^{(1)})\,e^{\varphi(\tau^{(1)})\cdot t^{(1)}}a_2(\theta^{(2)})\,e^{\chi(\theta^{(2)})\cdot t^{(1)}}E_0(e^{\theta^{(2)}\cdot t^{(2)}}|t^{(1)})$$

and as the next step it will be shown that this density does not depend on $\theta^{(2)}$. This means proving that

$$d(t^{(1)};\theta^{(2)}) = a_2(\theta^{(2)})\,e^{\chi(\theta^{(2)})\cdot t^{(1)}}E_0(e^{\theta^{(2)}\cdot t^{(2)}}|t^{(1)})$$

does not depend on $\theta^{(2)}$. Let $\Theta^{(2)}$ denote the domain of variation for $\theta^{(2)}$, let $\theta_0^{(2)}$ be an arbitrary but fixed element in $\Theta^{(2)}$ and let

$$\tilde{d}(t^{(1)}, \theta^{(2)}) = \frac{d(t^{(1)}, \theta^{(2)})}{d(t^{(1)}, \theta_0^{(2)})}.$$

With this notation

$$p(t^{(1)}; \tau^{(1)}) = a_1(\tau^{(1)})\, d(t^{(1)}, \theta_0^{(2)})\, e^{\varphi(\tau^{(1)})\cdot t^{(1)}}$$

is the density of $t^{(1)}$ corresponding to the measure in $\mathfrak{P}$ given by $(\tau^{(1)}, \theta_0^{(2)})$. For every $(\tau^{(1)}, \theta^{(2)}) \in \mathfrak{T}^{(1)} \times \Theta^{(2)}$ one obtains

$$\int \tilde{d}(t^{(1)}, \theta^{(2)}) p(t^{(1)}; \tau^{(1)})\, dP_0 t^{(1)} = 1,$$

that is

$$\int \tilde{d}(t^{(1)}, \theta^{(2)})\, dP_{(\tau^{(1)}, \theta_0^{(2)})} t^{(1)} = 1.$$

Since $\{P_{(\tau^{(1)}, \theta_0^{(2)})} t^{(1)} : \tau^{(1)} \in \mathfrak{T}^{(1)}\}$ is complete,

$$\tilde{d}(t^{(1)}, \theta^{(2)}) \equiv 1$$

and hence $d(t^{(1)}, \theta^{(2)})$ does not depend on $\theta^{(2)}$.

One may now conclude that $t^{(1)}$ is a cut with $\tau^{(1)}$ and $\theta^{(2)}$ as a corresponding pair of L-independent parameters, and on account of Lemma 10.2 one has $k^{(1)} = d$. ▶

Regular exponential families are open and connected and, since for any mixed parameter $(\tau^{(1)}, \theta^{(2)})$ corresponding to a regular exponential family the components $\tau^{(1)}$ and $\theta^{(2)}$ are variation independent (cf. Theorem 8.4), one has the following:

Corollary 10.4. *Suppose $\mathfrak{P}$ is regular. Then $t^{(1)}$ is a cut of size $d = k^{(1)}$ if and only if $\theta^{(1)}$ is of the form*

$$\theta^{(1)} = \varphi(\tau^{(1)}) + \chi(\theta^{(2)}).$$

Applications of Corollary 10.4 have been given in Section 9.2. As a further, immediate, consequence one has, provided $\mathfrak{P}$ is regular, that $t^{(1)}(x_1) + \cdots + t^{(1)}(x_n)$ is a cut with respect to $\mathfrak{P}^{(n)} = 1, 2, \ldots$ if and only if this is true for one n.

Next, a method for construction of exponential families with cuts, suggested by formula (4), will be mentioned.

The discussion will, for the sake of simple argumentation, be confined to the case where $\mathfrak{P}$ is regular and $k = 2$, and where $S^{(1)}$ either is of the form $\{0, 1, 2, \ldots, n\}$ for some positive integer n or is equal to N_0. But it will be obvious that the method indicated is applicable more generally.

Suppose, for the moment, that $t^{(1)}$ is a cut and that formula (4) holds. Taking $t_0^{(1)}$ to be 0 one sees that $c(\theta^{(2)}|t^{(1)})$ is of the form

$$c(\theta^{(2)}|t^{(1)}) = \zeta_0(\theta^{(2)})\zeta(\theta^{(2)})^{t^{(1)}} \tag{7}$$

where $\zeta_0(\cdot)$ is a Laplace transform. Presumably, in most cases of interest $\zeta(\cdot)$ will also be a Laplace transform. In that event, (7) shows that, under P_0 and for given $t^{(1)}$, the variate $t^{(2)}$ has the form

$$t^{(2)} = z_0 + z_1 + \cdots + z_{t^{(1)}} \tag{8}$$

$z_0, z_1, z_2, \ldots$ being independent random variables and $z_1, z_2, \ldots$ having identical distributions.

This leads to the idea of constructing a family $\mathfrak{P}$ in the following way. Take an arbitrary regular exponential family, of order one and with support $S^{(1)}$, to be the family of marginal distributions of $t^{(1)}$. Let

$$a_1(\alpha)b_1(t^{(1)})\,e^{\alpha t^{(1)}}, \qquad \alpha \in A \tag{9}$$

be the densities of this family, assume that $0 \in A$ and denote by c_1 the Laplace transform of the distribution of $t^{(1)}$ for $\alpha = 0$. Furthermore, take any two Laplace transforms (of one-dimensional distributions) ζ_0 and ζ, let B be the set $\{\beta: \zeta_0(\beta) < \infty \text{ and } \zeta(\beta) < \infty\}$ and suppose B is open. Define $\mathfrak{P}$ as the family

$$\mathfrak{P} = \{P_{(\alpha,\beta)}: \alpha \in A, \beta \in B\}$$

of distributions of t determined by the specification that under $P_{(\alpha,\beta)}$ the distribution of $t^{(1)}$ is given by (9) while the conditional distribution of $t^{(2)}$ is determined by (8) together with the requirement that the Laplace transform of z_0 is $\zeta_0(\cdot + \beta)/\zeta_0(\beta)$ and that the common distribution of $z_1, z_2, \ldots$ has Laplace transform $\zeta(\cdot + \beta)/\zeta(\beta)$.

Clearly, $t^{(1)}$ is a cut relative to $\mathfrak{P}$. Moreover—as desired—$\mathfrak{P}$ is exponential of order 2 and regular, with t as minimal canonical statistic.

To see this, let $\tilde{\mathfrak{P}}$ be the full exponential family in R^2 generated by P_0 (the element $P_{(0,0)}$ of $\mathfrak{P}$) and t, i.e. $\tilde{\mathfrak{P}}$ is the set of probability measures which are equivalent to P_0 and whose densities with respect to this measure are of the form $a(\theta)\exp(\theta \cdot t)$. Now, by the construction of $P_{(\alpha,\beta)}$ its Laplace transform $c_{(\alpha,\beta)}(\cdot)$ may be calculated thus

$$\begin{aligned}
c_{(\alpha,\beta)}(\theta) &= E_{(\alpha,\beta)}\, e^{\theta \cdot t} \\
&= E_{(\alpha,\beta)}\{e^{\theta^{(1)}t^{(1)}} E_{(\alpha,\beta)}(e^{\theta^{(2)}t^{(2)}}|t^{(1)})\} \\
&= E_{(\alpha,\beta)}\left[e^{\theta^{(1)}t^{(1)}} \frac{\zeta_0(\theta^{(2)} + \beta)}{\zeta_0(\beta)} \left(\frac{\zeta(\theta^{(2)} + \beta)}{\zeta(\beta)}\right)^{t^{(1)}}\right] \\
&= \frac{\zeta_0(\theta^{(2)} + \beta)}{\zeta_0(\beta)} \frac{c_1[\theta^{(1)} + \ln\{\zeta(\theta^{(2)} + \beta)/\zeta(\beta)\} + \alpha]}{c_1(\alpha)}.
\end{aligned} \tag{10}$$

Hence, in particular, the Laplace transform c of P_0 is given by

$$c(\theta) = \zeta_0(\theta^{(2)})c_1\{\theta^{(1)} + \ln \zeta(\theta^{(2)})\}. \tag{11}$$

On comparing (10) and (11) one sees that $P_{(\alpha,\beta)}$ is equal to the element of $\tilde{\mathfrak{P}}$ determined by the value $(\alpha - \ln \zeta(\beta), \beta)$ of the canonical parameter θ. Thus $\mathfrak{P} \subset \tilde{\mathfrak{P}}$ and it is now a simple matter to show that, in fact, $\tilde{\mathfrak{P}}$ is regular and $\mathfrak{P} = \tilde{\mathfrak{P}}$.

Example 10.5. Suppose $t^{(1)}$ follows a negative binomial distribution with point probabilities

$$\binom{\chi + t^{(1)} - 1}{t^{(1)}} \pi^{t^{(1)}}(1 - \pi)^{\chi},$$

χ being fixed and π varying in (0, 1). Let $z_0, z_1, z_2, \ldots$ be independent and such that z_0 has a gamma distribution with density

$$\frac{\delta^{\chi}}{\Gamma(\chi)} z^{\chi - 1} e^{-\delta z}$$

while $z_1, z_2, \ldots$ follow the exponential distribution

$$\delta e^{-\delta z},$$

the parameter δ having $(0, \infty)$ as domain of variation. Set $t^{(2)} = z_0 + z_1 + \cdots + z_{t^{(1)}}$.

Then $t^{(1)}$ is a cut in the exponential family of joint distributions of $t^{(1)}$ and $t^{(2)}$, with π and δ as a corresponding pair of L-independent parameters.

It may, incidentally, be noted that this example has the property that $t^{(2)}$ is also a cut, $(1 - \pi)\delta$ and $\pi\delta$ being L-independent. In fact, the marginal distribution of $t^{(2)}$ is the gamma distribution.

$$\frac{\{(1 - \pi)\delta\}^{\chi}}{\Gamma(\chi)} (t^{(2)})^{\chi - 1} e^{-(1 - \pi)\delta t^{(2)}},$$

and the conditional distribution of $t^{(1)}$ given $t^{(2)}$ is Poisson with mean value $\pi\delta t^{(2)}$.

▶

10.3 CUTS IN DISCRETE-TYPE EXPONENTIAL FAMILIES

It is assumed throughout this section not only that the elements of $\mathfrak{P}$ are discrete-type distributions but also that $\mathfrak{P}$ is full and linear. Thus x is minimal canonical.

Instances of cuts in such families were, in effect, given in Examples 3.2–3.6. These examples concerned the multinomial, multivariate Poisson, and negative multinomial families, all of which belong to the class of sum-symmetric power series families; this class is investigated separately in the last part of the present section.

The first part is devoted to the derivation of two necessary conditions for $t^{(1)}$ to be a cut. These drastically limit the class of possible cuts. In particular, if $\mathfrak{X}$ is a subset of N_0^k containing all the points $(\varepsilon_1,\ldots,\varepsilon_k)$ such that ε_i is either 0 or 1 $(i = 1,\ldots,k)$, then it appears, from the two conditions mentioned, that the only possible cuts are of the form

$$t^{(1)} = \left(\sum_{J_1} x_i, \ldots, \sum_{J_{k^{(1)}}} x_i\right), \tag{1}$$

$J_1,\ldots,J_{k^{(1)}}$ being disjoint subsets of $\{1,2,\ldots,k\}$. Basic special cases of (1) are $t^{(1)} = x_.(=x_1 + \ldots + x_k)$, and $t^{(1)} = x^{(1)}$ (for some partition $(x^{(1)}, x^{(2)})$ of x).

Theorem 10.5. *Suppose $\mathfrak{P}$ is regular. Let $S^{t^{(1)}}$ be the set of points in S having first component equal to a fixed value $t^{(1)}$.*

If there exists a point in the interior of the convex hull of S which has $t^{(1)}$ as first component and which does not belong to the convex hull of $S^{t^{(1)}}$ then the statistic $t^{(1)}$ is not a cut.

Proof. The family $\mathfrak{P}$ has an exponential representation of the form

$$\frac{dP_\theta}{dP_0} = a(\theta)\, e^{\theta . t}$$

For any $t \in R^k$, let the function $l(\theta) = \theta \cdot t + \ln a(\theta)$ be termed the log-likelihood function corresponding to t, even if t is not in S (just as in Chapter 9).

Now, let $\tilde{t} = (t^{(1)}, \tilde{t}^{(2)})$ be the point whose existence is hypothesized in the theorem. Since $\tilde{t}$ belongs to the interior of the convex hull of S the log-likelihood function corresponding to $\tilde{t}$ has a maximum (cf. Theorem 9.13), i.e. there exists a value $(\hat{\tau}^{(1)}, \theta^{(2)})$ of the mixed parameter such that, considering l as a function of $(\tau^{(1)}, \theta^{(2)})$, one has

$$l(\tau^{(1)}, \theta^{(2)}) \leq l(\hat{\tau}^{(1)}, \hat{\theta}^{(2)}) \tag{2}$$

for every $(\tau^{(1)}, \theta^{(2)})$.

Suppose $t^{(1)}$ is a cut. Then $\tau^{(1)}$ and $\theta^{(2)}$ are variation independent and

$$l(\tau^{(1)}, \theta^{(2)}) = l_1(\tau^{(1)}) + l_2(\theta^{(2)}) \tag{3}$$

where l_1 is the log-likelihood function for the marginal distribution of $t^{(1)}$ while l_2 is the log-likelihood function for the conditional distribution of $t^{(2)}$ given $t^{(1)}$.

Formulas (2) and (3) imply that l_2 has a maximum as $\theta^{(2)}$ varies over $\Theta^{(2)}$. Moreover, $\Theta^{(2)}$ is open.

Now, for any exponential family whose canonical parameter domain is open it is—whether the exponential representation considered is minimal or not—a necessary condition for the log-likelihood function to have a maximum that the value of the canonical statistic belongs to the relative interior of the convex support of the family. This is a simple consequence of the first assertion in

Theorem 9.13, and the well-known result that a local maximum of a concave function must, in fact, be global.

It follows that $\tilde{t}^{(2)}$ belongs to (the relative interior of) the convex hull of $S^{t^{(1)}}$, which is contrary to assumption. ▶

The use of Theorem 10.5 may be illustrated through the following remarks which pertain to the case $k = 2$.

Suppose all the points of $\mathfrak{X}$ are support points of the distributions of x (i.e. $\mathfrak{X}$ contains no superfluous points). Every minimal canonical statistic t is a (nonsingular) affine transformation of x and hence any $t^{(1)}$ corresponds to a system of parallel lines in the range space, R^2, of x, namely the lines on which $t^{(1)}$, viewed as an affine function on R^2, is constant. For $k = 2$ the result of Theorem 10.5 may therefore be paraphrased as follows. Consider an arbitrary system of parallel lines such that each line contains at least one point of S. In order for the $t^{(1)}$ determined by this system to be a cut it is necessary that for every one of the lines, the smallest line segment containing all points of S on the line be equal to the intersection between the line and the closure of the convex hull of S.

For instance, if $\mathfrak{X}$ is a subset of N_0^2 and if the points (0, 0), (0, 1), (1, 0), and (1, 1) all lie in $\mathfrak{X}$ then it is easily seen from a diagram such as Figure 10.1 that, besides x_1, x_2, and $x_.$, which are well known to be cuts in certain cases, the only possibility for a cut is $x_2 - x_1$. The latter may, however, be readily excluded by the next necessary condition.

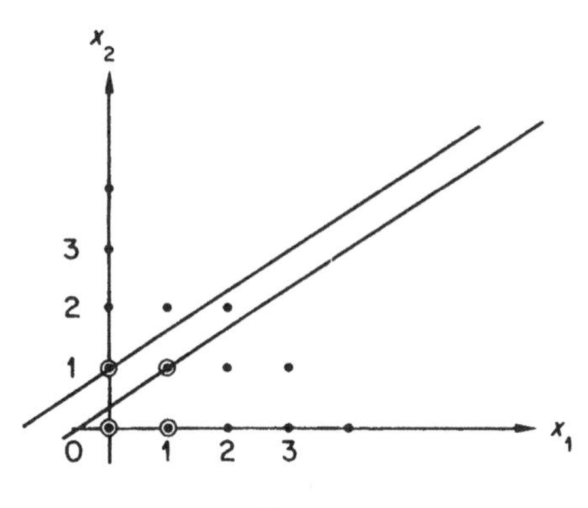

Figure 10.1

For simplicity, this second condition will be derived and stated for $k = 2$ only, but it will be clear that the derivation and result are extendable to general k. Again, let $\mathfrak{P}$ be regular and consider relation (5) of Section 10.2. Since $k = 2$, the range $\Theta^{(2)}$ of $\theta^{(2)}$ is an open interval of R. Assume that $\inf \Theta^{(2)} = -\infty$ and that for each $t^{(1)} \in S^{(1)}$, given by (3) of Section 10.2, the set of points $t^{(2)}$ for which

$(t^{(1)}, t^{(2)}) \in S$ is bounded below. These conditions are fulfilled, in particular, if $\mathfrak{X}$ is finite. Then, letting $\theta^{(2)} \to -\infty$ in (5) of Section 10.2, one finds that *for $t^{(1)}$ to be a cut the points $(t^{(1)}, \min t^{(2)})$, $t^{(1)} \in S^{(1)}$, must lie on a straight line*, the minimum being taken over the values $t^{(2)}$ with $(t^{(1)}, t^{(2)}) \in S$.

Example 10.6. The suppositions made to obtain the conclusion are satisfied in the situation illustrated by Figure 10.1 and for $t^{(1)} = x_2 - x_1$. Therefore $x_2 - x_1$ is not a cut. ▶

Example 10.7. Logistic dose-(binomial) response model. Figure 10.2 shows the set $\mathfrak{X}$ and its convex hull for the case where x equals the minimal canonical statistic (s, w) of the logistic, quantal response model with $n = 1$, $d = 5$ (cf. Example 9.14),

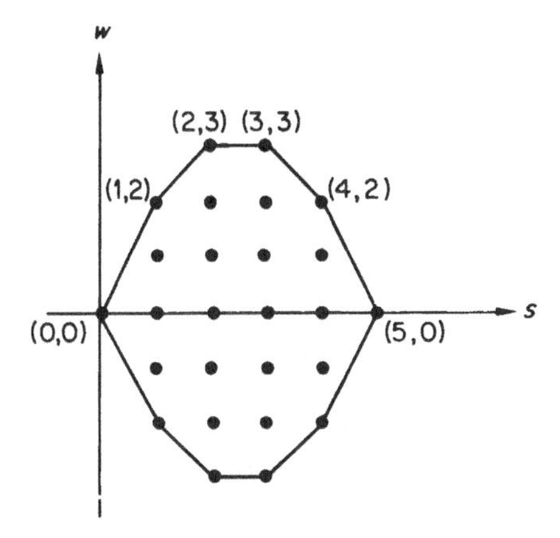

Figure 10.2

the doses being placed at $-2, -1, 0, 1, 2$. The second necessary condition above and a glance at Figure 10.2 immediately reveals that in this instance there are no cuts. ▶

In the remainder of this section $\mathfrak{X}$ is assumed to be a subset of N_0^k. Thus $\mathfrak{P}$ is a power series family and the point probabilities of $\mathfrak{P}$ may be written

$$p(x)\lambda^x/g(\lambda) \tag{4}$$

where $\lambda = (\lambda_1, \ldots, \lambda_k) = (e^{\theta_1}, \ldots, e^{\theta_k})$, $\lambda^x = \lambda_1^{x_1} \ldots \lambda_k^{x_k}$ and g is the generating function for the probabilities $p(x)$ of P_0.

Recall, from Example 8.4, that $\mathfrak{P}$ is a sum-symmetric power series family if and only if g depends on λ through $\lambda_. = \lambda_1 = \cdots + \lambda_k$ only. In this case g_0, defined by $g_0(\bar{\lambda})' = (g(\lambda)$, where $\bar{\lambda} = \lambda_./n$, is the generating function for the point probabilities

$p_0(x_.)$ of $x_.$ under P_0, and consequently

$$p(x) = p_0(x_.)\frac{x_.!}{x_1!\ldots x_k!}. \tag{5}$$

Hence the marginal distribution of $x_.$ is given by

$$p_0(x_.)\bar{\lambda}^{x_.}/g_0(\bar{\lambda}). \tag{6}$$

A probability distribution is called a *singular multinomial distribution* of order k and with numbering parameter n if it has support

$$\{(x_1,\ldots,x_k)\colon x_i \in N_0(i = 1,\ldots,k), x_1 + \cdots + x_k = n\}$$

and point probabilities

$$\frac{n!}{x_1!\cdots x_k!}\pi_1^{x_1}\ldots\pi_k^{x_k}$$

for some $\pi = (\pi_1,\ldots,\pi_k) \in \Pi_0$ where

$$\Pi_0 = \{(\pi_1,\ldots,\pi_k)\colon \pi_i > 0 \ \ (i = 1,\ldots,k), \pi_1 + \cdots + \pi_k = 1\}. \tag{7}$$

From (4) and (5) one finds:

Theorem 10.6. *Suppose $\mathfrak{P}$ is a sum-symmetric power series family, let $(x^{(1)},\ldots,x^{(m)})$ be a partition of x and let $k^{(i)}$ and $x_.^{(i)}$ denote, respectively, the dimension and the sum of the coordinates of $x^{(i)}(i = 1,\ldots,m)$.*

Then $x_.^{(*)} = (x_.^{(1)},\ldots,x_.^{(m)})$ is a cut. Moreover, the family of distributions of $x_.^{(*)}$ is a sum-symmetric power series family while the family of conditional distributions of x given $x_.^{(*)}$ is the product of the m singular multinomial families of orders $k^{(1)},\ldots,k^{(m)}$ and with trial parameters $x_.^{(1)},\ldots,x_.^{(m)}$. ▶

Conversely, one has:

Theorem 10.7 *If $x_.$ is a cut and the family of conditional distributions of x given $x_.$ is, for every value of $x_.$, the singular multinomial family of order k and with trial parameter $x_.$ then $\mathfrak{P}$ is sum-symmetric.*

Proof. Since $x_.$ is a cut, the class of marginal distributions of $x_.$ is a power series family having point probabilities

$$p_0(x_.)\lambda_0^{x_.}/g_0(\lambda_0) \tag{8}$$

for some p_0 and g_0. Thus the probability of x is

$$p_0(x_.)\frac{x_.!}{x_1!\cdots x_k!}(\lambda_0\pi)^x/g_0(\lambda_0)$$

which shows that $\mathfrak{P}$ is sum-symmetric. ▶

Theorem 10.8. *Suppose $\mathfrak{X}$ contains the origin and the unit vectors* $(1,0,\ldots,0),\ldots,(0,\ldots,0,1)$.

Then $\mathfrak{P}$ is a sum-symmetric power series family if and only if $x_.$ is a cut.

Proof. The only if assertion is a consequence of Theorem 10.6.

Suppose $x_.$ is a cut. By (4) and (8) the probability that $x_.$ equals 0, respectively 1, may be written

$$p_0(0)/g_0(\lambda_0) = p(0)/g(\lambda), \tag{9}$$

respectively

$$p_0(1)\lambda_0/g_0(\lambda_0) = \{\sum p(e_i)\lambda_i\}/g(\lambda) \tag{10}$$

where e_i denotes the ith unit vector. Combining (9) and (10) one obtains

$$\lambda_0 = c_1\lambda_1 + \cdots + c_k\lambda_k \tag{11}$$

for certain positive constants $c_1,\ldots,c_k$. Set

$$\tilde{\lambda} = (c_1\lambda_1,\ldots,c_k\lambda_k)$$
$$b(x) = p(x)c_1^{-x_1}\cdots c_k^{-x_k}$$
$$\tilde{g}(\tilde{\lambda}) = g(\lambda),$$

then the probability (4) of x can be expressed thus:

$$b(x)\tilde{\lambda}^x\tilde{g}(\tilde{\lambda}).$$

According to (9) and (11), the function $\tilde{g}$ depends on $\tilde{\lambda}$ only through $\tilde{\lambda}_.$. ▶

Note that if $\mathfrak{P}$ is a sum-symmetric power series family and if $(x^{(1)}, x^{(2)})$ is a partition of x then the family of conditional distributions of $x^{(2)}$ given $x^{(1)}$ is also a sum-symmetric power series family and it depends on $x^{(1)}$ through $x_.^{(1)}$, the sum of the coordinates of $x^{(1)}$, only.

Theorem 10.9. *Let $(x^{(1)}, x^{(2)})$ be a partition of x and suppose $\mathfrak{P}$ is a sum-symmetric power series family such that the probability that $x_.$ equals 1 is less than one. (In the case where this probability is one, $\mathfrak{P}$ is the singular multinomial family with trial parameter 1.)*

Then $x^{(1)}$ is a cut if and only if $\mathfrak{P}$ is either the multinomial, the multivariate Poisson or the negative multinomial family.

Proof. It is straightforward to check the if assertion.

Next, note that if the event that $x_. = 0$ or 1 has probability 1 then $\mathfrak{P}$ is the multinomial family with trial parameter 1. Henceforth it is therefore assumed that the probability of this event is less than 1.

Let $x^{(1)}$ be a cut and suppose first that the dimension of $x^{(1)}$ is one. For conciseness, let $p_n (n = 0,1,2,\ldots)$ denote the probability that $x^{(1)} = n$. The above assumption implies $p_n > 0$ for $n = 0$, 1, and 2, and since $x^{(1)}$ is a cut one has

$$p_0 p_2 / p_1^2 = c$$

where c is a constant (independent of the parameters). Generating function technique shows that

$$p_0 p_2 / p_1^2 = \tfrac{1}{2} g_0(\lambda_0) g_0''(\lambda_0) / \{g_0'(\lambda_0)\}^2,$$

and hence, for certain constants c' and c'',

$$g_0'(\lambda_0)/g_0(\lambda_0) = 1/(c' + c''\lambda_0).$$

It follows that g_0, which is the generating function for $x_.$ under P_0, corresponds to either a binomial ($c'' > 0$), Poisson ($c'' = 0$), or negative binomial ($c'' < 0$) distribution. This implies the desired result.

The general case, where the dimension of $x^{(1)}$ is not supposed to be one, may be thrown back on the one-dimensional case by remarking that the class of distributions of $(x_.^{(1)}, x^{(2)})$, where $x^{(1)}$ is the sum of the coordinates of $x^{(1)}$, is also a sum-symmetric power series family. ▶

10.4 S-ANCILLARITY AND EXPONENTIAL FAMILIES

Many examples of S-ancillarity in exponential families have been indicated in the foregoing, so the discussion here will be confined to some remarks on maximal S-ancillary statistics and to an example, concerning the correlation coefficient of the two-dimensional normal distribution, in which Corollary 10.3 and Theorem 10.2 are used to prove the non-existence of S-ancillary statistics.

Corollary 10.3 and Theorem 10.2 show, in particular, that if $\mathfrak{P}$ is open and convex and if there exists a cut, S-ancillary with respect to a parameter of interest ψ, whose size is the largest possible compatible with the rank of ψ then this cut is maximal S-ancillary. However, a cut of largest possible size need not exist and if it does not there may be no unique maximal S-ancillary statistic.

Example 10.8. For the two-by-two contingency table with the total fixed at n, a minimal canonical statistic is $(x_{11}, x_{.1}, x_{1.})$. Suppose ψ is the interaction parameter,

$$\psi = \ln \frac{p_{11} p_{22}}{p_{12} p_{21}}$$

Then $x_{.1}$ and $x_{1.}$ are both cuts, of size 1 and S-ancillary with respect to ψ. There exist no cuts of size 2 which are S-ancillary with respect to ψ, because in the contrary case $(x_{.1}, x_{1.})$ would, by Theorem 10.2, be a cut, which it is not (as is obvious from the fact that $x_{.1}$ and $x_{1.}$ are independent if and only if $\psi = 0$). Moreover, $x_{.1}$ and $x_{1.}$ are relatively maximal, as will be verified below, and hence there exists no unique maximal cut, S-ancillary with respect to ψ.

Suppose that $\tilde{T}^{(1)}$ is a cut, S-ancillary with respect to ψ and that, in the σ-algebra notation introduced in the beginning of Section 4.2,

$$\sigma(X_{.1}) \subseteq \sigma(\tilde{T}^{(1)})(=\sigma(\tilde{T}^{(1)}) \vee \varnothing). \tag{1}$$

By Lemma 10.1 there exist constants a, b, and c such that

$$\tilde{T}^{(1)} = aX_{.1} + bX_{1.} + c.$$

Obviously one may assume that $a = 1$ and $c = 0$, so

$$\tilde{T}^{(1)} = X_{.1} + bX_{1.}$$

In order to prove that $X_{.1}$ is relatively maximal it suffices to prove that $b = 0$.

Suppose first that the points

$$c_1 + bc_2,$$

where c_1 and c_2 belong to $\{0, 1, \ldots, n\}$, are all different. Then

$$\tilde{T}^{(1)}(x) = c_1 + bc_2$$

$$\Leftrightarrow \quad (X_{.1}(x), X_{1.}(x)) = (c_1, c_2)$$

which implies that the correspondence between $\tilde{T}^{(1)}$ and $(X_{.1}, X_{1.})$ is one-to-one Consequently $\tilde{T}^{(1)}$ is not a cut.

Now, assume there exist x and x' such that $(X_{.1}(x), X_{1.}(x)) \neq (X_{.1}(x'), X_{1.}(x'))$ and

$$X_{.1}(x) + bX_{1.}(x) = X_{.1}(x') + bX_{1.}(x')$$

which means that

$$\tilde{T}^{(1)}(x) = \tilde{T}^{(1)}(x').$$

Hence, on account of (1)

$$X_{.1}(x) = X_{.1}(x')$$

and thus

$$b(X_{1.}(x) - X_{1.}(x')) = 0.$$

It follows that $b = 0$.

By symmetry $X_{1.}$ must also be a relatively maximal cut, S-ancillary with respect to ψ. ▶

Example 10.9. Let (x_{i1}, x_{i2}), $i = 1, 2, \ldots, n$, be n independent and identically distributed two-dimensional normal variates with mean value (ξ_1, ξ_2) and variance matrix

$$\boldsymbol{\Sigma} = \begin{pmatrix} \sigma_1^2 & \rho\sigma_1\sigma_2 \\ \rho\sigma_1\sigma_2 & \sigma_2^2 \end{pmatrix}.$$

The family of joint distributions of these variates is regular exponential of order 5 and has an exponential representation with

$$t = (\sum x_{i1}^2, \sum x_{i1}, \sum x_{i1}x_{i2}, \sum x_{i2}, \sum x_{i2}^2)$$

as a minimal canonical statistic. The corresponding minimal canonical parameter is given by the following equations

$$\theta_1 = \frac{-1}{2(1-\rho^2)\sigma_1^2}$$

$$\theta_2 = \frac{\xi_1}{(1-\rho^2)\sigma_1^2} - \frac{\rho\xi_2}{(1-\rho^2)\sigma_1\sigma_2}$$

$$\theta_3 = \frac{\rho}{(1-\rho^2)\sigma_1\sigma_2}$$

$$\theta_4 = \frac{\xi_2}{(1-\rho^2)\sigma_2^2} - \frac{\rho\xi_1}{(1-\rho^2)\sigma_1\sigma_2}$$

$$\theta_5 = \frac{-1}{2(1-\rho^2)\sigma_2^2}.$$

The aim of the present example is to prove that there exists no cut, S-ancillary with respect to the correlation coefficient ρ.

One has

$$\rho = \psi(\theta) = \frac{\theta_3}{2\sqrt{(\theta_1\theta_5)}}$$

and hence ψ is of rank 3. If there exists a cut, S-ancillary with respect to ρ, its size must therefore be 1 or 2.

Suppose there is such a cut, of order 2. Theorem 10.2 implies that (t_2, t_4) is (equivalent to) this cut, i.e. that the family of marginal distributions of (t_2, t_4) is exponential of order 2. But $(t_2, t_4) = (\sum x_{i1}, \sum x_{i2})$ so a contradiction has obviously been reached.

Next, suppose there exists a cut of size 1, S-ancillary with respect to ψ. On account of Corollary 10.2 it causes no loss of generality to assume that this cut is a one-dimensional component $\tilde{t}^{(1)}$ of a minimal canonical statistic $\tilde{t}$. Let $\tilde{\theta} = (\tilde{\theta}^{(1)}, \tilde{\theta}^{(2)})$ denote the corresponding minimal canonical parameter. Since $\tilde{t}^{(1)}$ is S-ancillary with respect to ψ, the parameter ψ depends on $\tilde{\theta}$ only through $\tilde{\theta}^{(2)}$. Thus, on account of Lemma 10.1, $\tilde{t}^{(1)}$ is an affine transformation of (t_2, t_4). This implies that $\tilde{t}^{(1)}$ is normally distributed and hence the exponential family corresponding to the marginal distribution of $\tilde{t}^{(1)}$ is of order 2, which contradicts the fact that the size of $\tilde{t}^{(1)}$ is 1. ▶

10.5 M-ANCILLARITY AND EXPONENTIAL FAMILIES

A number of examples of M-ancillarity in exponential families have been given in Sections 4.1 and 4.4. Further examples are presented below. Moreover, a necessary condition for M-ancillarity in discrete type families will be established and illustrated. This condition is similar to the necessary condition for S-ancillarity provided by Theorem 10.5. It will be demonstrated that if the necessary condition to be presented is not fulfilled then, as a rule, some value (or values) of the statistic in question gives strong evidence against certain extreme values of the interest parameter. Except for Example 10.10, the families considered in this section are all of discrete type.

For each of the instances of M-ancillarity mentioned in the following the required property of universality is established by first proving that the marginal distribution of the relevant component $t^{(1)}$ is strongly unimodal and then invoking one of Corollaries 9.9 or 9.10.

Corollary 10.3 and Theorem 10.2 should be kept in mind in the following. They imply, in particular, maximality of all the M-ancillary statistics discussed.

The two final examples of the section contain instances of pointwise M-ancillarity.

Example 10.10. System reliability for components in series. Suppose a certain system consists of two components operating in series and having life lengths which follow independent, exponential distributions with mean values β_1 and β_2, respectively. The probability that the system fails before time t is $1 - \exp\{-\psi t\}$ where

$$\psi = \beta_1^{-1} + \beta_2^{-1}.$$

Suppose, moreover, that the life lengths of n_i items of the type used for the ith component, $i = 1, 2$, have been recorded in order to draw inference on ψ. The sums x_1 and x_2, say, of the observed lifelengths are gamma distributed with parameters (n_1, β_1) and (n_2, β_2).

The joint distribution of x_1 and x_2 is strongly unimodal and hence on account of Theorem 6.4 and Corollary 9.10, the difference $x_1 - x_2$ is M-ancillary with respect to ψ. (It may be noted that $x_1 - x_2$ is not G-ancillary.) An explicit expression for the conditional distribution given $x_1 - x_2$ may be found in Lentner and Buehler (1963). ▶

The kind of argument for M-ancillarity given in Example 10.10 goes to show the following general result. If $\mathfrak{P}$ is full, linear, of continuous type, and strongly unimodal then for any pair $t^{(1)}$ and $\theta^{(2)}$ the statistic $t^{(1)}$ is M-ancillary with respect to $\theta^{(2)}$. (Furthermore, under minor additional smoothness assumptions, conditions (iv) and (v) of Theorem 4.7 will be satisfied, cf. Theorem 9.27.)

Example 10.11. System reliability for components in parallel. With two components in parallel, functioning independently and failing with probabilities p_1

and p_2, the probability of failure of the system is $p_1 p_2$. Suppose p_1 and p_2 are small and that the components have been tested separately in, respectively, n_1 and n_2 Bernoulli trials, where n_1 and n_2 are comparatively large. Then the problem of drawing inference on $p_1 p_2$ from the observed numbers of failures x_1 and x_2 may be treated, approximately, as if x_i followed a Poisson distribution with mean value λ_i $(i = 1, 2)$ and as if the interest parameter was $\psi = \lambda_1 \lambda_2$.

In the formulation with Poisson variation, $x_1 - x_2$ is M-ancillary with respect to ψ, by Theorem 6.6 and Corollary 9.9. (As is well known, the distribution of $x_1 - x_2$, and hence also the distribution of x_1 given $x_2 - x_1$, is expressible in terms of the modified Bessel functions $I\nu$.) ▶

Examples 10.12. 2×2 *contingency tables.* If $x_{*\cdot}$ is given then $x_{\cdot 1}$ is M-ancillary with respect to the interaction parameter. In fact, the distribution of $x_{\cdot 1}$ is strongly unimodal because it is the convolution of the distributions of x_{11} and x_{21} which are binomial and hence strongly unimodal; now apply Corollary 9.9.

Suppose next that only the total $x_{\cdot\cdot}$ is given. The distribution of $(x_{\cdot 1}, x_{1\cdot})$ is then the $x_{\cdot\cdot}$-fold convolution of a distribution on $\{0, 1\}^2$ and this implies that it is strongly unimodal, see Pedersen (1975a). Consequently, $(x_{\cdot 1}, x_{1\cdot})$ is M-ancillary with respect to the interaction parameter. ▶

Example 10.13. $2 \times 2 \times 2$ *contingency tables.* Pedersen (1975b) established a number of cases of M-ancillarity in $2 \times 2 \times 2 \times 2$ tables. His results, some of which were obtained by quite intricate proofs (of strong unimodality), are summarized in Table 10.1, in which the second column indicates which interaction parameters are assumed to be zero as part of the model specification. There are, if $x_{\cdot\cdot\cdot}$ is given, one second order interaction

$$\theta_{123} = \ln \frac{p_{111} p_{221} p_{122} p_{212}}{p_{121} p_{211} p_{112} p_{222}}$$

Table 10.1. *Interest parameters and corresponding M-ancillary statistics, for various contingency table models*

Interest parameter	Submodel	$x_{\cdot **}$ given	$x_{\cdot\cdot *}$ given	$x_{\cdot\cdot\cdot}$ given
		M-ancillary statistic		
θ_{123}		$(x_{11\cdot}, x_{1\cdot 1}, x_{1\cdot\cdot})$		
θ_{12}	$\theta_{123} = 0$	$(x_{1\cdot 1}, x_{1\cdot})$		
$(\theta_{123}, \theta_{12})$		$(x_{1\cdot 1}, x_{1\cdot\cdot})$		
$(\theta_{12}, \theta_{13}, \theta_{23})$	$\theta_{123} = 0$	—	$(x_{1\cdot\cdot}, x_{\cdot 1\cdot})$	
θ_{12}	$\theta_{123} = \theta_{13} = \theta_{23} = 0$	—	$(x_{1\cdot\cdot}, x_{\cdot 1\cdot})$	$(x_{1\cdot\cdot}, x_{\cdot 1\cdot}, x_{\cdot\cdot 1})$

and three first order interactions

$$\theta_{12} = \ln \frac{p_{112}p_{222}}{p_{122}p_{212}}, \qquad \theta_{13} = \ln \frac{p_{121}p_{222}}{p_{122}p_{221}},$$

$$\theta_{23} = \ln \frac{p_{211}p_{222}}{p_{212}p_{221}}.$$

Table 10.1 gives the minimal canonical statistic corresponding to fixed values of the interest parameter, provided this statistic has been proved to be M-ancillary. An entry is empty if the question of M-ancillarity is, so far, undecided, and it contains a bar (–) if the question has no meaning in the model concerned.

It was furthermore observed by Pedersen (1975b) that for the table with given total $x_{\cdot\cdot\cdot}$ the set of marginals $(x_{1\cdot\cdot}, x_{\cdot 1\cdot}, x_{\cdot\cdot 1})$ is not M-ancillary with respect to $(\theta_{12}, \theta_{13}, \theta_{23}, \theta_{123})$, see Example 10.17 below. ▶

Example 10.14. Logistic dose-(binomial) response model. (Cf. Examples 9.14 and 10.7). The statistic $s = \sum a_i$, being a sum of independent binomial variates, has a strongly unimodal distribution and is hence M-ancillary with respect to β (whatever the number and placing of the doses and whatever the number of animals per dose).

With three doses placed at $-1, 0, 1$, the other component w equals $a_1 - a_{-1}$. It follows (using again Theorem 6.6 and Corollary 9.9) that w is M-ancillary with respect to α. However, this is an exceptional case; in general w is not M-ancillary, cf. Example 10.18. ▶

The following theorem and its corollary, together, are to a large extent analogous to Theorem 10.5.

Theorem 10.10. *Suppose $\mathfrak{P}$ is of discrete type and that Θ is of the form*

$$\Theta = \Theta^{(1)} \times R^{k(2)}. \tag{1}$$

Let $t^{(1)}$ be one of the possible values of $T^{(1)}$ and let $S^{t^{(1)}}$ be the set of points in S having first component equal to $t^{(1)}$.

If there exists a point in the convex hull of S which has $t^{(1)}$ as first component and which does not belong to the closure of the convex hull of $S^{t^{(1)}}$ then there exists a convex cone K in $R^{k(2)}$ such that for $\theta^{(2)} \in K$ and $|\theta^{(2)}| \to \infty$

$$P_\theta\{T^{(1)} = t^{(1)}\} \to 0 \qquad \textit{uniformly in } \theta^{(1)}.$$

Proof. $\mathfrak{P}$ has point probabilities

$$P_\theta\{T = t\} = a(\theta)b(t)e^{\theta \cdot t}$$

Set $C^{t^{(1)}} = \operatorname{cl} \operatorname{conv} S^{t^{(1)}}$ and suppose t_0 is a point such that $t_0^{(1)} = t^{(1)}$ and $t_0 \in \operatorname{conv} S \backslash C^{t^{(1)}}$. On account of Carathéodory's convexity theorem there exists an integer m with $1 \leq m \leq k+1$, points $t_1, \ldots, t_m \in S$ and positive numbers $\lambda_1, \ldots, \lambda_m$ such that $\lambda_1 + \cdots + \lambda_m = 1$ and

$$t_0 = \lambda_1 t_1 + \cdots + \lambda_m t_m. \tag{2}$$

Set

$$M(\theta) = \max\left\{\frac{P_\theta\{T^{(1)} = t_i^{(1)}\}}{P_\theta\{T^{(1)} = t^{(1)}\}} : i = 1, \ldots, m\right\}$$

and

$$d(s^{(1)}, \theta^{(2)}) = \sum_{s^{(2)} \in S_{s^{(1)}}} b(s^{(1)}, s^{(2)})\, e^{\theta^{(2)} \cdot s^{(2)}}.$$

One has

$$\frac{P_\theta\{T^{(1)} = t_i^{(1)}\}}{P_\theta\{T^{(1)} = t^{(1)}\}} = e^{\theta^{(1)} \cdot (t_i^{(1)} - t^{(1)})} \frac{d(t_i^{(1)}, \theta^{(2)})}{d(t^{(1)}, \theta^{(2)})}.$$

Now,

$$\begin{aligned} M(\theta) &\geq \prod_{i=1}^{m} \left(\frac{P_\theta\{T^{(1)} = t_i^{(1)}\}}{P_\theta\{T^{(1)} = t^{(1)}\}}\right)^{\lambda_i} \\ &= \exp\left\{\theta^{(1)} \cdot \left(\sum_{i=1}^{m} \lambda_i t_i^{(1)} - t^{(1)}\right)\right\} \prod_{i=1}^{m} \left(\frac{d(t_i^{(1)}, \theta^{(2)})}{d(t^{(1)}, \theta^{(2)})}\right)^{\lambda_i}. \end{aligned}$$

Equation (2) says, in particular, that

$$\sum_{i=1}^{m} \lambda_i t_i^{(1)} = t^{(1)}$$

and consequently

$$M(\theta) \geq \prod_{i=1}^{m} \left(\frac{d(t_i^{(1)}, \theta^{(2)})}{d(t^{(1)}, \theta^{(2)})}\right)^{\lambda_i}.$$

It is obviously that

$$d(t_i^{(1)}, \theta^{(2)}) \geq b(t_i)\, e^{\theta^{(2)} \cdot t_i^{(2)}}, \qquad i = 1, \ldots, m$$

and hence, using (2) again,

$$M(\theta) \geq \prod_{i=1}^{m} b(t_i)^{\lambda_i} \frac{e^{\theta^{(2)} \cdot t_0^{(2)}}}{d(t^{(1)}, \theta^{(2)})}.$$

It follows that

$$M(\theta) \geq c_0 \rho(\theta^{(2)})$$

where

$$c_0 = \frac{\prod_{i=1}^{m} b(t_i)^{\lambda_i}}{\sum_{s \in S_t^{(1)}} b(s)}$$

and

$$\rho(\theta^{(2)}) = \exp\{\theta^{(2)} \cdot t_0^{(2)} - \sup_{s^{(2)} \in S_{t^{(1)}}} \theta^{(2)} \cdot s^{(2)}\}.$$

Define K by

$$K = \{\theta^{(2)}: \inf_{s^{(2)} \in S_{t^{(1)}}} \theta^{(2)} \cdot (t_0^{(2)} - s^{(2)}) > 0\}.$$

The fact that $t_0 \notin C^{t^{(1)}}$ implies that K is non-empty, and obviously

$$\lambda K \subset K \qquad \text{for } \lambda > 0$$

and

$$\lambda K + (1 - \lambda)K \subset K \qquad \text{for } \lambda \in (0, 1),$$

so that K is a convex cone.

Now, if $\theta^{(2)} \in K$ and $|\theta^{(2)}| \to \infty$ then $M(\theta) \to \infty$ and this implies that $P_\theta\{T^{(1)} = t^{(1)}\} \to 0$. The convergence is uniform in $\theta^{(1)}$ as ρ does not depend on $\theta^{(1)}$. ▶

In cases where it is of interest to find K one may use that K is an intersection of halfspaces, namely

$$K = \bigcap_{s^{(2)} \in S_{t^{(1)}}} \{\theta^{(2)}: \theta^{(2)} \cdot (t_0^{(2)} - s^{(2)}) > 0\}.$$

Corollary 10.5. *Suppose S is finite and $\mathfrak{P}$ is full.*

If for one of the possible values $t^{(1)}$ of $T^{(1)}$ there exists a point in the convex hull of S which has first component $t^{(1)}$ and which does not belong to the closed convex hull of $S^{t^{(1)}}$ then $T^{(1)}$ is not M-ancillary with respect to $\theta^{(2)}$.

Proof. For $\theta^{(2)} \in K$ and $|\theta^{(2)}|$ large, $t^{(1)}$ is not a mode point for the family of distributions of $T^{(1)}$. ▶

In fact, for S finite, Theorem 10.10 implies not only that $t^{(1)}$ is not a mode point for distributions of $T^{(1)}$ with $\theta^{(2)} \in K$ and $|\theta^{(2)}|$ large, but that the event $\{T^{(1)} = t^{(1)}\}$, considered separately, affords strong evidence against such values of $\theta^{(2)}$ (both from the likelihood viewpoint and the plausibility viewpoint).

The illustrative remark concerning Theorem 10.5, made right after the proof of that result, applies *mutatis mutandis* to Corollary 10.5. In particular, suppose that $\mathfrak{P}$ is linear and regular, that $\mathfrak{X}$ is a finite subset of N_0^2, and that the points (0, 0), (0, 1), (1, 0), (1, 1) all have positive probability. Then the only statistics which can possibly be M-ancillary, with respect to the complementary parameter function which induces the same partition of $\mathfrak{P}$ as does the conditional distributions given the statistic, are $x_1, x_2, x_1 + x_2$, and $x_1 - x_2$. (Unlike the case in respect of S-ancillarity, the latter statistic $x_1 - x_2$ may, in fact, be M-ancillary.

Example 10.15. For the 2 × 2 contingency table with one margin fixed, shown on p. 36, the only statistics which are M-ancillary, with respect to their complementary parameter function, are $x_1, x_2, x_1 + x_2$, and $x_1 - x_2$ (the parameter function corresponding to $x_1 - x_2$ being $\ln(\{p_1 p_2\}/\{q_1 q_2\})$). ▶

Example 10.16. The 2 × 2 × 2 contingency table with given total n. For $n \geq 3$, the set of marginals $(x_{1..}, x_{.1.}, x_{..1})$ is not M-ancillary with respect to the set of first and second order interactions $(\theta_{12}, \theta_{13}, \theta_{23}, \theta_{123})$ (in the notation of Example 10.13).

To show that Corollary 10.5 applies, let $T^{(1)} = (X_{1..}, X_{.1.}, X_{..1})$, $T^{(2)} = (X_{111}, X_{.11}, X_{1.1}, X_{11.})$ and $t^{(1)} = (1, 1, 1)$. One has

$$S^{(1,1,1)} = \{(1, 1, 1, s^{(2)}) \colon s^{(2)} = (0, 0, 0, 0), (0, 0, 0, 1),$$
$$(0, 0, 1, 0), (0, 1, 0, 0), (1, 1, 1, 1)\}.$$

Moreover, $(0, 0, 0, 0, 0, 0, 0) \in S$, $(2, 2, 2, 0, 1, 1, 1) \in S$ and hence

$$(1, 1, 1, 0, \tfrac{1}{2}, \tfrac{1}{2}, \tfrac{1}{2}) \in \operatorname{conv} S.$$

But this latter point does not belong to $\operatorname{cl} \operatorname{conv} S^{(1,1,1)}$ and hence the supposition in Corollary 10.5 are fulfilled.

The above result was given by Pedersen (1975b). ▶

It was, furthermore, pointed out by Pedersen (1975b) that the result in Example 10.16 together with Theorem 10.10 simply implies that for every $n \geq 3$ there exists a strongly unimodal distribution on $\{0, 1\}^3$ such that its n-fold convolution is not strongly unimodal.

Example 10.17. Genotype distribution and selection. n individuals from a diploid population are classified according to genotype at a single, diallelic locus. Denote the genes by A and a, and let x_1, x_2, x_3 be the number of individuals of genotypes AA, Aa, aa, respectively. Supposing that the n individuals form a random sample from an infinite population in which the frequencies of the three genotypes are p_1, p_2, and p_3, the distribution of (x_1, x_2, x_3) is the trinomial

$$\frac{n!}{x_1! \, x_2! \, x_3!} p_1^{x_1} p_2^{x_2} p_3^{x_3}.$$

This expression may be written in the exponential form

$$a(\theta) p(z, x_2) \, e^{\theta_1 z + \theta_2 x_2}$$

where $z = 2x_1 + x_2$ is the number of A genes in the sample and where

$$\theta_1 = \frac{1}{2}\ln\frac{p_1}{p_3}$$

$$\theta_2 = \frac{1}{2}\ln\frac{p_2^2}{4p_1p_3}.$$

As discussed in Example 8.7, θ_2 is a meaningful indicator of deviation from Hardy–Weinberg distribution, especially when such deviation is caused by zygotic selection. Let $x = (z, x_2)$, then

$$\mathfrak{X} = \{(z, x_2): z = 0, 1, \ldots, 2n; x_2 = 0, 2, \ldots, \min\{z, 2n - z\} \text{ for } z \text{ even},$$
$$\times\ x_2 = 1, 3, \ldots, \min\{z, 2n - z\} \text{ for } z \text{ odd}\},$$

see Figure 10.3.

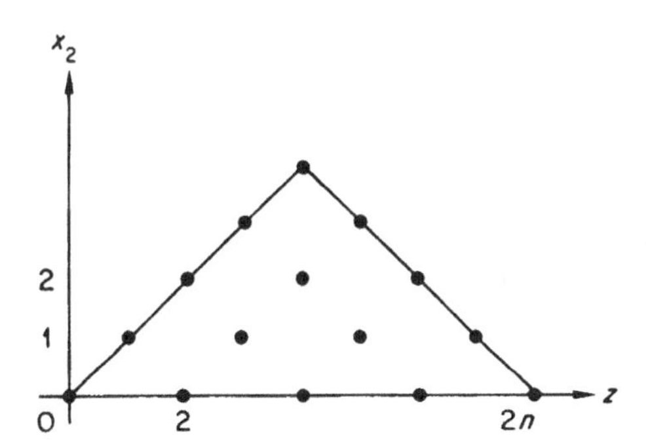

Figure 10.3 The set $\mathfrak{X}$ of values of $(z_1 x_2)$ and the convex hull C of $\mathfrak{X}_1$ for $n = 4$.

Each of the statistics x_1, x_2, and x_3 is M-ancillary (and, incidentally, also S-ancillary), x_1 with respect to $\ln(p_2/p_3)$, etc. It is evident from Figure 10.3 (and Corollary 10.5) that there is no other statistic which is M-ancillary with respect to its complementary parameter function.

In particular, z is not M-ancillary with respect to θ_2, and if an odd value of z is observed this, by itself, is strong evidence against negative, numerically large, values of θ_2. Thus, if θ_2 is the interest parameter and if one were to draw the inference solely in the conditional distribution of x_2 given z (which, of course, depends on θ_2 only) then one would, provided z was odd, be ignoring available information on θ_2. It seems, however, a reasonable conjecture that any even value of z is pointwise M-nonformative with respect to θ_2.

Should the interest be solely in the hypothesis of Hardy–Weinberg distribution then the parameter of interest is

$$\psi = \begin{cases} 0 & \text{for } \theta_2 = 0 \\ 1 & \text{for } \theta_2 \neq 0. \end{cases}$$

A discussion, from the plausibility viewpoint, of the amount of information which may be lost by drawing the inference about this parameter conditionally on z has been given in Barndorff-Nielsen (1977c). ▶

Example 10.18. Logistic dose-(binomial) response model. (Continuation of Example 10.14.) Let there be $d = 2j + 1$ doses placed at

$$-j, -j+1, \ldots, -1, 0, 1, \ldots, j-1, j$$

and let the number of animals be n at each dose. As mentioned in Example 10.14, the statistic s is M-ancillary with respect to β, and if $j = 1$ then w is M-ancillary with respect to α. But w is not M-ancillary for $j > 1$, because Corollary 10.5 works here for $T^{(1)} = w$, $t^{(1)} = 1$, as is apparent from a diagram such as Figure 10.2.

Suppose however, that $n = 1$, $j = 2$ and that w is observed to be 0. This observation together with the family of marginal distributions of w satisfies, as will now be verified, the requirement of the definition of pointwise M-nonformation with respect to α, and hence w is M-ancillary at 0 with respect to α.

In the case considered,

$$w = -2a_{-2} - a_{-1} + a_1 + 2a_2.$$

Thus, for $\beta = 0$ and with $p = (1 + e^{\alpha})^{-1}$ and $q = 1 - p$ one has that the distribution of w is symmetric and given by

w	$p(w; (\alpha, 0))$
± 3	p^2q^2
± 2	$pq^3 + p^3q$
± 1	$p^2q^2 + pq^3 + p^3q$
0	$p^4 + 2p^2q^2 + q^4$

For every value of p in $(0, 1)$ this distribution has mode at 0 so that the first condition for M-nonformation at 0 is satisfied. To see that the second condition is also fulfilled it suffices, by reason of continuity, to show that the subfamily of marginal distributions of w for which α equals 0 is universal. This subfamily is the full exponential family generated by the distribution of w for $\alpha = \beta = 0$. The latter distribution is strongly unimodal and hence (Corollary 9.9) the subfamily is universal. ▶

10.6 COMPLEMENT

It is not possible to extend Corollary 10.5 to hold for arbitrary regular families $\mathfrak{P}$ (of discrete type) as is shown by:

Example 10.19. Let $k = 2$ and let S consist of the origin and all points of the form $(\varepsilon^{(1)}/n, \varepsilon^{(2)}n)$ where $\varepsilon^{(1)} = \pm 1, \varepsilon^{(2)} = \pm 1$ and $n = 1, 2, \dots$. Take $\mathfrak{P}$ to be the family $\{P_\theta: \theta \in \Theta\}$ with

$$P_\theta\{0\} = a(\theta)\tfrac{1}{2}$$

$$P_\theta\{t\} = a(\theta)\frac{1}{2^{n+3}}e^{\theta \cdot t} \qquad \text{for } t = (\varepsilon^{(1)}/n, \varepsilon^{(2)}n)$$

$$c(\theta) = a(\theta)^{-1} = \frac{1}{2} + \frac{1}{8} \sum_{\varepsilon^{(1)}, \varepsilon^{(2)}} \sum_{n=1}^{\infty} 2^{-n} e^{\theta^{(1)}\varepsilon^{(1)}/n + \theta^{(2)}\varepsilon^{(2)}n}$$

and

$$\Theta = R \times (-\ln 2, \ln 2).$$

Clearly, $\mathfrak{P}$ is regular.

Here, for $t^{(1)} = 0$, any point $(0, t_0^{(2)})$ with $t_0^{(2)} \neq 0$ belongs to conv S but not to cl conv $S^{t^{(1)}} = \{0\}$. However

$$\frac{P_\theta\{T^{(1)} = \varepsilon^{(1)}/n\}}{P_\theta\{T^{(1)} = 0\}} = 2^{-n-2} e^{\theta^{(1)}\varepsilon^{(1)}/n}(e^{\theta^{(2)}n} + e^{-\theta^{(2)}n})$$

$$< \tfrac{1}{2} e^{|\theta^{(1)}|/n}$$

and thus for every $\theta^{(2)} \in (-\ln 2, \ \ln 2)$, $t^{(1)} = 0$ is a mode point for $P_{(0,\theta^{(2)})}t^{(1)}$. ▶

10.7 NOTES

When specialized to S-ancillarity the results in Section 10.1 coincide with material presented in Barndorff-Nielsen (1973a) and Barndorff-Nielsen and Blæsild (1975). The contents of Sections 10.2–10.4 stem from the same two papers and from Barndorff-Nielsen (1976a). However, Theorems 10.6–10.9 are but a paraphrasing of results due to Joshi and Patil (1970, 1971). Theorem 10.10 and Corollary 10.5 constitute an extension of the main conclusion in Barndorff-Nielsen and Kvist (1974).

As mentioned at the end of Section 4.1, the maximum likelihood estimator of an interest parameter ψ will in general not even be consistent if the number of incidental parameters tends to infinity together with the sample size n, but in important cases, primarily under exponential models, the incidental parameters may be eliminated by a suitable conditioning and the conditional maximum likelihood estimator is both consistent and asymptotically normal. Subject to fairly mild regularity conditions it is possible to show that the maximum plausibility estimate of ψ, whether conditional or unconditional (often these are the same, cf. Section 4.6), differs from the conditional maximum likelihood estimate by a quantity which is of the order of n^{-1} only—cf. Höglund (1974), and also Section 9.9.

References

Andersen, A. H. (1969). "Asymptotic results for exponential families." *Bull. Int. Statist. Inst.*, **43**, Book 2, 241–242.
Andersen, A. H. (1974). "Multidimensional contingency tables." *Scand. J. Statist.*, **1**, 115–127.
Andersen, E. B. (1970). "Asymptotic properties of conditional maximum-likelihood estimators." *J. Roy. Statist. Soc., B*, **32**, 283–301.
Andersen, E. B. (1971). "Asymptotic properties of conditional likelihood ratio tests." *J. Amer. Statist. Ass.*, **66**, 630–633.
Andersen, E. B. (1973). *Conditional inference and models for measuring*, Mentalhygiejnisk Forlag, Copenhagen.
Ando, A. and Kaufman, G. M. (1965). "Bayesian analysis of the independent multinormal process—neither mean nor precision known." *J. Amer. Statist. Ass.*, **60**, 347–358.
Anscombe, F. J. (1950). "Sampling theory of the negative binomial and logarithmic series distributions." *Biometrika*, **37**, 358–382.
Anscombe, F. J. (1953). "Contribution to the discussion of H. Hotelling 'New light on the correlation coefficient and its transforms.'" *J. R. Statist. Soc., B*, **15**, 193–232.
Anscombe, F. J. (1961). "Estimating a mixed-exponential response law." *J. Amer. Statist. Ass.*, **56**, 493–502.
Anscombe, F. J. (1964a). "Normal likelihood functions." *Ann. Inst. Statist. Math.*, **26**, 1–19.
Anscombe, F. J. (1964b). "Some remarks on Bayesian statistics." In G. L. Bryan and M. W. Shelly (Eds), *Human judgement and optimality*, Wiley, New York. Chap. 10.
Armitage, P. (1961). "Contribution to the discussion of C. A. B. Smith 'Consistency in statistical inference and decision.'" *J. Roy. Statist. Soc., B*, **23**, 1–37.
Bahadur, R. R. (1954). "Sufficiency and statistical decision functions." *Ann. Math. Statist.*, **25**, 423–462.
Barankin, E. W. (1951). "Concerning some inequalities in the theory of statistical estimation." *Skand. Aktuar.*, **34**, 35–40.
Barnard, G. A. (1949). "Statistical inference." *J. Roy. Statist. Soc., B*, **11**, 115–149.
Barnard, G. A. (1963a). "Some logical aspects of the fiducial argument." *J. Roy. Statist. Soc., B*, **25**, 111–114.
Barnard, G. A. (1963b). "The logic of least squares." *J. Roy. Statist. Soc., B*, **25**, 124–127.
Barnard, G. A. (1966). "Summary remarks." In N. L. Johnson and H. Smith, Jr. (Eds), *New Developments in Survey Sampling*, Wiley–Interscience, New York, 696–711.
Barnard, G. A. (1972). "The logic of statistical inference." *Brit. J. Phil. Sci.*, **23**, 123–132.
Barnard, G. A. (1974a). "On likelihood." *Conference Foundational Questions in Statistical Inference*. Memoirs No. 1, Dept. Theor. Statist., Aarhus Univ., 121–138.
Barnard, G. A. (1974b). "The foundations of statistics 1964–1974." *Bull. IMA*, **10**, 344–347.
Barnard, G. A. (1976). "Conditional inference is not inefficient." *Scand. J. Statist.*, **3**, 132–134.
Barnard, G. A., Jenkins, G. M., and Winsten, C. B. (1962). "Likelihood inference and time series. (With discussion)." *J. Roy. Statist. Soc., A*, **125**, 321–375.

Barndorff-Nielsen, O. (1969). "Lévy homeomorphic parametrization and exponential families." *Z. Wahrscheinlichkeitstheorie verw. Gebiete*, **12**, 56–58.

Barndorff-Nielsen, O. (1970). "Exponential families. Exact theory." Various Publications Series No. 19, Inst. Mathematics, Aarhus Univ. Reproduced as Part I, Barndorff-Nielsen (1973a).

Barndorff-Nielsen, O. (1973a). *Exponential families and conditioning*, Sc.D. thesis, University of Copenhagen.

Barndorff-Nielsen, O. (1973b). "Unimodality and exponential families." *Comm. Statist.*, **1**, 189–216.

Barndorff-Nielsen, O. (1973c). "On M-ancillarity." *Biometrika*, **60**, 447–455.

Barndorff-Nielsen, O. (1974). "On M-ancillarity and M-sufficiency." Contribution to the Kaleidoscope in *Conference on Foundational Questions in Statistical Inference*, Memoirs No. 1, Dept. Theor. Statist., Aarhus Univ., 308–311.

Barndorff-Nielsen, O. (1975). "Comments on paper by J. D. Kalbfleisch." *Biometrika*, **62**, 261–262.

Barndorff-Nielsen, O. (1976a). "Factorization of likelihood functions for full exponential families." *J. Roy. Statist. Soc.*, *B*, **38**, 37–44.

Barndorff-Nielsen, O. (1976b). "Plausibility inference. (With discussion.)" *J. R. Statist. Soc. B*, **38**, 103–131.

Barndorff-Nielsen, O. (1976c). "Nonformation." *Biometrika*, **63**, 567–571.

Barndorff-Nielsen, O. (1977a). "Exponentially decreasing distributions for the logarithm of particle size." *Proc. Roy. Soc. London, A*, **353**, 401–419.

Barndorff-Nielsen, O. (1977b). "Contribution to the discussion of D. V. Lindley 'The Bayesian approach'." To appear in *Scand. J. Statist.*

Barndorff-Nielsen, O. (1977c). "On conditional inference for deviation from Hardy–Weinberg distribution." F. B. Christiansen and T. Fenchel (Eds.), *Measuring Selection in Natural Populations*, dedicated to the memory of Ove Frydenberg, Springer, Berlin, 149–157.

Barndorff-Nielsen, O. and Blæsild, P. (1975). "S-ancillarity in exponential families." *Sankhyā*, *A*, **37**, 354–385.

Barndorff-Nielsen, O., Hoffman-Jørgensen, J., and Pedersen, K. (1976). "On the minimal sufficiency of the likelihood function." *Scand. J. Statist.*, **3**, 37–38.

Barndorff-Nielsen, O. and Kvist, H. K. (1974). "Note on exponential families and M-ancillarity." *Scand. J. Statist.*, **1**, 36–38.

Barndorff-Nielsen, O., and Pedersen, K. (1968). "Sufficient data reduction and exponential families." *Math. Scand.*, **22**, 197–202.

Bartlett, M. S. (1936). "Statistical information and properties of sufficiency." *Proc. Roy. Soc.*, *A*, **154**, 124–137.

Bartlett, M. S. (1937). "Properties of sufficiency and statistical tests." *Proc. Roy. Soc.*, *A*, **160**, 268–282.

Basu, D. (1955). "On statistics independent of a complete sufficient statistic." *Sankhyā*, **15**, 377–380.

Basu, D. (1958). "On statistics independent of a sufficient statistic." *Sankhyā*, **20**, 223–226.

Basu, D. (1959). "The family of ancillary statistics." *Sankhyā*, **21**, 247–256.

Basu, D. (1964). "Recovery of ancillary information." *Sankhyā*, *A*, **26**, 3–16.

Basu, D. (1967). "Problems relating to the existence of maximal and minimal elements in some families of statistics (sub-fields)." *Proc. Fifth Berkeley Symp. Math. Stat. Prob.*, Vol. I, 41–50.

Berg, S. (1974). "Factorial series distributions, with applications to capture–recapture problems." *Scand. J. Statist.*, **1**, 145–152.

Berg, S. (1977). "Certain properties of the multivariate factorial series distributions." *Scand. J. Statist.*, **4**, 25–30.

Berk, R. H. (1972). "Consistency and asymptotic normality of mle's for exponential models." *Ann. Math. Statist.*, **43**, 193–204.
Bildikar, S. and Patil, G. P. (1968). "Multivariate exponential-type distributions." *Ann. Math. Statist.*, **39**, 1316–1326.
Birnbaum, A. (1961). "On the foundations of statistical inference. I: Binary experiments." *Ann. Math. Statist.*, **32**, 414–435.
Birnbaum, A. (1962). "On the foundations of statistical inference. (With discussion)." *J. Amer. Statist. Ass.*, **57**, 269–326.
Birnbaum, A. (1969). "Concepts of statistical evidence." In S. Morgenbesser, P. Suppes, and M. White, (Eds), *Philosophy, Science and Method: Essays in Honor of Ernest Nagel*, St. Martin's Press, New York.
Birnbaum, A. (1970). "On Durbin's modified principle of conditionality." *J. Amer. Statist. Ass.*, **65**, 402–403.
Bochner, S. and Martin, W. T. (1948). *Several Complex Variables*, Princeton University Press.
Bolger, E. M. and Harkness, W. L. (1965). "Characterizations of some distributions by conditional moments." *Ann. Math. Statist.*, **36**, 703–705.
Bolshev, L. N. (1965). "On a characterization of the Poisson distribution and its statistical applications." *Theory Prob. Its Appl.*, **10**, 446–456.
Borell, C. (1975). "Convex set functions in *d*-space." *Period. Math. Hung.*, **6**, 111–136.
Borges, R. (1970). "Eine Approximation der Binomialverteilung durch die Normalverteilung der Ordnung $1/n$." *Z. Wahrscheinlichkeitstheorie verw. Gebiete*, **14**, 189–199.
Borges, R. (1971). "Derivation of normalizing transformations with an error of order $1/n$." *Sankhyā, A*, **33**, 441–460.
Box, G. E. P. and Tiao, G. C. (1973). *Bayesian Inference in Statistical Analysis*, Addison-Wesley, London.
Bradley, E. L. (1973). "The equivalence of maximum likelihood and weighted least squares estimates in the exponential family." *J. Amer. Statist. Ass.*, **68**, 199–200.
Brown, L. (1964). "Sufficient statistics in the case of independent random variables." *Ann. Math. Statist.*, **35**, 1456–1474.
Campbell, L. L. (1970). "Equivalence of Gauss's principle and minimum discrimination information estimation of probabilities." *Ann. Math. Statist.*, **41**, 1011–1015.
Chentsov, N. N. (1966). "A systematic theory of exponential families of distributions." *Theory Prob. Its Appl.*, **11**, 425–435.
Chentsov, N. N. (1972). *Statistical Decision Rules and Optimal Conclusions.* (In Russian). Nauka, Moscow.
Courant, R., and Hilbert, D. (1953). *Methods of Mathematical Physics*, Volume I, Interscience, New York.
Courant, R., and Hilbert, D. (1962). *Methods of Mathematical Physics*, Volume II, Interscience, New York.
Cox, D. R. (1958). "Some problems connected with statistical inference." *Ann. Math. Statist.*, **29**, 357–372.
Cox, D. R. (1972). "Regression models and life tables. (With discussion)." *J. R. Statist. Soc., B*, **34**, 187–220.
Cox, D. R. (1975). "Partial likelihood." *Biometrika*, **62**, 269–276.
Cox, D. R. and Lewis, P. A. W. (1966). *The statistical analysis of series of events*, Methuen, London.
Cox, D. R. and Snell, E. J. (1968). "A general definition of residuals. (With discussion)." *J. Roy. Statist. Soc., B*, **30**, 248–275.
Crain, B. (1974). "Estimation of distributions using orthogonal expansions." *Ann. Statist.*, **2**, 454–463.

Csörgö, M. and Seshadri, V. (1970). "On the problem of replacing composite hypotheses by equivalent simple ones." *Rev. Int. Statist. Inst.*, **38**, 351–368.

Darmois, G. (1935). "Sur les lois de probabilité a estimation exhaustive." *C.R. Acad. Sci. Paris*, **260**, 1265–1266.

Davidovič, Ju. S., Korenbljum, B. I., and Hacet, B. I. (1969). "A property of logarithmically concave functions." *Soviet Math. Dokl.*, **10**, 477–480.

Dawid, A. P. (1975). "On the concepts of sufficiency and ancillarity in the presence of nuisance parameters." *J. Roy. Statist. Soc., B*, **37**, 248–258.

Dawid, A. P., Stone, M., and Zidek, J. V. (1973). "Marginalization paradoxes in Bayesian and structural inference. (With discussion.)" *J. Roy. Statist. Soc., B*, **35**, 189–233.

Durbin, J. (1969). "Inferential aspects of the randomness of sample size in survey sampling." In N. L. Johnson and H. Smith, Jr. (Eds). *New Developments in Survey Sampling*. Wiley-Interscience, New York, 629–651.

Durbin, J. (1970). "On Birnbaum's theorem on the relation between sufficiency, conditionality, and likelihood." *J. Amer. Statist. Ass.*, **65**, 395–398.

Dynkin, E. B. (1951). "Necessary and sufficient statistics for a family of probability distributions." (In Russian.) An English translation is available in *Select. Trans. Math. Statist. Prob.*, **1**, (1961), 23–41.

Eaton, M, Morris, C., and Rubin, H. (1971). "On extreme stable laws and some applications." *J. Appl. Prob.*, **8**, 794–801.

Edwards, A. W. F. (1969). "Statistical methods in scientific inference." *Nature*, **222**, 1233–1237.

Edwards, A. W. F. (1974). "The history of likelihood." *Int. Statist. Rev.*, **42**, 9–15.

Efron, B. (1975). "Defining curvature of a statistical problem (with applications to second order efficiency). (With discussion.)" *Ann. Statist.*, **3**, 1189–1242.

Efron, B. and Truax, D. (1968). "Large deviations theory in exponential families." *Ann. Math. Statist.*, **39**, 1402–1424.

Erdélyi, A. *et al.* (1953). *Higher Transcendental Functions*, Vol. 2, McGraw-Hill, London.

Erlandsen, M. (1975). "On affine hypotheses in multivariate normal families." Research Report No. 10, Dept. Theor. Statist., Aarhus Univ.

Fekete, M. (1912). "Über ein Problem von Laguerre." *Rendiconti di Palermo*, **34**, 89–100, 110–120.

Feller, W. (1966). *An Introduction to Probability Theory and Its Applications*, Vol. II, Wiley, London.

Fenchel, W. (1953). "Convex Cones, Sets, and Functions." Lecture notes, Department of Mathematics, Princeton University.

Finucan, H. M. (1964). "The mode of a multinomial distribution." *Biometrika*, **51**, 513–517.

Fisher, R. A. (1920). "A mathematical examination of the methods of determining the accuracy of an observation by the mean error, and by the mean square error." *Monthly Not. Roy. Astr. Soc.*, **80**, 758–770.

Fisher, R. A. (1921). "The mathematical foundations of theoretical statistics." *Phil. Trans. Roy. Soc., A*, **222**, 309–368.

Fisher, R. A. (1925). "Theory of statistical estimation." *Proc. Camb. Phil. Soc.*, **22**, 700–725.

Fisher, R. A. (1934). "Two new properties of mathematical likelihood." *Proc. Royal Soc., A*, **144**, 285–307.

Fisher, R. A. (1935). "The logic of inductive inference." *J. Roy. Statist. Soc.*, **98**, 39–54.

Fisher, R. A. (1956). *Statistical Methods and Scientific Inference*, Oliver and Boyd, Edinburgh. (Third ed. 1973, Collier Macmillan, London.)

Fraser, D. A. S. (1956). "Sufficient statistics with nuisance parameters." *Ann. Math. Statist.*, **27**, 838–842.

Fraser, D. A. S. (1968). *The Structure of Inference*, Wiley, London.

Fraser, D. A. S. (1976). "Necessary analysis and adaptive inference. (With discussion.)" *J. Amer. Statist. Ass.*, **71**, 99–113.

Gart, J. J. and Pettigrew, H. M. (1970). "On the conditional moments of the k-statistics for the Poisson distribution." *Biometrika*, **57**, 661–664.

Gnedenko, B. V. and Kolmogorov, A. N. (1954). *Limit Distributions for Sums of Independent Random Variables* (translated from the Russian and annotated by K. L. Chung), Addison-Wesley, Cambridge, Mass.

Gradshteyn, I. S. and Ryzhik, I. M. (1965). *Tables of Integrals, Series, and Products.* Academic Press, London.

Haberman, S. J. (1974). "Log-linear models for frequency tables derived by indirect observation: maximum likelihood equations." *Ann. Statist.*, **2**, 911–924.

Halmos, P. R. and Savage, L. J. (1949). "Application of the Radon–Nikodym theorem to the theory of sufficient statistics." *Ann. Math. Statist.*, **20**, 225–241.

Hardy, G. H. and Wright, E. M. (1960). *An Introduction to the Theory of Numbers*, Fourth ed. Clarendon Press, Oxford.

Hipp, C. (1974). "Sufficient statistics and exponential families." *Ann. Statist.*, **2**, 1283–1292.

Hipp, C. (1975). "Note on the paper 'Transformation groups and sufficient statistics' by J. Pfanzagl." *Ann. Statist.*, **3**, 478–482.

Hoffmann-Jørgensen, J. (1970). "The Theory of Analytic Spaces." Various Publications Series No. 10, Institute of Mathematics, Aarhus University.

Holgate, P. (1970). "The modality of some compound Poisson distributions." *Biometrika*, **57**, 666–667.

Huzurbazar, V. S. (1950). "Probability distributions and orthogonal parameters." *Proc. Camb. Phil. Soc.*, **46**, 281–284.

Huzurbazar, V. S. (1956). "Sufficient statistics and orthogonal parameters." *Sankhyā*, **17**, 217–220.

Höglund, T. (1974). "The exact estimate—a method of statistical estimation." *Z. Wahrscheinlichkeitstheorie verw. Gebiete*, **29**, 257–271.

Ibragimov, I. A. (1956). "On the composition of unimodal distributions." *Theory Prob. Its Appl.*, **1**, 255–260.

Jeffreys, H. (1948). *Theory of Probability*, Second ed. Clarendon Press, Oxford.

Jensen, E. B. (1976). "Conditional plausibility inference in contingency tables." Research Report No. 17, Dept. Theor. Statist., Aarhus Univ.

Jensen, S. T. (1975). "Covariance hypotheses which are linear in both the covariance and the inverse covariance." Preprint 1, Inst. Math. Statist., Univ. Copenhagen. To appear in *Ann. Statist.*

Johansen, S. (1977. "Homomorphisms and general exponential families." In J. R. Barra, F. Brodeau, G. Romier, and B. van Cutsem. (Eds.), *Recent Developments in Statistics.* North-Holland, Amsterdam, 489–499.

Jones, P. C. T. and Mollison, J. E. (1948). "A technique for the quantitative estimation of soil micro-organisms." *J. Gen. Microbiology*, **2**, 54–69.

Joshi, S. W. and Patil, G. P. (1970). "A class of statistical models for multiple counts." *Random Counts in Scientific Work*, Vol. 2, Pennsylvania State University Press, 189–203.

Joshi, S. W. and Patil, G. P. (1971). "Certain structural properties of the sum-symmetric power series distributions." *Sankhyā*, *A*, **33**, 175–184.

Kalbfleisch, J. D. (1975). "Sufficiency and conditionality. (With discussion.)" *Biometrika*, **62**, 251–268.

Kalbfleisch, J. D. and Prentice, R. L. (1973). "Marginal likelihoods based on Cox's regression and life model." *Biometrika*, **60**, 267–278.

Kalbfleisch, J. D. and Sprott, D. A. (1970). "Applications of likelihood methods to models involving large numbers of parameters. (With discussion.)" *J. Roy. Statist. Soc., B*, **32**, 175–208.

Kalbfleisch, J. D. and Sprott, D. A. (1973). "Marginal and conditional likelihoods." *Sankhyā, A*, **35**, 311–328.

Kalbfleisch, J. G. and Sprott, D. A. (1974). "Inferences about hit number in a virological model." *Biometrics*, **30**, 199–208.

Kamke, E. (1930). *Differentialgleichungen Reeller Funktionen*, Akademische Verlagsgesellschaft, Leipzig.

Kamke, E. (1974). *Differentialgleichungen. Lösungsmethoden und Lösungen.* Third unaltered edition. Chelsea, Bronx.

Kawata, T. (1972). *Fourier Analysis in Probability Theory*, Academic Press, London.

Keilson, J. (1965). *Green's Function Methods in Probability Theory*, Griffin, London.

Keilson, J. and Gerber, H. (1971). "Some results for discrete unimodality." *J. Amer. Statist. Ass.*, **66**, 386–389.

Khinchin, A. I. (1949). *Mathematical Foundations of Statistical Mechanics*, Dover, New York.

Koopman, L. H. (1936). "On distributions admitting a sufficient statistic." *Trans. Amer. Math. Soc.*, **39**, 399–409.

Kosambi, D. D. (1949). "Characteristic properties of series distributions." *Proc. Nat. Inst. Sci. India, A*, **15**, 109–113.

Kullback, S. (1959). *Information Theory and Statistics*, Wiley, New York.

Landers, D. and Rogge, L. (1972a). "Existence of sufficient regular conditional probabilities." *Manuscripta Math.* **7**, 197–204.

Landers, D. and Rogge, L. (1972b). "Minimal sufficient σ-fields and minimal sufficient statistics. Two counterexamples." *Ann. Math. Statist.*, **43**, 2045–2049.

Lauritzen, S. (1975). "General exponential models for discrete observations." *Scand. J. Statist.*, **2**, 23–33.

Lehmann, E. L. (1959). *Testing Statistical Hypotheses*, Wiley, New York.

Lentner, M. M. and Buehler, R. J. (1963). "Some inferences about gamma parameters with an application to a reliability problem." *J. Amer. Statist. Ass.*, **58**, 670–677.

Lindley, D. (1958). "Fiducial distributions and Bayes' theorem." *J. Roy. Statist. Soc., B*, **20**, 102–107.

Lindley, D. V. (1971). "Bayesian Statistics, a Review." *Reg. Conf. Ser. Appl. Math.*, **2**, SIAM, Philadelphia.

Linnik, Yu. V. (1964). *Decomposition of probability distributions*, Oliver and Boyd, Edinburgh.

Linnik, Yu. V. (1968). *Statistical Problems with Nuisance Parameters*, Amer. Math. Soc. (Translation Series), New York.

Mardia, K. V. (1972). *Statistics of Directional Data*, Academic Press, London.

Mardia, K. V. (1975). "Statistics of directional data. (With discussion.)" *J. Roy. Statist. Soc., B*, **37**, 349–393.

Marshall, A. W. and Olkin, I. (1974). "Majorization in multivariate distributions." *Ann. Statist.*, **2**, 1189–1200.

Martin-Löf, P. (1970). "Statistiska Modeller." Lecture notes (in Swedish), University of Stockholm.

Martin-Löf, P. (1974a). "The notion of redundancy and its use as a quantitative measure of the discrepancy between a statistical hypothesis and a set of observational data. (With discussion.)" *Conference on Foundational Questions in Statistical Inference*, Memoirs

No. 1, Dept. Theor. Statist., Aarhus Univ., 1–42. (Has also appeared in *Scand. J. Statist.*, **1**, (1974), 3–18.)

Martin-Löf, P. (1974b). "Exact tests, confidence regions and estimates. (With discussion)." *Conference Foundational Questions in Statistical Inference*, Memoirs No. 1, Dept. Theor. Statist., Aarhus Univ., 121–138.

Martin-Löf, P. (1974c). "Repetitive structures and the relation between canonical and microcanonical distributions in statistics and statistical mechanics. (With discussion)." *Conference on Foundational Questions in Statistical Inference*, Memoirs No. 1, Dept. Theor. Statist., Aarhus Univ., 271–294.

Martin-Löf, P. (1975). "Reply to Sverdrup's polemical article: 'Tests without power.'" *Scand. J. Statist.*, **2**, 161–165.

Mathiasen, P. E. (1977). "Prediction Functions." Research Report No. 24, Dept. Theor. Statist., Aarhus Univ.

Menon, M. V. (1966). "Characterization theorems for some univariate probability distributions." *J. Roy. Statist. Soc., B*, **28**, 143–145.

Nachbin, L. (1965). *The Haar Integral*, Van Nostrand, New York.

Nelder, J. A. and Wedderburn, R. W. M. (1972). "Generalized linear models." *J. Roy. Statist. Soc., A*, **135**, 370–384.

Neyman, J. and Pearson, E. S. (1933). "On the problem of the most efficient test of statistical hypotheses." *Phil. Trans. Roy. Soc., A*, **231**, 289–337.

Neyman, J. and Pearson, E. S. (1936). "Sufficient statistics and uniformly most powerful tests of statistical hypotheses." *Stat. Res. Mem.*, **1**, 113–137.

Neyman, J. and Scott, E. (1948). "Consistent estimates based on partially consistent observations." *Econometrika*, **16**, 1–32.

Patil, G. P. (1968). "On sampling with replacement from populations with multiple characters." *Sankhyā, B*, **30**, 354–366.

Patil, G. P. and Seshadri, V. (1964). "Characterization theorems for some univariate probability distributions." *J. Roy. Statistist. Soc., B*, **26**, 286–292.

Pedersen, J. G. (1975a). "On strong unimodality of two-dimensional discrete distributions with applications to M-ancillarity." *Scand. J. Statist.*, **2**, 99–102.

Pedersen, J. G. (1975b). "On strong unimodality and M-ancillarity with applications to contingency tables." *Scand. J. Statist.*, **2**, 127–137.

Pedersen, J. G. (1976). "Fiducial inference." Memoirs No. 2, Dept. Theor. Statist., Aarhus University. To appear in *Int. Statist. Rev.*

Petersen, C. G. J. (1896). "The yearly immigration of young plaice into the Limfjord from the German Sea, etc." *Rept. Danish Biol. Sta. for 1895*, **6**, 1–48.

Pfanzagl, J. (1968). "A characterization of the one parameter exponential family by existence of uniformly most powerful tests." *Sankhyā, A*, **30**, 147–156.

Pfanzagl, J. (1972). "Transformation groups and sufficient statistics." *Ann. Math. Statist.*, **43**, 553–568.

Pitcher, T. S. (1957). "Sets of measures not admitting necessary and sufficient statistics or subfields." *Ann. Math. Statist.*, **26**, 267–268.

Pitman, E. J. G. (1936). "Sufficient statistics and intrinsic accuracy." *Proc. Camb. Phil. Soc.*, **32**, 567–579.

Prékopa, A. (1971). "Logarithmic concave measures with application to stochastic programming." *Acta. Sci. Math.*, **32**, 301–316.

Prékopa, A. (1973) "On logarithmic concave measures and functions." *Acta Sci. Math.*, **34**, 335–343.

Prohorov, Yu. V. (1966). "Some characterization problems in statistics." *Proc. Fifth Berkeley Symp. Math. Statist. Prob.*, **1**, 341–349.

Puri, P. S. (1968). "Some further results on the birth- and death-process and its integral." *Proc. Camb. Phil. Soc.*, **64**, 141–154.

Raiffa, H. and Schlaifer, R. (1961). *Applied Statistical Decision Theory*, Division of Research, Harvard Business School, Boston, Massachusetts.

Rao, C. R. (1973). *Linear Statistical Inference and Its Applications*, Second ed. Wiley, London.

Rasch, G. (1960). *Probabilistic Models for some Intelligence and Attainment Tests*, Studies in Mathematical Psychology I. Danish Inst. Educational Research Copenhagen.

Rasch, G. (1961). "On general laws and the meaning of measurement in psychology." *Proc. Fourth Berkeley Symp. Math. Stat. Prob.*, Vol. 5, 321–333.

Rasch, G. (1968). "A mathematical theory of objectivity and its consequences for model construction." Paper read at European Meeting on Statistics, Econometrics, and Management Science, Amsterdam, 2–7 September, 1968.

Rasch, G. (1974). "On specific distributions for testing of hypotheses." *Conference on Foundational Questions in Statistical Inference*, Memoirs No. 1, Dept. Theor. Statist., Aarhus Univ., 101–110.

Roberts, G. E. and Kaufman, H. (1966). *Table of Laplace Transforms*, Saunders, London.

Rockafellar, R. T. (1970). *Convex Analysis*, Princeton University Press.

Sandved, E. (1967). "A principle for conditioning on an ancillary statistic." *Skand. Aktuar.*, **50**, 39–47.

Sandved, E. (1972). "Ancillary statistics in models without and with nuisance parameters." *Skand. Aktuar.*, **55**, 81–91.

Savage, L. J. (1954). *The Foundations of Statistics*, Wiley, New York.

Savage, L. J. (1970). "Comments on a weakened principle of conditionality." *J. Amer. Statist. Ass.*, **65**, 399–401.

Sherman, S. (1955). "A theorem on convex sets with applications." *Ann. Math. Statist.*, **26**, 763–767.

Silverstone, H. (1957). "Estimating the logistic curve." *J. Amer. Statist. Ass.*, **52**, 567–577.

Soler, J. L. (1977). "*Infinite dimensional type statistical spaces*. (Generalized exponential families.)" In J. R. Barra, F. Brodeau, G. Romier and B. van Cutsem (Eds.), *Recent Developments in Statistics*, North-Holland, Amsterdam.

Sprott, D. A. (1973). "Normal likelihoods and their relation to large sample theory of estimation." *Biometrika*, **60**, 457–465.

Sundberg, R. (1974). "Maximum likelihood theory for incomplete data from an exponential family." *Scand. J. Statist.*, **1**, 49–58.

Sverdrup, E. (1965). "Estimates and test procedures in connection with stochastic models for deaths, recoveries, and transfers between different states of health." *Skand. Aktuar.*, **48**, 184–211.

Sverdrup, E. (1966). "The present state of the decision theory and the Neyman-Pearson theory." *Rev. Int. Statist. Inst.*, **34**, 309–333.

Tweedie, M. C. K. (1946). "The regression of the sample variance on the sample mean." *J. London Math. Soc.*, **21**, 22–28.

Tweedie, M. C. K. (1947). "Functions of a statistical variate with given means, with special reference to Laplacian distributions." *Proc. Camb. Phil. Soc.*, **43**, 41–49.

Tweedie, M. C. K. (1957). "Statistical properties of inverse Gausian distributions." *Ann. Math. Statist.*, **28**, 362–377.

Valentine, F. A. (1964). *Convex Sets*. McGraw-Hill, New York.

Washio, Y., Morimoto, H., and Ikeda, N. (1956). "Unbiased estimation based on sufficient statistics." *Bull. Math. Statist.*, **6**, 69–94.

Wedderburn, R. W. M. (1974). "Quasi-likelihood functions, generalized linear models, and the Gauss–Newton method." *Biometrika*, **61**, 439–447.

Welch, B. L. (1939). "On confidence limits and sufficiency, with particular reference to parameters of location." *Ann. Math. Statist.*, **10**, 58–69.

Wijsman, R. A. (1973). "On the attainment of the Cramér–Rao lower bound." *Ann. Statist.*, **1**, 538–542.

Wilson, E. B. and Hilferty, M. M. (1931). "The distribution of chi-square." *Proc. Nat. Acad. Sci.*, **17**, 684–688.

Witting, H. (1966). *Mathematische Statistik*, B. G. Teubner, Stuttgart.

Zinger, A. A. and Linnik, Yu. V. (1964). "A characteristic property of the normal distribution." *Theory Prob. Its Appl.*, **9**, 624–626.

Author Index

Andersen, A. H. 154, 190, 221
Andersen, E. B. 38, 221
Ando, A. 132, 221
Anscombe, F. J. 30, 180, 185, 221
Armitage, P. 65, 221

Bahadur, R. R. 69, 221
Barankin, E. W. 137, 221
Barnard, G. A. 1, 2, 7, 17, 20, 21, 52, 53, 69, 70, 221
Barndorff-Nielsen, O. 1, 2, 31, 65, 69 70, 93, 98, 107, 136, 190, 218, 219, 222
Bartlett, M. S. 68, 137, 222
Basu, D. 45, 46, 69, 222
Berg, S. 135, 222
Berk, R. H. 190, 223
Bildikar, S. 115, 190, 223
Birnbaum, A. 1, 2, 45, 65, 223
Blæsild, P. 219, 222
Bochner, S. 106, 223
Bolger, E. M. 64, 223
Bolshev, L. N. 64, 223
Borell, C. 95, 223
Borges, R. 180, 223
Box, G. E. P. 30, 180, 223
Bradley, E. L. 185, 223
Brown, L. 136, 223
Buehler, R. J. 211, 223

Campbell, L. L. 137, 223
Chentsov, N. N. 136, 137, 152, 190, 223
Courant, R. 88, 223
Cox, D. R. 52, 54, 65, 69, 180, 183, 223
Crain, B. 137, 223
Csörgö, M. 63, 64, 224

Darmois, G. 136, 224
Davidovič, Ju. S. 97, 224
Dawid, A. P. 8, 56, 66, 224
Durbin, J. 65, 66, 224
Dynkin, E. B. 136, 224

Eaton, M. 104, 224
Edwards, A. W. F. 17, 27, 224
Efron, B. 68, 173, 190, 224
Erdélyi, A. 224
Erlandsen, M., 151, 152, 224

Fekete, M. 99, 224
Feller, W. 117, 136, 224
Fenchel, W. 82, 224
Finuncan, H. M. 16, 61, 62, 224
Fisher, R. A. 8, 31, 34, 35, 36, 37, 68, 136 224
Fraser, D. A. S. 8, 70, 225

Gart, J. J. 134, 225
Gerber, H. 99, 226
Gnedenko, B. V. 96, 101, 225
Gradshteyn, I. S. 133, 225

Haberman, S. J. 177, 225
Hacet, B. I. 97, 224
Halmos, P. R. 38, 69, 225
Hardy, G. H. 95, 225
Harkness, W. L. 64, 223
Hilbert, D. 88, 223
Hilferty, M. M. 180, 229
Hipp, C. 8, 136, 225
Hoffmann-Jørgensen, J. 42, 69, 222, 225
Höglund, T. 17, 190, 219, 225
Holgate, P. 101, 225
Huzurbazar, V. S. 31, 225

Ibragimov, I. A. 97, 225
Ikeda, N. 134, 228

Jeffreys, H. 30, 225
Jenkins, G. M. 221
Jensen, E. B. 61, 62, 70, 225
Jensen, S. T. 152, 225
Johansen, S. 137, 225
Jones, P. C. T. 128, 225
Joshi, S. W. 118, 219, 225

Kalbfleisch, J. D. 51, 54, 65, 225, 226
Kamke, E. 88, 226
Kaufman, G. M. 132, 221
Kaufman, H. 133, 228
Kawata, T. 107, 226
Keilson, J. 99, 136, 226
Khintichin, A. I. 137, 226
Kolmogorov, A. N. 96, 101, 225
Koopman, L. H. 136, 226
Korenbljum, B. I. 97, 224
Kosambi, D. D. 115, 226
Kullback, S. 189, 226
Kvist, H. K. 219, 222

Landers, D. 69, 226
Lauritzen, S. 137, 226
Lehmann, E. L. 68, 136, 226
Lentner, M. M. 201, 226
Lewis, P. A. W. 52, 183, 223
Lindley, D. V. 8, 136, 226
Linnik, Yu. V. 63, 67, 69, 226

Mardia, K, V. 114, 226
Marshall, A. W. 101, 129, 226
Martin, W. T. 106, 223
Martin-Löf, P. 137, 177, 190, 226, 227
Mathiasen, P. E. 31, 190, 227
Menon, M. V. 64, 227
Mollison, J. E. 128, 225
Morimoto, H. 134, 228
Morris, C. 104, 224

Nachbin, L. 5, 227
Nelder, J. A. 137, 227
Neyman, J. 68, 137, 227

Olkin, I. 101, 226

Patil, G. P. 64, 115, 118, 190, 219, 223, 225
Pearson, E. S. 68, 137, 227
Pedersen, J. G. 70, 99, 212, 213, 216, 227
Pedersen, K. 69, 136, 222
Petersen, C. G. J. 189, 227
Pettigrew, H. M. 133, 225
Pfanzagl, J. 8, 137, 227
Pitcher, T. S. 69, 227
Pitman, E. J. G. 136, 227
Prékopa, A. 95, 227
Prentice, R. L. 54, 226
Prohorov, Yu. V 64, 227
Puri, P. S. 64, 228

Raiffa, H. 132, 136, 228
Rao, C. R. 128, 228
Rasch, G. 64, 67, 69, 228
Roberts, G. E. 133, 228
Rockafeller, R. T. 73, 74, 75, 76, 78, 79, 80, 82, 83, 85, 86, 87, 88, 89, 90, 143, 166, 169, 172, 187, 188, 228
Rogge, L. 69, 226
Rubin, H. 104, 224
Ryzhik, I. M. 133, 225

Sandved, E. 70, 228
Savage, L. J. 38, 65, 69, 189, 228
Schlaifer, R. 132, 136, 228
Scott, E. 227
Seshadri, V. 63, 64, 224, 227
Sherman, S. 97, 228
Silverstone, H. 158, 228
Snell, E. J. 180, 223
Soler, J. L. 137, 228
Sprott, D. A. 30, 51, 226, 228
Stone, M. 8, 224
Sundberg, R. 177, 228
Sverdrup, E. 29, 36, 70, 228

Tiao, G. C. 30, 180, 223
Truax, D. 190, 224
Tweedie, M. C. K. 134, 184, 228

Valentine, F. A. 73, 228

Washio, Y. 134, 228
Wedderburn, R. W. M. 137, 179, 185, 227
Welch, B. L. 68, 229
Wijsman, R. A. 137, 229
Wilson, E. B. 180, 229
Winsten, C. B. 221
Witting, H. 136, 229
Wright, E. M. 95, 225

Zidek, J. V. 8, 224
Zinger, A. A. 63, 229

Subject Index

Affine hull 3
Affine hypothesis or subfamily of exponential family 125–126, 151, 177
Affine independence 3, 112–113
Affine mapping 76, 90, 204
Affine subset 3
Affine support 90
Ancillarity 2, 9, 33–37, 48–50, 57, 64–65, 68–70, 109, (*see also* Ancillary statistic; Exponential family)
  B- 35–36, 38, 43–46, 49, 56, 62, 65–68, 192–193
  conditional B- 45–46, 57–58, 62, 192
  G- 37, 49, 52, 65
  M- 11, 37, 49, 52, 56, 58, 65, 70, 211–218
  pointwise M- 217–218
  principle of 34, 67
  quasi-ancillarity 57–58, 70, 191–196
  S- 36, 49–52, 56, 58, 65–66, 70, 208–210, 219
Ancillary statistic 2, 9, 33, 36, 69, 109 (*see also* Ancillarity)
  maximal 57–58
  maximal B- 43, 45, 58, 66
  maximal M- 58, 211
  maximal quasi- 58, 195
  maximal S- 58, 208
  relatively maximal B- 43, 45, 66
  relatively maximal S- 208–209
Anderson 97

Barnard 7
Barrier cone 74
Basu's theorem 46, 183
Behrens–Fisher problem 69
Bernoulli trials 173
Bessel functions 98, 113, 129, 212
Beta distribution 142
Binomial distribution 14–15, 24, 27, 36, 135, 173, 179–180, 182
  convolution of 99
Binomial experiments 31
Birnbaum's theorem 2, 65
Birth and death process 29, 64
Boltzmann's law 137

Canonical parametrization 113
  continuity of 120
  minimal 113
Canoncial statistic 113
  minimal 113
Capture–recapture 135, 189
Carathéodory's theorem 73
Cauchy distribution 43
c-discrete distribution, *see* Distributions
Censored exponential lifetimes 131
Censored family of exponential family 130–131
χ^2 distribution 63, 184
  non-central 129
Completeness 45–46
  bounded 45–46, 56
  of exponential families 118
Completion of exponential family 154–158
  of convex exponential family 163–164
Concave function 22–23, 77
  polyhedral 169
Conditional B-ancillarity, *see* Ancillarity
Conditional family of exponential family 130
Conditional inference 52, 62, 70
Conditional mean value 133–134
Conditionality axiom 65
Confidence sets 16
Conjugate exponential family 131–133, 149
Contigency tables 16, 28, 61–62, 119, 154 167, 212
  2×2 36, 44, 48, 63, 208–209, 212, 216
  $2 \times 2 \times 2$ 216
Continuous-type distribution, *see* Distributions

Convex duality 2, 19, 71, 140
Convex exponential family 117
Convex function 76–95, (*see also* Subdifferential of; Subgradient of; Supremum of collection of)
  closure of 78
  conjugate 80–84, 88–89
  convex hull of 79
  effective domain of 77
  epigraph of 76
  essentially smooth 87–88
  essentially strictly 87
  partial conjugate of 83, 89
  polyhedral 79, 172
  steep 86–87
Convex sets 73–76
  dimension of 73
  polyhedral 89
  relative boundary of 73
  relative interior of 73
Convex support, *see* Distributions
Correlation coefficient 210
Correlation matrix 37, 53, 119
Corresponding pair of L-indpendent parameters 50
Cramér–Rao lower bound 137
Cumulant transform 105–106, 114, 142, 197
Cumulants 106, 114, 134
Cuts 50–52, 64
  in exponential families 128, 196–208
  proper 50

Derived exponential families 125
Differential theory, convex 84–89
Dimension of convex set, *see* Convex sets
Direction of R^k 74
Dirichlet distribution 98, 132
Discrete-type distribution, *see* Distribution
Discrimination information, *see* Information
Dispersion index 134
Distinguish 12
Distributions, affine support of 5
  conditional singular 5–6
  convex support of 90
  of c-discrete type 6
  of continuous type 6
  of discrete type 6
  singular 5
Dose-response model, *see* Logistic dose-(binomial) response model

Duality 1, 7, 9, 19–20, 22, 109
  convex 19, 71, 107, 139
  sample-hypothesis 19–20, 139

Efron 173
Epigraph of a convex function, *see* Convex function
Essentially smooth convex function, *see* Convex function
Essentially strictly convex function, *see* Convex function
Estimation, *see* Least squares; Maximum b-lods estimation; Maximum likelihood estimation; Maximum plausibility estimation
Euclidean 5
Exponential distribution 63, 68, 202 211
Exponential family (or model) 1–2, 7–8, 23, 30, 37–38, 109–150, (*see also* Affine hypothesis or subfamily of; Censored family of; Completeness of; Completion of; Conditional family of; Conjugate; Cuts; Derived; Infinite divisibility in; Linear hypothesis or subfamily of; Marginal family of; Maximum likelihood estimation; Minimal representation; Mixed parametrization; Regular; Standard representation)
  ancillarity in 191–196, 208–219
  asymptotic properties 190, 219
  connected 117
  convex 117
  full 116, 125
  generalized 137
  generated by probability measure and statistic 115–116
  independence in 147–150, 182–184
  interior 196
  linear 19, 113, 215
  open 117
  order of 112
  partly 37–38, 185
  steep, 117, 126, 142, 152–153, 172
  steep, example of non- 152–153
  (strong) unimodality and (strict) universality in 164–168
Exponential representation of densities of exponential family 111
  minimal 112–113
Extreme point of convex set 75

Face 75
  proper 75
Factorial series families 134–136, 188–189
Fenchel's inequality 80
Fiducial inference 70
Fisher 8. 17, 31, 34, 36, 68, 70
Fisher information function or matrix 150, 177, 182–183, 189–190
Fisher's prediction function 31
Fisher's scoring method 185
Fourier–Laplace transform 105

Gamma distribution 8, 53, 97, 107, 132, 146, 149, 151, 166, 180, 182–183, 202, 211
Gauss 137
Gauss–Newton method 185
Geiger counter 27
Generated family, of probability functions 6
  of probability measures 6
Genotype distribution 44, 54, 122, 177, 216
Geometric distribution 167, 169
Group family 6–8, 52, 67
  transitive group family 6, 16

Hardy–Weinberg distribution 122–123, 177, 216, 218
Hartog's theorem 106
Hazard function or rate 53, 131
Hölder's inequality 100–101, 103
Hyperbolic distribution 97
  multivariate 97
Hypergeometric distribution 189
  multivariate 62, 100, 135, 167
Hypothesis aspect 19

Ibragimov 97
Incidental parameter 33, 37
Independence 26, (*see also* Exponential family)
  approximate L- 30
  conditional 46, 119, 147
  infinitesimal L- 30–31, 182–185
  likelihood (L-) 19, 26–31, 51, 64, 148–150, 196–198
  plausibility (Π-) 26
  stochastic 19, 26–27, 31, 119, 149–150
  under a function 26
  variation 26, 28, 60–61, 122–123, 170, 198–199
Independent experiments 21
Indicator function of a convex set 82
Indicator of a set 1–2
Infimal convolution of convex functions 83
Infinite divisibility in exponential family 136
Information, *see* Fisher information function or matrix
  discrimination 189–190
  statistical 1–2, 33–34, 37–38, 49, 215, 217–218
Interest, parameter of 33
Interior exponential family, *see* Exponential family
Inverse Gaussian distribution 117, 184
  generalized 97
Inversion formula 21–22

Jensen's inequality 91

Khintichin 101–102
Kronecker's theorem 95
Kuhn–Tucker conditions 187
Kurtosis 119

Laplace distribution 168
Laplace transforms 100, 103–108, 114, 133, 184, 201–202
  effective domain of 103
Least squares estimation 185
Legendre transformation 87–89, 101
Legendre type 88–89, 142
Level sets 78
Lifetime data, *see* Regression analysis
Likelihood 1–2, 11, 15–17, 20, 215, (*see also* Likelihood function; Maximum likelihood estimation; Prediction)
  axiom 65
  conditional 65
  equation 152, 160, 176–177
  independence, *see* Independence
  marginal 65
  ratio test, *see* Test
Likelihood function 1, 7, 11–17, 19, 21–22, 27, 29–30, 41, 43, 54, 62, (*see also* Likelihood)
  log- 9, 20–21, 23, 30, 139, 203–204

Likelihood function (*continued*)
log-likelihood functions in exponential families 139–143, 150–164, 175–180, 203–204
normed 13–14, 16
sup-log- 140
Lineality space 74–75
Linear hypothesis or subfamily of exponential family 126, 153–154
Linear regression analysis, *see* Regression analysis
Location-scale model 35–36, 62–63
Lods functions 1–2, 7, 9, 13, 19–23, 30, 109, 181
b- 13, 21–23
equivalence of 21–22
f- 13, 21–23
linear 23
normed 21
Logarithmic distribution 127
multivariate 118
Log-concavity, of a function 93
of a probability measure 95
Log-convexity 100–101
Logistic dose-(binomial) response model 156–158, 161, 205, 213, 218

Marginal family of an exponential family 127–129
Markov kernel 5, 38–39, 45–46
Markov process 29–30
Martin-Löf 177
Maximal invariants 52
Maximum b-lods estimation 22–23
Maximum entropy 137
Maximum likelihood 8
Maximum likelihood estimation 13, 15, 22, 27, 30–31, 37, 62, 189
conditional 37–38, 58–59, 219
in exponential families 137, 150–164, 175–177, 183–188, 190
of sub-parameters 62, 156, 158
Maximum likelihood prediction 24, 172
Maximum plausibility estimation 13, 15, 17, 22, 58–62, 189, 219
conditional 58–62, 219
in exponential families 168–170, 188, 190
Maximum plausibility prediction 24, 171–173
Mean value, mapping 121, 188
parametrization 121
Minimal sufficient σ-algebra 69
Minimum discrimination information 137
von Mises–Fisher distribution 53, 113–114, 130
Mixed parametrization 121–122, 148–149, 183
Mixture experiment 34–35, 69
Mode, mapping 17, 168, 188
point 11, 16, 24–25, 48–49, 60–61
point of family of conditional probability functions 12
point of family of probability functions 11, 13, 15, 59
point of probability function 11
size, constant 11–13
Model control 62, 133
Model function 7, 23
Multinomial distribution 26–27, 61, 100, 107, 118, 132, 207
Multiple recapture census 135

Negative binomial distribution 15, 28, 56, 100, 128, 185, 202
Negative multinomial distribution 28, 100, 107, 118, 207
Neyman–Pearson 68
Nonformation 1–2, 33–35, 37–38, 46–48, 50, 56, 65, 70
B- 35, 47–48
G- 35, 47, 49, 52, 55
M-B- 65
M- 35, 47–49, 217
pointwise 46
pointwise B- 47–48, 66
pointwise M- 47–49, 218
pointwise S- 47–49
principle of 34–35
S- 35–36, 47–49
Normal cone 73–74
Normal cone mapping 74
Normal distribution 8, 52, 62–63, 97 119, 130, 132, 193
multivariate 7, 28, 37, 97, 104 107, 116, 119, 122, 126, 146, 151–154, 209

Normal vector to a convex set 73

Ods function 1, 22, 62
b- 22–23
f- 22–23
normed 22
Orbit 4, 52
Order of an exponential family, *see* Exponential family
Orderings, b- 20
f- 20
Open exponential family, *see* Exponential family
Open kernel 117–118
Orthogonal parameters 30–31, 182–185

Parameter function 6
Parametrization 6, (*see also* Canonical; Mean value; Mixed)
Pareto distribution 152
Partially observed exponential situation 176–177
Perfect fit 48
Plausibility 1–2, 6, 11, 17, 21, 48, 215, 218, (*see also* Maximum plausibility estimation; Plausibility function Prediction)
ratio test, *see* Test
Plausibility function 1, 7, 11–16, 15, 19, 21–22, 58–60, 62, 169, (*see also* Plausibility)
conditional 58–60
log- 9, 20, 21–23, 140
log-plausibility function in exponential families 143–144, 168–170
normed 13–14, 16
sup-log- 140
Poisson distribution 27–28, 57, 100–101, 107, 115, 118, 132–134, 142, 149, 180 195–196, 202, 207, 212
unimodality of mixtures of 101
Poisson process 48
with log-linear trend 52
Poisson regression, *see* Regression analysis
Polar of a convex cone 74
Polyhedral concave function, *see* Concave functions
Polyhedral convex function, *see* Convex function
Polyhedral convex set, *see* Convex sets
Polytope 76
Power series family 118, 205, (*see also* Sum-symmetric)
Precision in (conditional) inference 34–36
of a (multivariate) normal distribution 7
of a von Mises–Fisher distribution 113
Prediction 11, (*see also* Maximum likelihood prediction; Maximum plausibility prediction; Prediction function)
Prediction function 1, 20, 23, 31, (*see also* Prediction)
likelihood prediction 24–25, 31
likelihood prediction, for exponential families 170–172
plausibility prediction 24–25, 31
plausibility prediction, for exponential families 171–173
Probability 20, (*see also* Probability function)
Probability function 11–13, 19, 21–22, 25, 30, (*see also* Probability)
log- 9, 20, 23, 139
log-probability functions in exponential families 164–168, 177–182
Probability measure, of c-discrete type 6
of continuous type 6
of discrete type 6
Product exponential family 127

Quasi-ancillarity, *see* Ancillarity
Quasi-convex function 77, 96, 98
Quasi-sufficiency, *see* Sufficiency

Rank of a mapping 191
Realizable, sample points 6
values of a statistic 6
Recession, cone 74, 82
function 79, 83
Recombination 44, 54–55
Reference set 70
Regression analysis, linear 36
of lifetime data 53
Poisson 51
Regular exponential family 116–117, 126, 114, 149, 153, 200, 203–204, 215
example of non- 117
Relative boundary of convex set, *see* Convex sets
Relative interior of convex set, *see* Convex sets
Relatively maximal ancillary statistic, *see* Ancillary statistic

Resultant density for a von Mises–Fisher distribution 128–129
Resultant length density for a von Mises–Fisher distribution 129
Rockafeller 89

Sample aspect 19
Schur concavity 101
Separable σ-algebra 40
Separate inference 2, 9, 33–34, 36, 68
Separating hyperplane 73
 properly 73
 strongly 73
Separation, inferential 2, 7, 9, 11, 33, 37–38, 46, 56, 69
Singular distribution, *see* Distributions
Singular multinomial distribution 206
Size of a quasi-ancillary statistic 192
Skewness 119, 179
Skitović–Darmois theorem 67
Stable distributions 104, 117, 142, 184
Standard representation for an exponential family 115
 minimal 115
Statistical field 5
Statistical information, *see* Information
Statistical mechanics 8, 137
Steep convex function, *see* Convex function
Steep exponential family, *see* Exponential family
Stopping rule 14, 65
Subdifferential of a convex function 84
Subgradient of a convex function 84
Sufficiency 1–2, 8, 9, 33, 48–50, 52, 57, 64–65, 68–70, 109, 133, 136–137, 152, 191, (*see also* Sufficient statistic)
 axiom 65
 B- 35, 37–38, 43, 45, 49, 65–66, 69
 G- 37, 52–55, 67, 70
 M- 37, 52, 55, 70
 principle of 35
 quasi- 57–58, 70
 S- 48–51, 56, 66, 70
Sufficient statistic 9, 33, 35, 38, 69, 191, (*see also* Sufficiency)
 minimal 12, 15, 22, 56–58, 69, 111, 126
 minimal B- 38, 40–45, 58
Sum-symmetric power series family 118, 205–208
Support function of a convex set 83
Support of a probability measure 90
Supremum of collection of convex functions 80
System reliability 211–212

Test, conditional 68
 likelihood ratio 140, 190
 plausibility ratio 140
 similar 68–69
Transformations 4–6, 30–31
 group of 4–6, 52–54
 normalizing 30, 177–182
 spread-stabilizing 30, 177–182
 transitive class of 4–5, 11, 13, 47
 unitary class of 4–5, 52
 variance-stabilizing 30, 179
Transitive class of transformations, *see* Transformations
Trinomial distribution 99, 195
Truncated family of exponential family 130–131

Unimodality 2, 11, 49, 71, (*see also* Exponential family)
 of continuous type distributions 96–98, 101
 of discrete type distributions 98–100
 strong 49, 107, 144, 211–212, 216, 218
 strong unimodality of continuous type distributions 96–98, 101
 strong unimodality of discrete type distributions 98–100
Uniqueness in statistical inference 56–57, 65
 non- 2, 37, 65–68
Unitary class of transformations, *see* Transformations
Universality 11–13, 25, 47, 49, 60, 63, 169, 172, 211, (*see also* Exponential family)
 strict 11, 60

Variation independence, *see* Independence

Wedderburn 179
Wishart distribution 98, 107, 128, 149–150

Printed and bound by CPI Group (UK) Ltd, Croydon, CR0 4YY

27/10/2024

14580350-0001